D0845585

Toward an

EVOLUTIONARY BIOLOGY

of

LANGUAGE

Toward an EVOLUTIONARY BIOLOGY *of* LANGUAGE

PHILIP LIEBERMAN

THE BELKNAP PRESS OF HARVARD UNIVERSITY PRESS
Cambridge, Massachusetts · London, England
2006

Printed in the United States of America

Portions of Chapters 3, 6, and 7 appeared in different form in Lieberman, *The Biology and Evolution of Language,* Harvard University Press, 1984.

Library of Congress Cataloging-in-Publication Data

Lieberman, Philip.
Toward an evolutionary biology of language / Philip Lieberman.
p. cm.
Includes bibliographical references and index.
ISBN 0-674-02184-3 (cloth : alk. paper)
1. Biolinguistics. 2. Evolution.
3. Language and languages—Origin. I. Title.

P132.L533 2006
401—dc22 2005059101

To the memory of Elizabeth Ann Bates

Contents

Preface

WITH THE PASSING OF TIME, the preface to a book that records research that is in progress must grow longer. It is apparent that human language is unique and that some of its biologic bases are species-specific. However, I hope that it will become clear that various attributes of language are present or present in reduced degree in other species. I also hope to show that neural mechanisms specific to humans that confer linguistic ability also regulate other aspects of behavior by virtue of their evolutionary history. My own research on the evolution of language and this book owe much to Edmund S. Crelin. I still remember Ed's words in 1970 as he looked at a cast of the Neanderthal La Chapelle-aux-Saints fossil: "He's a big baby." We then started to attempt to track the evolution of human speech, one of the critical elements of human language. Bill Laughlin in 1972 called my attention to the radiographs that track the development of the larynx, pharynx, tongue, and mouth from birth to age twenty years. In 1999 my most demanding critic, my son Daniel, with Robert McCarthy, reevaluated these radiographs and provided a starting point for new quantitative studies of the evolution of the anatomy that allows us to produce the full range of human speech. They found that our vocal anatomy gradually develops, reaching its adult-like form between ages six to eight years. Mc-

Carthy and his colleagues have since been able to quantify the appraisal of Neanderthal speech capabilities by techniques that place limits on the possible vocal anatomy of Neanderthals and other long-dead hominids. Jeffrey Laitman's work on the evolution of human vocal anatomy also must be acknowledged.

None of these studies would have made much sense if the Haskins Laboratories research group directed by Franklin Cooper and Alvin Liberman had not already shown that speech has a special role in human language. The high information transfer rate of speech allows humans to overcome the limits of short-term memory. The pioneering studies of Alan and Trixie Gardner revealed some of the latent linguistic capabilities of chimpanzees; studies of apes in their natural habitats revealed some of the aspects of behavior that differentiate us from them. Chris Boehm deserves special thanks for his advice on what chimpanzees can do.

My studies on the neural bases of human language and thought at Brown University would have been impossible without the support and insights of Joseph Friedman, who is as close to a saint as is humanly possible to his patients afflicted with Parkinson's disease. The list of former and present students who contributed the fresh views and energy of youth to the body of evidence on which this book rests includes Terry Nearey, James Atkinson, Phil Morse, Peter Blackwell, Patricia Keating, Robert Buhr, Karen Landahl, Bill Katz, William Ryalls, Brad Seebach, Shari Baum, Chiu-Yu Tseng, Judith Parker, Molly Mack, Joan Sereno, Edward Kako, Liane Feldman, Emily Pickett, Techumseh Fitch, Thanasi Protopappas, Jennifer Adeylot, Jesse Hochstadt, Angie Morey, Mara Larson, Beverly Young, and Sandra Mather. Over the years, Sheila Blumstein, James Anderson, William Warren, and Mike Tarr patiently listened as I tried out ideas, sometimes differing but always generous with their time and insights. At a distance, Fareneh Vargha-Khadem and her colleagues and Simon Fisher kept me current on their ground-breaking studies of the FOXP2 gene. The list could go on and I must apologize for any omissions. Funding from the National Institute of Health and NASA under grant NCC9-58 with the National Space Biomedical

Research Institute greatly facilitated the research discussed in this book.

Ernst Mayr, whom I met only once, provided much of my education through his books. Although I only encountered their thoughts through the medium of paper and ink, Victor Negus and Karl Lashley were mentors. And the master, Charles Darwin, must be acknowledged in any work touching on evolution. At MIT, where I spent both my undergraduate and graduate years, Morris Halle's generous spirit was evident. The young Noam Chomsky created a soaring space in which everything seemed possible. Ken Stevens and Roman Jakobson brought us back to earth with theories that could be tested and facts. Arend Bouhuys guided me through the mysteries of physiology in that distant period. In the period leading to this book, I owe special thanks to Fred Dick and Daniel Lieberman for their insights and corrections concerning brain and body, respectively. My wife Marcia provided more than encouragement; her questions sharpened the hypotheses that form part of this book. Elizabeth Knoll, who has guided most of my books at Harvard University Press, again provided sound editorial advice. Elizabeth Collins had to unscramble text mangled by a runaway word processor and my being blind to typos. Her meticulous and sensible decisions yielded a readable text.

And yet another Elizabeth played a critical role in this work. In the prime of life and full of energy, she noted that my 1984 book, *The Biology and Evolution of Language,* had been ten to twenty years ahead of its time; but she also observed that almost twenty years had passed. And apart from her insights on the nature of the processes by which we attain language, she provided a signal example of courage during her final days. To her, Elizabeth Ann Bates, is this book dedicated.

Toward an EVOLUTIONARY BIOLOGY *of* LANGUAGE

CHAPTER 1

The Mark of Evolution

Nothing in biology makes sense except in the light of evolution.

—THEODOSIUS DOBZHANSKY (1973)

THE QUESTIONS that are addressed in this book concern the nature and evolution of the biologic bases of human language. That human beings are biologically predisposed to acquire and use language is indisputable. However, the path blazed by linguists that other scholars often follow ignores the principles and findings of biology. Moreover, theories that seem to reflect nineteenth-century phrenology constrain the interpretation of many current studies on how brains work.

I will attempt to show that linguists, who have treated syntax as though it were the central feature of human language, have failed to take account of some of the basic principles of evolutionary biology. It is certainly the case that only humans can produce long, complex sentences and can acquire and use tens of thousands of words. However, "voluntary" speech is absent in our ape cousins, separated from our common ancestor by a mere 6 to 7 million years. Linguists and cognitive scientists also tend to underestimate the central role of human speech in language. Speech makes it possible to exchange information at rates exceeding any other vocal signal. Voluntary speech production also entails having the neural "reiterative" capacity that many linguists now associate with syntax. Fully human speech capacity involves having a species-specific tongue and brain that reflect both the continuity and the tinkerer's logic that mark biologic evolution.

However, the main focus of this book is on an aspect of the tinkerer's logic that has a more general role. Neural structures that initially evolved to facilitate motor control, perhaps starting with walking, appear to have been modified to yield the unique qualities of human language—voluntary speech production and syntactic processes that allow us to create an unbounded number of novel words and sentences. Evolution has produced a brain, the human brain, that can freely "reiterate," that is, reorder and modify a finite set of individual motor "gestures" involving our tongue, lips, and larynx to produce a potentially unbounded number of words. That same brain appears to make use of similar neural processes to reiterate a finite set of "syntactic rules" and words to create a potentially unbounded set of novel sentences. Reiteration accounts for our being able to form complex sentences such as "The boy who was wearing a small red hat was hugged by the tall girl." Reiteration yields the ability to insert adjectives, adverbs, relative clauses, prepositional clauses, and other types of clauses into sentences.

Current research identifying the brain's "center" of religion, fear, or language disregards evidence from hundreds of studies that show that most complex behaviors are regulated by neural circuits linking activity in many parts of the brain, including subcortical structures such as the basal ganglia and hippocampus. The Broca-Wernicke language organ theory is simply wrong. The basal ganglia, which support neural circuits implicated in syntax, also confer cognitive flexibility, allowing individuals to adapt to changing circumstances. They also continue to play a part in regulating motor control in walking, running, talking, dancing, and so on. Current research also shows that disturbances such as schizophrenia, disinhibition, and obsessive-compulsive disorder, as well as verbal apraxia (difficulty in controlling tongue, lip, and larynx gestures), derive from disruption of neural circuits linking the cortex and the basal ganglia. Indeed, advances in molecular genetics have identified the FOXP2 "language gene" involved in verbal apraxia. FOXP2 is not a language gene. However, it provides new insights on the evolution of the human brain and language.

There are other aspects of the neural bases of language and thought that I shall also note. The human brain's memory can both temporarily retain long sequences of concepts, coded in words so that we may comprehend the meaning of a sentence, and hold an almost-unbounded store of information over the course of a lifetime. And there is a balance between nature and nurture—evidence from a wide range of studies rules out our having genetically transmitted innate knowledge of the details of syntax. The human neural system in its totality is the basis for our singular cognitive ability, allowing us to form and draw on an almost limitless conceptual base and adapt our behavior and thoughts to changing circumstances.

The Alternative View

The alternative "standard" view cannot be dismissed out of hand. Expressed in the theories of linguists and philosophers such as Noam Chomsky (1986), Steven Pinker (1994, 1998), and Jerome Fodor (1983), this view states that the neural basis of human language is a "module" devoted to language and language alone, and this module is distinct from the mechanisms that regulate other aspects of human or animal behavior. Modular theories implicitly claim that the functional architecture of the human brain is similar to that of a conventional digital computer in which a discrete set of devices controls a printer, another the display, another the keyboard, and so on.

Young Frankenstein's Tap Dance and Syntax

Although Chomsky now concedes that human language must have an evolutionary past, he still maintains that some key element is specific to language. At one time (Chomsky 1976) words were the key element, but syntax has been Chomsky's focus for decades. The current hypothetical neural mechanism that makes language possessing syntax possible is a narrow faculty of language (FLN) that allows us to produce novel sentences. According to Hauser,

Chomsky, and Fitch (2002), the FLN yields recursion, the process that allows a phrase to be embedded within a similar phrase to form complex sentences. Recursion is a theory-specific form of reiteration that reflects the syntactic theory first proposed in Chomsky (1957) that "generates" (i.e., "forms") complex sentences. In generative grammars, phrases such as relative clauses are initially represented by a hypothetical sentence node "S" that is embedded within a phrase of the carrier sentence. Subsequent theory-specific syntactic rules then rewrite parts of the embedded sentence to form the desired relative clause.

For example, the relative clause in the sentence "*The man who was wearing a red hat was old*" is hypothetically derived from a string of words that could have formed the sentence "*The man was wearing a red hat,*" which is embedded within the first phrase (the man) of the carrier sentence "*The man was old.*" I return to this issue in Chapter 7. The hypothetical FLN plays no part in regulating any other aspect of human behavior. It constitutes a module of the brain devoted to syntax. The difference between a human brain and a chimpanzee brain is that syntactic language is a human "optional feature." Presumably, if we were able to plug an FLN module into a chimpanzee's brain, the ape would possess human language.

In discussing how this hypothetical unique FLN may have evolved, Hauser, Chomsky, and Fitch take a step forward when they suggest that studies of animal communication may shed light on this question. However, they neglect to consider motor control. A scene in Mel Brooks's movie *Young Frankenstein,* in which Dr. Frankenstein and the monster tap dance, speaks to this question; it captures the intuition that creative motor control and cognition are related. In the movie, Frankenstein's "proof" that the monster possesses a human brain is a complex tap dance he and the monster perform. This convinces the savants of the Academy of Sciences that the creature is human, though, as is traditionally the case, he still has a large screw protruding from his forehead. The script writers and director were not far off the mark. Human morphology places limits on the movements that a person can execute, yielding a large but finite number of steps that a dancer can potentially make. Creative dance involves reordering and reiterating steps selected from this finite set. Thus, paraphrasing Descartes, the movie's message is, "I dance, therefore I am." The motto that I propose here is, "I walk, run, and talk—therefore I am."

Universal Grammar

According to Chomsky, human syntactic ability derives from a localized, innate, species-specific organ of the human brain dubbed the Universal Grammar (UG), which presumably incorporates the hypothetical FLN. No one would dispute the fact that children have an innate propensity to acquire complex "rules" or syntax, or that human language is singularly creative. We can potentially form an infinite number of sentences. However, Chomsky's claim is that all of the syntactic rules and principles of every human language that was, is, or will be spoken are genetically coded in the UG. Moreover, the UG is identical for each and every "normal" human being—all of the hypothetical syntactic rules and principles must be present in every human brain (Chomsky 1980a, b, 1986, 1995). In short, the Chomskian claim is that every "normal" human being starts life with an identical store of knowledge concerning syntax. When exposed to a fragment of a particular language, the UG triggers a detailed representation of the language's syntax. In computerese, the UG containing the detailed syntax of every human language is preloaded.

Biologic evidence does not support Chomsky's claim. In these pages I present evidence from independent studies that link the neural bases of syntactic ability to brain mechanisms that regulate motor control and other aspects of human and animal behavior, such as coping with life's changing circumstances and mood. Innate human capacities clearly set boundaries on behavior. We can't fly, breathe under water, or comprehend infinitely long sentences. However, solid biologic evidence rules out any version of innate Universal Grammar.

As Charles Darwin (1859) pointed out, evolution is opportunistic and has a "historic" logic of its own making. Existing structures and systems are adapted to serve new ends, often maintaining their original functions as well. Once a new behavior is in place, natural selection may then modify a structure to enhance that aspect of life; but some, or all, of the demands of the starting point may persist. The chance events and evolutionary changes that yielded human linguistic and cognitive ability are apparent in the anatomy and physiology of the human brain and body. For example, although we differ in many ways from chimpanzees, we too must eat. The shape of the human tongue differs from a chimpanzee's because it has been adapted to facilitate speech communication; but our tongues must still serve to swallow food, and the neural control sequence that coordinates tongue and larynx movements in swallowing is quite similar for chimpanzees and humans (Palmer et al. 1992). I present data that demonstrates that the brain mechanisms that yield human syntax ability also have evolutionary antecedents outside the domain of language. The subcortical basal ganglia structures of the human brain that are critical elements of the neural systems that allow us to comprehend the meaning or to form a sentence also continue to support neural circuits that regulate motor control as well as aspects of cognition, mood, and much else. The evolutionary record of the changes that yielded human language is evident in the morphology and physiology of the brain and body; disputes concerning the evolution of language follow from different readings of the text.

Uncertainty arises because the text has become obscured; the species who possessed intermediate stages of language are extinct. The evolutionary record is clearer when similar aspects of behavior in living, related species can be compared. Because virtually all living animals possess sight, the evolution of biologic bases of vision has been traced back to insects; some of the genes that determine the morphology of the human eye occur in fruit flies (Carroll, Grenier, and Weatherbee 2001). The comparative method that makes this possible was part of Charles Darwin's toolkit. As Ernst Mayr (1982) pointed out, Darwin introduced the research paradigm that has guided biology since 1859—the gradual exploration in related species of links between behavior and biology. The evolutionary path can then be discerned by tracking morphologic changes having behavioral consequences. But the comparative method generally entails having access to a line of living animals that demonstrably are related and that all manifest a similar observable behavior, such as vision. If all goes well, the evolution of the biologic substrate that regulates that behavior can then be discerned.

Despite the fact that we humans are the only living species that possess language, the situation is not hopeless. As I hope to demonstrate in the pages that follow, the present anatomy and physiology of the human brain and body reveal its evolutionary history, which, in turn, provides insights on the nature of the biologic bases of human cognition, language, and other aspects of human behavior. Some of the evolutionary-biologic connections may seem at first to be bizarre, such as the relation of the neural structures implicated in language to disorders such as schizophrenia, Parkinson's disease, and obsessive-compulsive disorder; to abilities such as virtuoso piano performances; and to everyday activities such as walking, dancing, or driving your car.

Roadmap to the Book

Chapter 1

This first chapter presents an overview. As the title of this book suggests, we can only move toward an evolutionary biology of language. A definitive account must await this book's successors. I also review some aspects of evolutionary biology that may be unfamiliar to some readers.

Chapter 2

Chapter 2 takes note of comparative studies that identify the "primitive" and "derived" features of human language. Primitive features are attributes found in related species that derive from a common ancestor. For example, most animals, including human beings, have five digits on each foot and hand. This presumably reflects the five digits of our common ancestor. In contrast, the hoofs of horses represent a "derived" feature, a unitary digit that reflects the particular line of descent of horses. We are related to early hominids who descended from an ancestral species similar in many ways to present-day apes. But apes don't walk and one derived human feature shared by all hominids is bipedal, upright locomotion and the supporting anatomy for this mode of locomotion. Therefore, we can trace the evolution of walking by means of anatomic studies.

Comparative studies of the linguistic abilities of living apes (see Chapter 2) show that they have limited lexical and syntactic ability. The anatomic similarities between apes and humans have been evident since the end of the seventeenth century (Tyson 1699); chimpanzees are the living species most closely related to human beings. Genetic analyses indicate that the last common ancestor of chimpanzees and humans lived 6 to 7 million years ago. About 96 percent of chimpanzee and human genes were present in the common ancestor. There is about a 1.2 percent difference between these genes in humans and chimpanzees due to gradual "drift," making the total difference about 5 percent (Britten 2002). For comparison, the human-mouse difference is about 41 percent, with the last common ancestor living about 75 million years ago (Mouse Genome Sequencing Consortium 2002). Thus, though chimpanzees are not living replicas of our common ancestor, they present a reasonable model for assessing the capabilities of the earliest hominids. Studies of chimpanzees that have been taught to use manual sign language (Gardner and Gardner 1969, 1973, 1984, 1994) or other manual systems (Savage-Rumbaugh, Rumbaugh, and McDonald 1985; Savage-Rumbaugh et al. 1986) show that they can communicate and think using words. Chimpanzees listening to spoken English can comprehend distinctions in meaning conveyed by simple syntax (Savage-Rumbaugh and Rumbaugh 1993).

If the common ancestor of chimpanzees and humans possessed the same, or superior, mental facilities as living chimpanzees, we must conclude that limited lexical, syntactic, and vocal abilities are primitives of human language. Chimpanzees and other apes do not appear to have complex syntactic abilities, nor do they acquire almost limitless vocabularies. However, what chimpanzees and other apes cannot do is talk! Chimpanzees and other primates appear to be capable of communicating referential information by means of a limited number of "fixed" calls (Slocombe and Zuberbuhler 2005a, b). However, they cannot readily coin new vocal words and speak. Following the principles of evolutionary biology, we must conclude that talking—voluntary, almost limitless, speech production—is a derived feature of human language.

Chapter 3

Chapter 3 discusses the singular contribution of human speech. Speech is a key element of human language, yielding a high data transmission rate. The rate at which information is transmitted by means of the "speech code" (Liberman et al. 1967) makes complex human language possible. Speech is the default medium of human language; written language was invented in only the last 10,000 years and still is not universal. Indeed children acquire most aspects of their native language before literacy. Without speech or alternative systems such as manual sign language, complex language would not be possible. We would forget the beginning of a complex sentence before reaching its end. And manual sign languages, which can take the place of speech, are recent inventions, dating back to the end of the eighteenth century.

Moreover, neurophysiologic studies suggest that manual sign languages make use of the neural structures that regulate speech, perhaps with the addition of activity in right hemisphere cortical areas that process visual signals (Newman et al. 2002). Gestural communication may have played a greater role in the early stages of hominid evolution (Hewes 1973; Lieberman 1984; Corballis 2002) and gestures retain a role in communication by hearing persons, typically accompanying speech in normal conversations (McNeill 1985). It is clear that manual gestures can suffice to transmit language, but it is equally clear that for hearing persons, speech has the advantage of keeping your hands free and not fixing vision on the signer.

The signal advantage that human speech has over all other acoustic signals is its speed. If you were to tape record and transcribe any normal conversation, you would need about 20 to 30 alphabetic characters or "phonemic" segments per second to write down the words. (In linguistic theories, a phoneme is essentially a sound contrast that differentiates words, such as /p/ versus /b/ in the words "pat" and "bat.") The high data transmission rate of speech might seem mysterious because the rate at which phonemes are transmitted during normal discourse exceeds the fusion frequency of the human auditory system. When nonspeech sounds, such as clicks or drum beats, are generated, the individual sounds merge into an indistinguishable buzz at rates exceeding 15 sounds per second. Yet we "hear" the sounds of speech, the phonemes, at rates of 20 to 30 per second. Research that started in the 1950s shows that a complex process of encoding and decoding accomplishes this feat (Liberman et al. 1967).

As we speak, the acoustic parameters that convey individual sounds are melded into syllabic-like units, transmitted at a slow rate (not exceeding seven segments per second). As we perceive speech, the phonemes are recovered by a neural decoding process that makes use of implicit knowledge of the constraints of speech production. If we were limited to the slow nonspeech transmission rate, we would run afoul of the memory limits of the human brain and forget the beginning of a long sentence before we heard its final words. Many attempts have been made to use other sounds to code the letters of the alphabet. They all are no better than the dots and dashes of Morse code. As is the case for the clicks that convey Morse code, a listener would have to bend all of his or her attention to the sequence of sounds. The meaning of the sentence is lost. This limit apparently follows from the fact, discovered by George Miller in his 1956 study, that we can hold 7±2 items in our roughly two-second-long short-term memory span.

Speech has a less obvious role when we attempt to comprehend the meaning of a sentence, or the fragments of sentences that often mark conversations. A sort of silent speech, which I discuss in Chapter 3, plays a part in our comprehending the meanings of words and syntax (which, to a great degree, cannot be untangled) by maintaining words in the neural computational space termed "verbal working memory" (a specialized short-term memory) in which sentence comprehension takes place (Baddeley 1986). In short, speech is an essential component of human linguistic ability. Recent studies show that speech sounds even have a significant advantage over the gestures of American Sign Language, which appears to be limited to a short-term memory span of five, plus or minus one, items (Boutla et al. 2004).

Chapter 4

Chapter 4 discusses some of the neural bases of human language in the broader framework of current studies of how biologic brains may work. It is in no way an attempt to present a comprehensive account of "how the brain works to make language possible." That goal may someday be achieved, but it is not yet within our grasp.

The chapter starts with some basic background information that will make the discussion of experimental findings accessible to nonspecialists. I note recent developments that suggest that the brain's "dictionary" entails access to the store of real-world knowledge that defines the meaning of words. Neural circuits linking cortex to the hippocampus appear to play an essential part in this process. But the primary focus in this chapter is on the creativeness of human language and cognition. The ability to reiterate or reprogram a finite set of automatic acts appears to be one of the keys to both articulate human speech and syntax, as well as to the flexibility and creativeness of human thought processes. Converging evidence from studies of aphasia, Parkinson's disease, oxygen deficits and other insults to the human brain, imaging studies of neurologically intact human subjects, and behavioral and neurophysiologic studies of other species point to neural circuits linking areas of the cortex with subcortical basal ganglia, thereby providing this reiterative capacity. Coupled with a vast memory store that can access the referents of an almost limitless number of words, the result is fully human linguistic ability. The take-home message is that the evolutionary root of this creative ability is motor control.

In this approach I am following in the footsteps of Karl Lashley, one of the pioneers of modern neuroscience. In 1951, Lashley pointed out the continuity between the organization of serial order in motor and cognitive acts. Anticipating theoretical linguists and mathematicians, Lashley recognized the complexity of language, particularly in "syntactic coordination." But Lashley did not believe that the syntax of language was a singular event; he noted similarities in other domains and the probable evolutionary antecedent of syntax—motor control. As Lashley pointed out:

> Temporal integration is not found exclusively in language; the coordination of leg movements in insects, the song of birds, the control of trotting and pacing in a gaited horse, the rat running the maze, the architect designing a house, and the carpenter sawing a board present a problem of sequences of action . . . each of which similarly requires syntactic organization. (Lashley 1951, p. 113)

The neural and behavioral evidence that would have fleshed out Lashley's claim was not available fifty years ago, but subsequent studies show that he was correct. I present evidence that the evolution of the neural bases for adaptive motor control was the preadaptive basis of human syntactic ability and cognitive flexibility. The evidence derives from hundreds of independent experiments and observations by many research groups.

This body of evidence also shows that the traditional Broca-Wernicke theory for the neural bases of human language is incorrect. Language is not centered in these hypothetical language areas of the cortex of the human brain. In fact, as we shall see, language can be retained after these areas are destroyed.

Chapter 5

Chapter 5 continues on the same track. The evolution of the neural mechanism initially adapted for motor control appears to be the key to the evolutionary process that yields our ability to create a potentially infinite number of actions or thought processes from a finite number of stereotyped motor sequences, words, or thoughts.

Linguists may argue that syntax differs from motor control processes outside the supposed domain of linguistics and does not simply involve changing the order in which words occur. This is true. Words always have specific grammatical functions, constrained by semantic and syntactic criteria specified in the brain's neural "dictionary." For example, understanding the meaning of a statement such as "the old man saw the boy who was gangly fall down" entails knowing that the word "who" refers to the boy. The syntactic structure of the sentence and the grammatical function of the word "who" restrict its reference. But similar constraints mark motor control. The act of walking, for example, involves positioning your foot to achieve heel strike at the precise moment of contact with a floorboard or the irregular surface of a country path. Your heel must flex at the proper moment; and that moment is different depending on whether you are walking or running, your posture, and whether the path is stony or smooth. There is a "syntax" of walking.

The discovery of the FOXP2 gene provides the potential for dating the evolution of the human brain and fully human language. FOXP2, the so-called language gene, is a regulatory gene that governs the embryonic development of the subcortical neural structures that yield the reiterative capacity necessary to produce articulate speech as well as syntax and cognitive flexibility. An anomalous version of this gene is responsible for the orofacial apraxia, speech, and linguistic and cognitive deficits of afflicted members of a large extended family (Vargha-Khadem et al. 1995, 1998, 2005; Lieberman 2000, 2002; Watkins et al. 2002; Lai et al. 2001, 2003). The chimpanzee version of this gene, foxp2, differs from the human version, and an analysis of the evolution of FOXP2 indicates that it achieved its human form within the past 100,000 years (Enard et al. 2002).

In short, the premise that is explored and documented in Chapters 4 and 5 is that cognitive flexibility and syntactic ability devolve from adaptive motor control. And it will become apparent that some aspects of human linguistic and cognitive ability in language are present in reduced degree in other species. Therefore, I do not adopt the hermetic stance taken in many discussions of the evolution of human language—which is to assume that language is a module disjoint from other aspects of human and animal behavior. As we shall see, it is possible that neural adaptations that facilitated walking and running may have triggered the evolution of the neural circuits that now confer human linguistic ability. Moreover, the evolution of the neural bases of human cognitive and linguistic ability most likely extends far back into the depths of time, well before the first hominids appeared some 6 or 7 million years ago. Some of these neural structures, the basal ganglia, can be traced back to amphibians who appeared hundreds of millions of years ago, and they are evident in contemporary frogs. These ancestors of present-day frogs obviously did not possess language; their basal ganglia most likely regulated adaptive motor acts.

Chapter 5 also briefly reviews theories for the evolution of the human brain and language. It points out the distinction between factors that may have directed natural selection for language and the resulting biologic mechanisms. Factors such as the lateralization of the brain (which also occurs in frogs) cannot explain human cognitive capacities. Other factors, such as the absolute size of the brain, most likely entered into the evolution of the human brain.

Chapter 6

Chapter 6 returns to the question of the evolution of the anatomic bases of human speech. Some aspects of speech reflect the continuity of evolution and there is a need for careful quantitative studies of vocal communication in other species. The airway above the larynx, the supralaryngeal vocal tract (SVT), shapes the sounds produced by the larynx to form different vowels and consonants.

The evolution of the human tongue is the key element in the formation of the species-specific human SVT, which allows us to produce the full range of human sounds, including the supervowel [i] (the vowel of the word "see"). Victor Negus (1949) pointed out the fact that apes and human newborns have a very different tongue from adult humans; they inherently cannot produce the full range of human speech sounds (Lieberman 1968a; Lieberman et al. 1972;

Lieberman, Crelin, and Klatt 1972; Lieberman, Klatt, and Wilson 1969; Carre, Lindblom, and MacNeilage 1995). A certain degree of confusion and controversy has developed concerning the anatomic basis for producing the sounds of human speech. Similar confusion surrounds the significance of being able to produce the sound [i]—a sound that enhances the process of speech production and perception that provides its high data transmission rate. Misinterpretations of the linguistic abilities of extinct Neanderthal hominids are common. Speech was almost certainly present in Neanderthal hominids, but they were a species who probably lacked tongues that could have produced the full range of human speech.

Chapter 7

Chapter 7 presents some suggestions for advancing linguistic research. A better understanding of the mechanisms of sound change and phonology could be developed if researchers were to take the wise counsel of Roman Jakobson, one of the foremost scholars of the last century, whose linguistic studies took into account findings on how speech is actually produced. The sounds of speech, for example, cannot be uniquely related to invariant motor commands, tongue positions, or other articulatory gestures, as linguists have often assumed. A mixed system involving both auditory and articulatory factors instead appears to characterize the sound pattern that conveys words. Appropriate consideration of the physiology of speech production and perception factors could lead to insights on the nature of phonologic processes.

Moreover, there is a lesson in the study of evolution for linguistic research on syntax. The studies discussed in Chapters 4 and 5 demonstrate that the neural structures implicated in motor learning and motor control also regulate cognitive processes such as those involved in comprehending the syntactic distinctions that convey the meaning of a sentence. It cannot be the case that the physiology of the neural systems of the brain differs for cognitive, linguistic, or motor control tasks. Although some aspects of motor control such as breathing are innately specified (Bouhuys 1974; Langlois, Baken, and Wilder 1980), most motor acts, even ones that we execute without conscious thought, are learned. Hence, on these grounds alone, we can dismiss the supposition that an innate Universal Grammar exists in the brains of all human beings that specifies the detailed syntax of every language that ever was, or will be, spoken to the end of time.

Moreover, though many aspects of the physiology and the operations of the neural network that sequences motor acts and thought patterns are unknown, it is clear that they do not resemble the sequential, algorithmic operations typically used to describe syntactic processes. Many linguists study phonology, the sound pattern of language, but the methodology employed in these studies usually mirrors that used to describe syntax. Sequential, abstract algorithms, or rules, are devised that attempt to describe utterances. Linguistic studies of meaning, called semantics, have also tended to employ the abstract rule-governed model provided by syntax. As Croft (1991) points out, this model has yielded few insights into the nature of language. No comprehensive description of the syntax of any natural language has been achieved, despite intense efforts by hundreds of linguists over the course of more than four decades. The findings of many independent experiments have led to some understanding of the neural bases of motor control, which cannot be described by serial algorithmic processes. These insights could inform linguistic theories that attempt to describe the "rules" that people use to comprehend sentences. Biologic brains simply do not make use of the sequential, algorithmic operations that typify most contemporary linguistic research.

Chapter 8

In closing we return to the start. We are only at the threshold of knowledge on how biologic brains work. I present some open questions and suggestions for research that may advance the state of human knowledge.

The primary argument of this book is that the biologic bases of linguistic as well as cognitive ability cannot be studied in isolation from other aspects of human behavior or the behavior of other species.

Charles Darwin's Toolbox

Virtually everyone working on the biologic bases and evolution of human language would acknowledge that we are following the footsteps of Charles Darwin. However, it is easy to underestimate the scope of Darwin's achievements.

Although Darwin would doubtless have been amazed at the techniques that have allowed us to obtain data and insights on the processes involved in evolution, ranging from studies of genes to research on tongue movements in speech and swallowing, the tools that he used to cut through the layers of time are still sharp. Darwin (1859) took account of four related factors that could yield "new" species: (1) "the struggle for existence," (2) variation, the feedstock for natural selection, (3) natural selection, which enhances the probability of an individual's having more surviving progeny, and (4) the fortuitous chance events that channel evolution into new directions—a process termed preadaptation by Mayr (1982) and, with a somewhat different focus, exaptation (Gould 1977; Gould and Lewontin 1979).

Variation, Natural Selection, and the Struggle for Existence

Darwin's "struggle for existence" recognizes the fact that life is uncertain and precarious. He was influenced by the political philosopher Thomas Malthus, who in the early years of the nineteenth century predicted an impending disaster. The population of Europe, which had remained static for centuries (Darnton 1985), was rapidly expanding. Malthus raised the alarm: food supplies would not be sufficient to feed everyone and only the fittest would survive. Malthus failed to anticipate advances in agriculture that would outpace population growth; but Darwin realized that, in the state of nature, resources are finite. For most species, and that included virtually all humans until comparatively recent times, life is chancy. Until the introduction of modern sanitation in the nineteenth century, few children survived though many were born. Survival still is uncertain for much of the world's human population in this, the twenty-first century. The struggle for existence also manifests itself in the fact that most species that have lived on earth are extinct. An estimated 500 million species are extinct, compared to a few million living species (Jacob 1977).

Natural selection, which follows from the struggle for existence, is the evolutionary mechanism most often equated with Darwinian theory. Darwin, in a presciently modern note, said that:

> A struggle for existence inevitably follows from the high rate at which all organic beings tend to increase. Every being, which during its natural lifetime produces several eggs or seeds, must suffer destruction during some period of its life, and during some season or occasional year, otherwise, on the principle of geometrical increase, its numbers would so quickly become so inordinately great that no country could support the product. Hence, as more individuals are produced than can possibly survive, there must in every case be a struggle for existence, either one individual with another of the same species, or with individuals of distinct species, or the physical conditions of life. It is the doctrine of Malthus applied with manifold force to the whole animal and vegetable kingdoms; for in this case there can be no prudent restraint from marriage. There is no exception to the rule that every organic being naturally increases at so high a rate, that if not destroyed, the earth would soon be covered by the progeny of a single pair. Even slow-breeding man has doubled in twenty-five years, and at this rate, in a few thousand years, there would literally not be standing room for his progeny. (1859, p. 63)

The Darwinian struggle for existence is not Tennyson's vision of "Nature red in tooth and claw" or the neo-Darwinian views endorsed by nineteenth-century captains of industry or present-day CEOs to justify the accumulation of wealth and influence. Darwin's words are clear:

> I should premise that I use the term Struggle for Existence in a large and metaphorical sense, including dependence of one being on another, and including (which is more important) not only the life of the individual, but success in leaving progeny. (1859, p. 62)

Darwin noted the subtle relationships that hold between animals, plants and human activity, anticipating current ecological models that take account of interactions between the environment and human endeavors. Darwin most likely would have cited recent proposals concerning the evolutionary consequences of mutual altruism (Axelrod and Hamilton 1981), group cohesion (Dunbar 1993), and speech as a reproductive isolating mechanism (Barbujani 1991; Barbujani and Sokal 1990, 1991; Lieberman 1992) as factors that bear on the struggle for existence. Many factors enter into the equation that yields success in the struggle for existence and biologic fitness—an individual's reproductive success.

Natural selection is a filtering mechanism; its feedstock is the ever-present genetic variation that marks the individuals that compose a species. Darwin was unaware of how variation occurs at the genetic level, but he was clear concerning the role of variation:

> Owing to this struggle for life, any variation, however slight and from whatever cause proceeding, if it be in any degree profitable to an individual of any species, in its infinitely complex relations to other organic beings and to external nature, will tend to the preservation of that individual, and will generally be inherited by its offspring. The offspring, also, will thus have a better chance of surviving, for of the many individuals of a species which are periodically born but a few number can survive. I have called this principle, by which each slight variation, if useful, is preserved, by the term of Natural Selection. (1859, p. 61)

The Darwinian model thus is, as Ernst Mayr (1982) notes, a population model. Natural selection can work if, and only if, variation is present in the individuals that comprise a species. What's real is the pool of variation in individuals that differ somewhat but that comprise an isolate capable and willing to mate and produce progeny. As Ernst Mayr pointed out in his introduction to the facsimile edition of *On the Origin of Species* (1859, 1964), the Darwinian model is the antithesis of Platonic, essentialistic models such as Chomsky's (1976, 1980a, b, 1986, 1995) "knowledge of language," which is possessed by an idealized speaker-hearer.

Linguistic research generally focuses on idealized utterances (usually written texts) produced by hypothetical, disembodied speaker-hearers that reflect competence rather than actual utterances. To Chomsky and like-minded linguists, phenomena that deviate and are not consistent with a current theory are "production effects" or part of the "peripheral" grammar. In short, variation is irrelevant noise, generally ignored by theoretical linguists.

Darwin's paradigm for biologic research rests on neither pure inference nor deduction. Data derived from initial observations or experiments provides the basis for a theory, whose predictions are then tested against data from subsequent observations and experiments (Mayr 1982). The theory then is modified in the light of these subsequent data. In one of his experiments, Darwin tested his theories concerning variation and selection using pigeons. He conducted a reverse-engineering exercise in which he "debreeded" pigeons. Pigeon fanciers have over the ages bred domesticated breeds that, to anyone used to the common rock pigeons that today roost on buildings, would seem to be different species. Darwin's pigeon experiment started with breeds of domesticated birds that so differed that

> if shown to an ornithologist, and he were told that they were wild birds, would certainly, I think, be ranked as well-defined species. Moreover, I do not believe that any ornithologist would place the English carrier, the short-faced tumbler, the runt, the

> barb, pouter, and fantail in the same genus; more especially as in each of these breeds several truly-inherited sub-breeds, or species as he might have called them, could be shown him.
>
> Great as the differences are between the breeds of pigeons, I am fully convinced that the common opinion of naturalists is correct, namely, all have descended from the common rock pigeon (*Columba livia*). (1859, pp. 22, 23)

Darwin reasoned that the different breeds had been produced by artificial selection acting on the pool of variation that existed in the ancestral rock-pigeon stock. He believed that breeders, over generations of pigeons, had narrowed the degree of genetic variation in particular breeds by continually mating birds who exhibited a particular variation. Breeding long-beaked males with long-beaked females, pigeons who tumbled in flight with other tumbling birds, and so on had produced "true-breeding" varieties—males and females in whom the degree of genetic variation was so limited that their offspring reliably had long beaks, tumbling in flight, pout, and so on. Darwin simply allowed fancy true-breeding pigeons to mate with different true-breeding pigeons, such as pouters with runts and tumblers with homing pigeons. The result of mixing the genetic stew was to restore the ancestral pool of genetic variation, yielding rock pigeons.

Chance-Preadaptation-Exaptation

As is the case for other aspects of life, chance plays a role in biologic evolution. The "proximate" logic of evolution can be seen in the annals of human history. The present English royal family descends from a petty German princeling rather than from the Norman conquerors of Saxon England because of events in 1689. After Charles II fled, Parliament decided that a weak, pliable monarch who would not interfere with their decisions would be a suitable "ruler." A German who could not even speak English ultimately became king.

Chance is an essential element in the Darwinian model. Darwin (1859) stressed the gradual role of natural selection. As Darwin states time and again:

> Natural selection can act only by the preservation and accumulation of infinitesimally small inherited modifications, each profitable to the preserved being . . .
>
> Natural selection can act only by taking advantage of slight successive variations; she can never take a leap, but must advance by the shortest and slowest steps. (1859, pp. 95, 194)

But Darwin was not a fool. Leaps in evolution do occur; we are not highly developed fish swimming in the briny deep. Darwin knew that he had to account for transitions such as fish evolving into terrestrial, air-breathing animals. Gradual natural selection could act to perfect the behavioral attributes of fish, but fish do not breathe air so how could lungs have come into being? Darwin's solution, based on biologic fact, was the role of the environment and chance events—he observed the evolution of the air-breathing lung fish of South America and noted that:

> [an] organ might be modified for some other and quite distinct purpose . . . The illustration of the swimbladder in fishes is a good one, because it shows us clearly the highly important fact that an organ originally constructed for one purpose, namely flotation, may be converted into one for a wholly different purpose, namely respiration. (1859, p. 190)

The evolution of the lungs was triggered by chance events. Lungfish, fish that could transfer air back from their gasping mouths to their bloodstream, survived when they were stranded in dry river beds. An ecological change in habitat, transitory terrestrial life, resulted in natural selection favoring the survival and reproduction of the individual fish who could force air back through their swim bladders, oxygenating their blood. The "preadaptive" basis of the lung was the swim bladder.

Although some controversy now exists as to whether swimbladders first developed from lungs or the other way round, the principle holds. Since Darwin's time, it has become apparent that this process plays a crucial role in the evolution of new modes of behavior.[1] The hinge bones of the reptilian jaw, for example, serve to increase the sensitivity of the mammalian middle ear. That's why an earache can result from grinding your teeth; the pain pathways still preserve the wiring diagram that existed before the time of the therapsids, the mammal-like reptiles that were the ancestors of all mammalian species some 200 million years ago. Mammals must maintain contact with infants. Hence all young mammals produce isolation cries, and the increased sensitivity of the mammalian auditory system that results from the mechanical sound-amplifying system bones of the middle ear clearly increases biologic fitness. Milk glands, likewise, derive from sweat glands (Long 1969).

Therefore, as later studies note (Gould and Eldridge 1977), the course of evolution is not a slow, steady progression. Ecological changes initiated by changes in climate and the "infinitely complex relations of organic beings" create new opportunities. Anatomic structures and neural mechanisms that evolved to serve one function just happen to be useful in adapting to changed circumstances. They serve as branch points, opportunities for natural selection to modify "old" organs to carry out a new function that enhances biologic fitness in a changing world. In other instances, dramatic changes can arise by mutations acting on regulatory genes that act on the biologic substrate that regulates an existing behavior (Alberch 1989; Arthur 2002). Mutations that enhanced existing patterns of hominid behavior may have played a critical role in shaping human cognitive and linguistic ability. However, we have to remember that the process of modification itself can be gradual when measured in geological time. As Carroll and his colleagues, when reviewing the contributions of molecular biology to the study of evolution, note:

> Dramatic morphological changes, such as the fin-to-limb transition in vertebrates . . . involved many regulatory genetic, developmental, and anatomic modifications that could not and did not evolve instantaneously. Instead, these structures were sculpted by regulatory evolution over millions of years. (Carroll, Grenier, and Weatherbee 2001, p. 191)

We shall see that both preadaptation and genetic mechanisms that are expressed in ontogenetic development played a role in shaping the anatomy involved in speech production, as well as the neural mechanisms that yield fully human linguistic and cognitive ability. The anatomy of the human face, mouth, tongue, and upper airway involved in breathing and eating has been modified to enable us to produce sounds that reduce errors in speech perception. The cost has been increased risk of choking to death on food. The changes from the common ancestral anatomy of apes and humans, evident in present-day chimpanzees, produce these negative effects on biologic fitness. Darwin, again, appears to have first noted the effects on choking:

> The strange fact that every particle of food and drink which we swallow has to pass over the orifice of the trachea, with some risk of falling into the lungs, notwithstanding the beautiful contrivance by which the glottis is closed. (1859, p. 191)

Embryology and Evo-Devo

To Darwin, ontogenetic development was another window on the course of evolution. During his pigeon experiments Darwin, ever the careful observer,

> compared young pigeons of various breeds, within twelve hours after being hatched; I carefully measured the proportions (but will not here give details) of the beak, width of mouth, length of nostril and of eyelid, size of feet and length of leg, in the wild

> stock, in pouters, runts, barbs, dragons, carriers, and tumblers . . . when the nestling birds of these several breeds were placed in a row, though most of them could be distinguished from each other, yet their proportional differences in the above specified several points were incomparably less than in the full grown birds. (1859, p. 445)

Darwin concludes that the morphology of the embryo

> reveals the structure of its progenitor. In two groups of animal, however much they may at present differ from each other in structure and habits, if they pass through the same or similar embryonic stages, we may feel assured that they have both descended from the same or nearly similar parents, and are therefore in that degree closely related. Thus, community in embryonic structure reveals community of descent. (1859, p. 449)

Haeckel's (1866, 1896) "law"—that ontogeny recapitulates phylogeny, which overstates this finding—parallels Darwin's observations. Von Baer (1828, c.f. Wimsatt 1985) had earlier pointed out that features appearing earlier in fetal life appear to reflect characteristics shared by species related to the organism in question. The features that differentiate a species from ancestral and related living species tend to be expressed later. The value placed on inferences on evolution derived from ontogeny has fluctuated. As Gould's 1977 book *Ontogeny and Phylogeny* noted, controversy still surrounds this issue. However, apart from the merits of Von Baer's or Haeckel's laws, insights on evolution can be gained through the study of ontogenetic development. Few scholars, for example, argue with the premise that the embryonic skulls of dinosaurs shed light on the evolution of the cranial features of sauropod dinosaurs 80 million years ago (Chiappe, Salgado, and Coria 2001).

Advances in molecular genetics may perhaps convince all but the most entrenched skeptics that insights on human evolution can be gained by studying development. A new scientific field, evolutionary development (evo-devo), traces the effects of genes regulating development and speciation. As Carroll, Grenier, and Weatherbee (2001) point out in their overview of recent molecular genetic studies, many human genes can be traced back to fruit flies. But it is clear that we are not large flies, and many of the significant differences between flies and people appear to involve genes that regulate ontogenetic development. Although the genes of extinct animals generally are not available for study, molecular biologists can reconstruct the evolution of gene families by studying animals considered to be living fossils. Living species that are close to the most deeply branching members of an evolutionary tree may serve to indicate the genetic "toolkit" that was present before the branch point and the evolution of successor species.

Some of the critical biologic attributes that confer human linguistic and cognitive ability may derive from changes in genetic regulatory mechanisms that occurred 6 or 7 million, 2 million, and 500,000 years ago, and during the past 100,000 years. The 6 or 7 million-year date reflects the time at which human beings and living apes, particularly chimpanzees, shared a common ancestor (Sarich 1974; Stringer 1992; Wood 1992; Wood and Collard 1999). The genetic mechanisms that may have been involved in this initial divergence that led to present-day apes and humans are unknown, but chimpanzees, in a meaningful sense, are living fossils. Chimpanzees retain many of the skeletal and cranial features of the earliest-known fossil hominids. Genes such as ASPM may have been factors in the enlargement of the brain in early members of the genus Homo some 2 million years ago (Zhang 2003; Evans et al. 2004). The 500,000-year date is the time at which analysis of DNA recovered from Neanderthal fossils indicates divergence from the lineage that ultimately resulted in anatomically modern human beings (Krings et al. 1997; Ovchinnikov et al. 2000; Adcock et al. 2001).

As already noted, two of the biologic attributes that allow us to produce human speech—specialized anatomy and a brain capable of executing the complex muscular maneuvers that are necessary to talk—are discussed in detail in this book. These attributes are absent in living apes and most likely were absent or present in reduced degree in some archaic hominids. The 100,000-year date is probably close to the outer bound for the appearance of modern humans (Wood and Collard 1999; Templeton 2002; Clark et al. 2003; White et al. 2003). The human face restructures over the course of the first years of life from the morphology present in Homo erectus and Neanderthals. This process and the subsequent restructuring of the tongue may reflect changes in regulatory genes. The result is that humans have a species-specific SVT that is capable of producing all speech sounds (Negus 1949; Lieberman and Crelin 1971; Fitch and Giedd 1999; Lieberman and McCarthy 1999; Lieberman et al. 2001; Vorperian et al. 2005). I return to discuss, in some detail, both speech-producing anatomy and the aspects of the brain that are regulated by the FOXP2 gene.[2] Darwin would have applauded the use of the techniques of molecular genetics to yield insights on human evolution.

CHAPTER 2

Primitive and Derived Features of Language

THEORIES CONCERNING the evolution of language often start with the premise that human language is disjoint from the communications system of other species. It is also the case that contemporary scholars usually have little contact with animals other than their pets. In contrast, Charles Darwin, an English country gentleman, observed and interacted with a wide range of animals on a daily basis. Lacking contact with animals, one can suppose that they are incapable of transmitting any referential information and that their communications primarily convey emotional states and a limited repertoire of genetically specified calls, gestures, or facial expressions that serve to signal food, the presence of danger, potential mates, and so on. And although linguists have long realized that human linguistic ability involves the interplay of different components that hinge on different biologic capacities, there is a tendency to treat an animal's communication system as a simple unitary system.

This chapter identifies some of the aspects of human linguistic ability that may be present in other species. This exercise is necessary if we are to identify the missing elements that may characterize human language. A full treatment of animal communication is not my intent. As the previous chapter noted, human linguistic ability devolves from three basic capacities: (1) lexical capacity, the ability to learn and use words, (2) morphologic and syntactic processes that entail conveying meaning by systematic local modifications of words or of words that form sentences, and (3) phonetic and phonologic processes that produce and modify the sounds of speech that convey words and sentences. Other factors, such as turn-taking and social interaction, enter into human communication, but these elements are generally considered to be the core properties of human linguistic ability. I hope to demonstrate that some of these components are present in reduced form in other species, including some far removed from humans. In short, the differences between human linguistic ability and the observed language skills of other living animals, though real, are not total. The conclusions that I hope will emerge are (1) that hominid "protolanguage" never existed and (2) that speech, an almost limitless memory capacity coded by words, and complex syntax are among the "derived" features of human language whose evolution we must account for.

In some ways the problem before us is similar to that which would be faced by a Martian scholar in 2006 attempting to trace the "evolution" (invention and development) of automobiles. The scholar might discover that some of the bits and pieces, such as wheels, could be traced back to the chariots described by Homer; that internal combustion engines were present in 1880 and pneumatic tires in nineteenth-century bicycles; but that nothing resembling a car existed until the very end of the nineteenth century. But the Martian would first have to identify the basic components of a contemporary car to trace the gradual development of pneumatic tires in bicycles, diesel and spark-ignited engines in boats, spring systems in Napoleon's carriages, effective brakes in mid-nineteenth-century railroad cars, electric motors, and so on—components that, when put together in a functional system, resulted in the sudden emergence of cars.

Words

What Words Convey

A language without words doesn't exist. Words convey concepts. Some words, usually "technical" specialized words, refer to specific things or actions or classes of things and actions. E-mail is one example. But the meaning of a word almost never is precisely equivalent to a thing, a set of things, or a property of a set of things. When pushed, the meaning conveyed by most words becomes imprecise, fuzzy, and subject to the influence of context, culture, and time.

However, equating the meaning of a word with a specific thing or action is an exercise that pervades the linguistic and philosophic literature. Jonathan Swift parodied Liebniz's (1949 translation, p. 387) presumed equivalence between words and objects in *Gulliver's Travels.* The learned scholars of the Academy of Laputa attempted to correct the inherent imprecision of words. The meaning of a word never is precisely the same for any two people. The solution thus was, as Swift noted,

> a Scheme for abolishing all Words whatsoever . . . since Words are only names for Things, it would be more convenient for all Men to carry about them, such Things as were necessary to the particular Business they are to discourse on . . . many of the most Learned and Wise adhere to the new Scheme of expressing themselves by Things, which hath only this Inconvenience attending it; that if a Man's Business be very great, and of various Kinds, he must be obliged in Proportion to carry a greater Bundle of Things upon his Back, unless he can afford one or two strong Servants to attend him. I have often beheld two of these Sages almost sinking under the Weight of their Packs, like Peddlers among us, who when they meet in the Streets, would lay down their Loads, open their Sacks, and hold Conversation for an Hour together; then put up their Implements, help each other to resume their Burthens, and take their Leave. (1726; 1970, p. 158)

As Swift realized, words are not simply labels for things or actions; the learned sages would have had problems even if they had had trucks rather than sacks. Futile attempts to define words in terms of the properties of things persist, though the learned exercises involve predicate logic instead of sacks. Bertrand Russell, for example, spent a good part of his productive life in a vain attempt to refine the Laputan solution. Russell's autobiography (1967) notes the problems that beset him when he attempted to show that words can be defined in terms of logical operations that ultimately refer to things. The problem is that words, even "simple" words that we "know" refer to material objects, cannot be defined by reference to material objects. Consider a simple word like "table." We, of course, are not referring to a particular table in our mental dictionary when we use this word. But if we think about the way in which we would use "table" in even a few sentences, it is not possible to define that word, as Russell attempted to do, in terms of some procedure that attempts to capture the essence of tableness by partitioning the universe into a set of objects that *are* tables and a set of objects that are *not* tables. The problem is inseparable from any attempt to relate words directly to things. As Jacob Bronowski notes:

> you cannot make a single general statement about anything in the world which really is wholly delimited, wholly unambiguous, and divides the world into two pieces . . . ; you cannot say anything about a table or chair which does not leave you open to the challenge, "Well, I am using this chair as a table . . . it is now a table or chair." . . . the world does not consist simply of an endless array of objects and the word "table" was not invented in order to bisect the universe into tables and non-tables. And if that is true of "table," it is true of "honor" and it is true of "gravitation," and it is true, of course, of "mass" and "energy" and everything else. (1978, pp. 106, 107)

We know when we are talking about a real table. A table has legs; we can sit down before a table, put things on a table, and recognize one when we see one. We, moreover, can generally tell when something is not a table. It should therefore be possible to devise a simple definition of a table. We could adopt Russell's solution and simply state that the word "table" refers to the class of things that people consider to be tables and does not refer to the class of things that are not considered to be tables. Suppose we did this and enumerated a long list of the objects that were tables and the objects that were not tables. The list could refer to the meaning of the word "table" only in a particular setting. If we are sitting and eating dinner at a table, the set of possible tables does not include chairs. But as Bronowski pointed out, the set of objects that are tables must include chairs if the setting is a buffet dinner at which I say, "I'll use this chair as a table."

You may be thinking that this is a silly discussion; we all know what we mean when we use the word "table." We know that the word "table" has a floating set of references; but we also know that there is some limit to its range. The problem derives from using discrete formal logic to capture both the precision and the fuzziness conveyed by a word. We can use the word "table" to refer to all manner of things that have some property of tableness in a particular setting. The degree to which we each ascribe the property of tableness to something in any setting will vary. And the quality of tableness will change for each of us. What seems to be a table at some time in some place may not be a table to us at another time in another place.

The fact that language is inherently ambiguous and uncertain can lead to problems, but it is also the source of the power of language. The ambiguity and uncertainty of the linguistic system most likely matches the richness of our inner psychological conceptual framework. As Bronowski points out, the concept conveyed by the word "atom" would have been different for a nuclear physicist and a taxi driver in 1982. The concept, moreover, would have been understood differently by nuclear physicists in 1922 and in 2002—and it will probably be understood very differently yet again by a physicist in 2100. The total context of a word always plays a part in specifying its meaning. Words and sentences are almost always ambiguous; they are never precise unless we take into account the context of a communication. This lack of precision is not a deficiency of language; it is rather an aspect of language that mirrors human thought. As Cassirer (1944) and Bronowski (1971) noted, the inherent imprecision of any linguistic statement keeps the frontier of human language open. We must always creatively interpret an utterance: the new interpretation always has the potential of achieving new insights. Thought is not static; language mirrors thought, whether artfully in the brilliant word games of Nabokov or crudely in political rhetoric. A linguistic communication always is inherently ambiguous to some degree and must be interpreted.

The inherent ambiguity of human language is furthermore shared by all systems of mathematical logic that involve self-reference and arithmetic enumeration. As Bronowski (1971) noted, the theorems of Godel and Tarski show that all systems of mathematical logic that approach the power of human language are inherently ambiguous. Croft (1991) again showed that the attempts of logicians, including Russell, Chomsky, and Montague (1974), to develop formal systems of logic that are unambiguous probably are inherently flawed. But linguist-logicians have not thrown in the towel. Here we see the inherent ambiguity of language—the meaning conveyed by this phrase depends on you and I having some common contextual knowledge. If you haven't a clue about the meaning of "throwing in the towel," you haven't any direct or indirect knowledge of boxing.

The Neural Bases of Words

A word seems instead to invoke a set of memory traces that reflect the direct and indirect events of a person's life and the conceptual inferences that may be drawn from these memory traces. When I hear, read, or think of the word "elephant," I can relate it to an image in my mind's eye of an elephant or any number of the properties of an elephant. Neuroimaging experiments support the contention that I am "seeing" an elephant or some property of an elephant when I read or hear that word. Martin et al. (1995a, b) used positron emission tomography (PET) to map brain activity in various parts of the cortex while subjects listened to and thought about a word. PET can track metabolic activity in particular regions of the brain; the technique (discussed briefly in Chapter 4) is cumbersome and has comparatively low spatial resolution, but it does provide objective data on the neural bases of thinking. Martin and his colleagues found that primary visual cortical areas associated with the perception of shape and color are activated when we think of or read the name of an animal such as an elephant. These neuroimaging experiments reveal that the same regions of "sensory" temporal cortex that are the neural bases of perceiving an elephant also serve as stores of the "semantic" information that characterizes the word "elephant" in the brain's dictionary. The same cortical areas are active when a person sees an object, when the word that refers to the object is read or heard, or when a person thinks about the concept conveyed by the word. These effects are not limited to native speakers of English. Martin and Chao (2001) review independent neuroimaging studies for subjects speaking different languages that show similar results.

The subjects in these experiments, which were conducted in the suburbs of Washington, D.C., most likely had never seen an elephant outside of a zoo. The cortical motor areas of the brain of an Indian mahout, who works with elephants, would also probably be activated by the word "elephant." The imaging studies of Martin and his colleagues reinforce the premise that the knowledge "coded" in a word involves accessing the neuroanatomic structures and circuits that constitute the means by which we attain and/or make use of the knowledge coded by that word. The subjects in these experiments were asked to either name the color associated with an object or word (e.g., *yellow* for a pencil) or state the action associated with the word or object (e.g., *write* for a pencil). The parts of the brain involved in the perception of the objects or actions that the words conveyed were activated.

As Martin et al. (1995a) note, "Generation of color words selectively activated a region in the ventral temporal lobe just anterior to the area involved in the perception of color, whereas generation of action words activated a region in the left temporal gyrus just anterior to the area involved in the perception of motion." Their PET data show that primary motor cortex, implicated in manual control, is activated when we think of the way we use an object. Subjects who were asked to name pictures of tools also activated areas of motor cortex. The fact that areas of the brain that are in use when you use a hammer are activated when you think of a hammer or read the word "hammer" suggests that the brain's "dictionary" codes the uses of a hammer as well as its image. In other words, the real-world uses of a hammer are part of the brain's representation of the word "hammer."

It is significant that the areas of sensory cortex activated in these experiments are multisensory. Other neural circuits supported in these regions of cortex are implicated in tactile sensation and audition (Ungerleider 1995). Cross-modal effects clearly exist. Rizzolatti et al. (1996) found activation in "mirror neurons" in premotor cortex when monkeys *viewed* actions. These mirror neurons fire motor regions of the cortex when monkeys view actions or hear sounds that these actions produce (Rizzolatti and Arbib 1998; Kohler et al. 2002). For example, mirror neurons fire when a monkey sees hands tearing a piece of paper or hears the resulting sound. There may be no clear distinction between the neural mechanisms involved in storing nonlinguistic concepts in our mind-brain and those implicated in perception. Neurophysiologic data, for example, show that the structures of the brain that we use for first seeing an event or object are part of our memory. Brodmann's area 17, an area of cortex associated with early stages of visual perception, is activated when subjects are asked to image simple patterns (Kosslyn et al. 1999). Independent experiments show other regions of temporal cortex that are activated when people recognize complex objects and faces or recall memories of visualized events.

Complex neural circuits involving activity in both frontal and posterior regions of the brain appear to be involved in both forming and retrieving the neural memory "traces" that constitute the meaning of a word. As we shall see in Chapter 4, neural circuits that involve connections from prefrontal regions of the cortex and the subcortical hippocampus appear to be critical elements of the neural system that accesses the meaning of words from the brain's dictionary. Damage to prefrontal-to-hippocampal circuits may be the basis of Alzheimer's dementia in which semantic knowledge is lost (Kirchoff et al. 2000; Velanova et al. 2003; Wheeler and Buckner 2003).

The associations connected with a word also go far beyond general characteristics. Words evoke specific personal memories as well as general knowledge. When I hear or think of the word "bicycle," I can recall my father holding me upright as I first attempted to ride. The brain's dictionary and the store of words therein constitute a recollection of times past that is beyond the descriptive powers of formal logic. How can the meaning of "love" be described by means of predicate logic? That limit also presently exists for neuroimaging techniques. What can we say about the neural bases of words such as "love" or "honor"? That also goes for the words that convey emotion. It's clear that human beings do not utter primal cries when they are angry, happy, annoyed, and so on. We express our emotions by means of language. Do the neural bases of these utterances differ from referential speech? We do not know.

Nor have neuroimaging techniques been used to study what human words "mean" to animals. It would be instructive to determine the neural activation patterns of dogs who understand words such as "out" and "biscuit." All of the dogs that have been part of my life learned the meaning of these words without benefit of any formal instruction. My family had to spell out the word "out" to avoid having one of our 40 kg Briards racing to the door. The circus dog studied by Warden and Warner (1928) understood 50 English words. A contemporary German border collie raised in a German-speaking home understands more than 200 words (Kaminski, Call, and Fischer 2004). The collie, a pet living with a young family, has learned the names of children's toys. Moreover, as is the case with a human child, he learns the name of a new object after a single exposure and follows simple verbal instructions, such as bringing an object to a particular person. I would predict activation patterns similar to those noted for humans. The chimpanzee language experiments that I discuss in this chapter demonstrate beyond reasonable doubt that apes comprehend spoken words and sentences. Are cortical areas that are homologous to those activated in humans also activated in dogs and chimpanzees that understand speech? I think the answer is "yes": it has been apparent for many years that particular neural structures in animals' brains are activated when they hear conspecific vocal signals (e.g., Peterson et al. 1978; Poremba et al. 2004). After discussions of the acoustic parameters that play a role in human speech, later chapters provide more detail on speech perception by animals.

In humans, it also is clear that the sounds that convey words automatically activate the neural systems that are involved in talking. A recent fMRI study shows that, when we listen to a word, parts of motor cortex that are involved in speech production become active (Wilson et al. 2004); these motor areas also are activated when people look at someone talking (Watkins, Strafella, and Paus 2003). Visual and auditory information are integrated when we listen to a person whose face we can see. Speech is better perceived when we can see the face of the person who is talking (Massaro and Cohen 1995; Driver 1996; Lindblom 1996). The McGurk "effect" in which "perception" of a [g] consonant shifts to [d] when you see the face of the person speaking a [b], is an extreme example (McGurk and MacDonald 1976). Recent preliminary findings reported by Rizzolatti (at a 2003 Munich Max Planck meeting) may explain these effects: mirror neurons in the monkey homologue of Broca's area become active when a monkey sees the lips of another monkey while listening to it vocalizing. The McGurk effect may reflect cross-modal excitation of neural mechanisms "tuned" to the "quantal" consonants (cf. Chapters 3 and 4).

Neuroimaging data from studies of sentence comprehension, which also are discussed in Chapter 4, show that, although we can think in terms of words without any overt indication of speech, we internally model them in a form of silent speech. This process entails activating the neural structures that regulate overt speech. We must convey our thoughts by means of speech, orthography representing speech, or manual systems such as one of the sign languages that have been crafted in the last 200 years or so. And as is the case for the dictionaries that we are accustomed to using, the sound pattern, the word's "spelling," seems to be the dictionary entry point.

Accessing Words

The findings of Damasio et al. (1996) reinforce the view that words are accessed by neural circuits that link the word's sound pattern to conceptual knowledge. Discrete cortical areas seem to partition the conceptual space. Deficits in naming were studied in twenty-nine patients who had focal brain damage that resulted in their not being able to name photographs from three categories—the faces of well-known people, animals, and tools—though they knew what they represented. These deficits were established by asking the subjects to name a set of photographs and to describe each photograph as best they could. Their responses were compared with the responses of normal controls matched for age and education. In total, twenty-nine brain-damaged subjects were studied who couldn't name particular types of photographs.

Seven subjects were impaired solely on photographs of persons; two on persons and animals; five only on animals; five on animals and tools; seven only on tools; and four on persons, animals, and tools. All of these subjects had cortical and underlying subcortical lesions localized to three adjoining regions of the posterior region of the brain (along the temporal pole and regions inferior to Wernicke's area; c.f. Chapter 4). The naming deficits roughly correlated with damage to each of the three adjoining cortical areas and the underlying subcortical structures in this region. Damasio and her colleagues found that functional magnetic resonance imaging (fMRI) activation of these same regions occurred when these photographs were shown to nine neurologically intact subjects. Variations occur from subject to subject, which Damasio and her colleagues consider the result of different life histories; the detailed circuitry for the meaning and sound pattern of each word, in their view, is acquired rather than genetically specified.

The Arbitrary Nature of the Phonetic Code

The sounds that convey the word "dog" bear no relationship whatsoever to a dog's vocalizations, anatomy, or any aspect of a dog's behavior. As both Saussure (1959) and Hockett (1960) noted, the arbitrary relationship between sounds and meaning is one of the signal properties of language. A similar arbitrary relationship between the *phonetic-manual-facial* signal that constitutes a signed word and its meaning holds for most aspects of manual sign language. Some of the gestures that make up American Sign Language (ASL) had an iconic, mimetic origin but have shifted over time toward arbitrary patterns (Stokoe 1978). However, the psychological reality of the sound pattern that conveys a word is quite apparent. Systematic speech production errors form one source of data that reveal both the psychological reality and the internal structure of the sound pattern (Jakobson, Fant, and Halle 1952; Chomsky and Halle 1968). For example, the "stop" consonants [p] and [b] of the words "pad" and "bad" are more likely to be mispronounced and confused than [p] versus the stop consonant [d] of the word "dad." This error pattern reflects the fact that [p] and [b] are formed by lip movements that yield more similar acoustic signals than those produced by the tongue gesture that is necessary to produce a [d]. As pointed out by Roman Jakobson (1940), who was arguably one of the most creative linguists of the past century, the sequence in which children acquire the sounds of language is influenced by the phonetic contrasts of the particular languages to which a child is exposed.

Acoustic studies show that infants start to attempt to produce the vowels of English in the first years of life (Buhr 1980). Surprisingly, adult competence is not achieved until children are older than fourteen and well into puberty (Lee, Potamianos, and Narayanan 1999). Speech motor control clearly is an exceedingly complex process. Speech perception is shaped by the linguistic environment early on in life. By age six months, infants living in Swedish- and English-speaking homes differentially respond to sounds in terms of the vowel systems of the two languages (Kuhl et al. 1992; see also Chapter 6). Moreover, as noted above, the neural process by which the meaning of a sentence is recovered appears to involve a form of silent speech, in which neural mechanisms that also play a part in producing overt speech maintain words in a short-term memory system that integrates knowledge of syntax, word meaning, and context, to derive the meaning of a sentence.

Animal Words

One feature of language is the fact, often noted, that humans possess words. Whether animals have that capacity has been a subject of debate for centuries.

What Do Animals "Say"?

A common view concerning animal communication is that their vocalizations, gestures, and facial expressions always have fixed, simple referents. This may be the case for phylogenetically simple animals. The alarm call of a frog (Bogert 1960) or the mating signal of a cricket (Hoy and Paul 1973) appears to have the same meaning, independent of the context in which it occurs. Following this line of reasoning, many ethological studies of communication—for example, Smith (1977)—have been focused on discovering signals that elicit *consistent* responses in one animal or group of animals whenever they are produced by another animal. The underlying assumption was that animals are automatons that produce a particular signal whenever the appropriate stimulus occurs. Smith, whose 1977 study reflected the prevailing contemporary model and otherwise cannot be faulted, noted that "in natural circumstances it often appears as if most animals perform their displays whenever an appropriate set of circumstances arises, whether or not the displays appear to be necessary or effective, as if the displays were more or less automatic responses. Further, in natural circumstances each participant does not usually appear to perceive itself as the target of another's signaling, as conversing humans do" (Smith 1977, p. 265).

Although Charles Darwin started biology on the road to discovering how we came to be, he wasn't always on the right path. The fallback position to invoking innate mechanisms to explain behavior that characterizes Chomskian linguistics and some aspects of evolutionary psychology (Pinker 1994, 1998, 2002), as well as the traditional ethological model, can be traced to Darwin's 1872 book *The Expression of the Emotions in Man and Animals.* According to Darwin, the communications of animals are innately determined, following from "the principles of actions due to the constitution of the Nervous System, independently from the first of the Will, and independently to a certain extent of Habit" (Darwin 1872, p. 29).

Darwin categorically states that the communicative signals of animals "differ little from reflex actions" (1872, p. 48) triggered by specific stimuli. The premise that each signal has a specific, "invariant" meaning is expressed in Darwin's descriptions of animal communication. Darwin also takes the same stance when he discusses the expression of emotion in human beings. Particular facial expressions are innate and have a single, fixed meaning regardless of context. Particular muscles have single fixed purposes, with the "grief muscles" of the eye as an example (p. 189). In an account that reads like a parody of the "forbidden experiment"—raising a child in the absence of language to determine the "innate first language"—Darwin claims that shrugging one's shoulders is an innate characteristic of Frenchmen and Italians but not of true Englishmen. In Darwin's words:

> Englishmen are much less demonstrative than the men of most other European nations, and they shrug their shoulders far less frequently and energetically than Frenchmen and Italians do . . . I have never seen very young English children shrug their shoulders, but the following case was observed with care by a medical professor and careful observer . . . The father of this gentleman

> was a Parisian, and his mother a Scotch lady. His wife is of British extraction on both sides, and my informant does not believe that she ever shrugged her shoulders in her life. His children have been reared in England, and the nursemaid is a thorough Englishwoman, who has never been seen to shrug her shoulders. Now, his eldest daughter was observed to shrug her shoulders at the age of between sixteen and seventeen months; her mother exclaiming at the time, "Look at the little French girl shrugging her shoulders!" . . . This little girl, it may be added, resembles her Parisian grandfather in countenance to an almost absurd degree. (Darwin, 1872, pp. 264–265)

We may infer the expression of the child's French genetic disposition for shrugging, given her living in a shrug-free environment.

The Darwinian model thus entails establishing an invariant relation between a signal and a particular behavioral pattern. The signal's meaning is demonstrated when it consistently appears in a particular context and elicits an invariant behavioral response. Thus Smith (1977) states that the Venezuelan *Cebus nigrivittatus* monkey vocalization that sounds like a "huh-yip" to human observers is not a food-sharing call because it occurs both when monkeys feed and when they do not feed. Smith concludes that its "semantic" referent, therefore, must be something common to both situations. In the analytic framework that structures Smith's interpretation, the call can be either a food-sharing call or a group cohesion call. It cannot have both functions. The monkey call would be a food-sharing call in this ethological model if, and only if, it were solely produced when the monkeys fed. Smith concludes that the call's "functions probably have to do with group cohesion" (1977, p. 311). The primary reflex model for animal communication does not take into account the possibility that the monkeys' "huh-yip" call may have different meanings in different contexts.

Recent ethological studies fortunately take a broader view. For example, Zuberbuhler (2002) shows that Diana monkeys respond differentially to the alarm calls of Campbell's monkeys (a different species) depending on the particular sequence of calls emitted by the Campbell's monkeys. However, the traditional ethological model is alive and well—animals are automatons that automatically broadcast specified signals, triggered by particular events or classes of events. Evidence to the contrary generally is interpreted as "noise" that appropriate statistical analysis will eliminate. For example, Cheney and Seyfarth (1980, 1990) show that different vervet monkey alarm calls signal the presence of predatory hawks or snakes in the context of "normal" monkey life. When it was found that the calls served to alarm monkeys to other predators in different situations (the "hawk" call serving to signal distant threats), the validity of their study was questioned. But the context-dependent meaning of vervet monkey calls can instead be viewed as evidence for wordlike behavior on the part of the monkeys. The monkeys may have generalized the referents of the call. The context in which the call occurs specifies the predator.

Phonetic Limits on Animal Vocalizations

Moreover, monkeys do not have the option of producing an expansive repertoire of vocal calls. Field observations of chimpanzees (Goodall 1986) and acoustic analyses of the vocalizations of nonhuman primates (e.g., Lieberman 1968a, 1975; Lieberman, Crelin, and Klatt 1972; Hauser 1996) suggest that nonhuman primates lack the neural capacity that would allow them to form novel vocalizations. Acoustic analyses (see Chapter 3) suggest that vocal phonetic ability of nonhuman primates is largely confined to a fixed repertoire of calls. (The limits appear to be largely neural, not anatomic, contrary to misreadings of these papers; for further discussion of the acoustic characteristics of animal communication, see Chapters 3, 6, and 7.) In this respect, the phonetic limitations of nonhuman primates may, itself, have resulted in primates developing communications systems in which a signal has multiple, context-dependent referents. In other words, one of the characteristics of human speech is evident in their vocal communications—the disjuncture between sounds and meaning noted by linguists. A sound does *not* have a fixed unitary reference.

Species living in a relatively fixed environment may be able to compete in the Darwinian struggle for existence using a set of innate, communicative signals that elicit single, fixed behavioral responses. But if the environment should change, resulting in new opportunities and challenges, animals that lack the ability to produce "new" vocalizations could instead adapt existing calls to new situations. The words of human language continually take on "new" referents or have multiple referents. Contemporary Russian soldiers use the word "archers" for riflemen. The word "hey" can signify "stop it" in many contexts, or it can serve to simply start an e-mail or open a conversation. We don't know what the monkey "huh-yip" call observed by Smith means. In the context of a moving band of monkeys, it may serve to maintain cohesion; in the context of a subgroup of monkeys feeding, it might attract other monkeys to the food source. But this is all speculation—controlled playback experiments would be needed to establish the call's meaning.

Apes Using Human Language

A Martian observer arriving here 100 years ago would not have been able to determine that human beings have the biologic capacity to place the wheels of an airplane (without being able to see the wheels) moving at 150 miles an hour onto a narrow strip of concrete and bring the plane to a safe halt. If a Martian observer had arrived here 20,000 years ago, no evidence would have suggested that human beings were capable of writing or reading. The Martian observer would not have been able to determine whether the absence of writing reflected biologic limitations or the cultural context. It is now apparent that most human beings have the neural capacities that enable literacy because "normal" children from any cultural context can learn to read if instructed properly. Barring conditions such as dyslexia, success in achieving literacy in human beings whose ancestors were illiterate demonstrates that humans generally have the biologic capacity for reading and writing.

The ape-language studies of the last century were aimed at determining whether chimpanzees, gorillas, and orangutans had the biologic capacity to acquire some aspects of human language. These experiments have, in my view, been misrepresented by adherents of the Cartesian claim (dating to 1646) that all aspects of language belong to man and man alone. It is strange that Chomsky, who places such emphasis on innate, genetically transmitted human capacities for language, adopted this position (Chomsky 1966, 1986). Virtually all of the genetically specified aspects of human anatomy and physiology have antecedents in other species. Finding reduced linguistic capacities in other closely related species would reinforce the case for innate human linguistic features. Moreover, the ape-language studies were not intended to prove that there was no difference between the linguistic and cognitive capacities of humans and apes. The major finding was that apes have the biologic capacity to acquire and productively use human words and syntax at a reduced level equal to that of most three-year-old children.

Attempts to teach apes human language date back at least to the time of La Mettrie (1747), who claimed that an ape could be taught to talk using methods that had been developed to teach deaf children language. The ape, La Mettrie claimed, would then be "a perfect little gentleman." Success in this endeavor was deferred until 1965 in a project initiated by Alan and Beatrix Gardner. The Gardners took account of the failure of many attempts to teach chimpanzees to talk. Later acoustic analyses and neurophysiologic studies showed that chimpanzees lack the biologic capacities necessary to produce human speech, but the Gardners' primary motivation to use ASL was the disappointing results of the attempt by Hayes and Hayes (1951) to teach a chimpanzee to talk. The failure of attempts to get chimpanzees to talk suggested that they might be better adapted for communication using manual gestures or other visual cues.

The procedure that the Gardners adapted was direct and simple.

Human children acquire speech and language when they are raised in a normal linguistic and social environment. The Gardners thus reasoned that the biologic differences relevant to language between chimpanzees and human beings could be assessed if chimpanzees were raised in similar conditions. Hayes and Hayes (1951) had, in fact, attempted to raise an infant chimpanzee in a normal human environment with spoken English as the mode of communication. The Gardners instead used ASL, the gestural language that is often used by deaf people in North America. The Gardners raised Washoe, a female chimpanzee that was about 10 months old at the start of the project in 1966, in a setting in which she had close interaction with humans who communicated with her using ASL. Washoe lived with the Gardners and their research assistants until 1970.

During that period she was treated like a "normal" child, except that she was in contact with an adult during all of her waking hours. Washoe was diapered, clothed, and wore shoes. As she matured, she learned to use cups and spoons and to help clear the table after dinner. She learned to use the toilet, played with toys, broke some, and, like young children, observed language in its ordinary uses. Washoe continually observed ASL conversations between adults and communications in ASL directed toward her, although the adults did not necessarily expect her to understand what was being signed in the early stages of the chimpanzee-rearing project. Practice sessions, analogous to those in which parents attempt to "teach" young children how to pronounce words more distinctly, also occurred. In short, Washoe learned ASL words as might a human child. Washoe acquired about fifty ASL words in the first year of cross-fostering.

The initial Washoe project suffered from the deficiency that some of her human companions were not proficient signers of ASL. Washoe's exposure to ASL also did not start until she was 10 months old. To address these issues, an expanded project was initiated in which four other chimpanzees were raised in an ASL environment almost from birth. Starting in November 1972, the newborn chimpanzees—Moja, Pili, Tatu, and Dar—were placed in four adoptive "families" at the Gardners' laboratory in the suburbs of Reno, Nevada. The four newborns arrived in Reno within a few days of birth. Each chimpanzee was placed in a "family," and one or more human members of these families maintained contact with the chimpanzees from morning until bedtime. Each family included one or more persons proficient in ASL. The family groups were stable throughout the project.

Chimpanzee Words

The Gardners documented the procedures they employed to assess the vocabularies of the chimpanzees, their training procedures, and their analyses of the linguistic skills of ASL-using chimpanzees compared to those of children acquiring speech or ASL (if they were deaf). The analyses presented in Gardner and Gardner (1969, 1971) show that Washoe's progress during the first three years of ASL exposure was qualitatively similar to that of the children studied by Brown (1973) and Bloom (1970) in contemporary studies. The chimpanzee utterances that were made in a number of twenty-minute sessions were transcribed using procedures modeled on the techniques developed in these child language studies. A team of two persons was necessary. One person whispered a spoken transcription of the signing, together with notes on the context, into a cassette recorder, while the other person performed "the usual roles of teacher, caretaker, playmate and interlocutor" (Gardner and Gardner 1978, p. 57).

The findings for the expanded project (Gardner, Gardner, and Van Cantfort 1989) included comparisons of the number of different utterances, the number of utterances per hour, and the grammatical categorization of the early vocabulary of chimpanzees versus those of children. The relationship between questions posed to the chimpanzees and their replies was also noted. The ASL outputs of the chimpanzees were within the range reported for the early stages of language acquisition for the English-speaking children studied by Brown (1973) and Bloom (1970) and videotaped children using ASL. The conclusion was that the chimpanzees' utterances were similar to those of children in the earliest stages of language acquisition. Locatives (words signifying locations) such as "in," "out," and "there" occurred frequently, as did appropriate responses to Wh- questions—that is, sentences such as "Where is John?"

Washoe's expressive vocabulary after fifty-one months included 132 signs and met a conservative criterion of reliable usage. The sign first had to be reported on three independent occasions by three different observers. It then had to occur spontaneously and appropriately at least once on fifteen consecutive days. The four young chimpanzees who were exposed to ASL almost from birth initially had a rate of word acquisition similar to that of children. However, the rate of word acquisition soon leveled out. The upper limit of a chimpanzee's productive vocabulary appears to be about 150 words. Passive vocabularies appear to be far greater, but no systematic studies in chimpanzees similar to those exploring the ability of dogs to recognize words have yet been reported.

A Vocabulary Test for Chimpanzees

In a study published in 1984, the chimpanzees' productive vocabulary was tested using procedures that eliminated any possible effects of inadvertent cueing and minimized observer errors (Gardner and Gardner 1984). Because young chimpanzees' ASL signing, like young children's speech, is imperfect, it was necessary to guard against the observers "seeing" more than the chimpanzees were producing. It is common for human observers to "hear" more than young children actually say. In a project that traced the acquisition of speech by children, my colleagues and I found that even trained phoneticians frequently "heard" children produce words that subsequent acoustic analysis revealed were not spoken (Lieberman 1980).

Human observers commonly hear words and sounds that are not conveyed by the acoustic signal; the contextual and pragmatic context fills in gaps in communication. The effects have been documented in studies employing acoustic analyses of speech to compare what actually was said in a conversation with what adult listeners thought they heard. Context augments the acoustic signal (Pollack and Pickett 1963; Lieberman 1963b; Pitt and Samuel 1995; Samuel 1996, 1997, 2001; Shockey 2003). The Gardners' chimpanzee vocabulary test guarded against this phenomenon by using two observers who could not see the projected images (avoiding the effects of extralinguistic context) or each other. The observers each independently recorded the chimpanzees' ASL signs. The two observers' transcriptions of the chimpanzees' ASL signs for 760 trials agreed 90 percent of the time. The chimpanzees were tested one by one, as they sat alone in a chamber looking out at a screen on which color slides of objects that the chimpanzees had never seen before were projected. The color slides included multiple pictures of objects that appeared to be in the chimpanzees' vocabularies, such as trees, apples, meat, shoes, and so on. Each slide represented a category rather than a singular object. The images of trees, for example, included many trees in different locations, pictured from different angles. The chimpanzee's task was to name each pictured object and the two observers independently transcribed each sign that each chimpanzee produced.

Biuniqueness

About 150 different words were reliably signed. The errors that the chimpanzees made were as significant as their correct responses. Errors reflected either "semantic" misidentification—for example, between different fruits—and "phonetic" misidentification between ASL signs that had similar hand shapes or movements. The chimpanzee word error pattern thus demonstrates the psychological reality of the two levels that distinguish the linguistic elements that we call words—their semantic properties and their arbitrary phonetic properties. Charles Hockett in 1960 coined the term "biuniqueness" to represent the two levels of representation that mark the words of human language: the phonetic representations of words—the sounds that are represented in our brains—and their meanings. Chimpanzees raised in an environment that involves productive linguistic interchanges with humans clearly manifest biuniqueness. In an independent study, Savage-Rumbaugh and her colleagues (1986) showed a similar 150-word upper limit for the productive vocabularies of pygmy chimpanzees *(Pan paniscus).*

The Chimpanzee Language Debate

In a number of widely publicized publications, Terrace and his colleagues (Terrace et al. 1979) converted a flawed experiment into a successful attack on the Gardners' findings. Because Terrace's claims are often taken at face value, some detail may be helpful in forming a view on this issue.

Terrace at a 1975 conference on the evolution of language declared that he could teach an ape language using the operant-conditioning techniques proposed by B. F. Skinner (1957). At one time, many psychologists believed that the key to learning virtually any skill was to have a person or creature attempt some act and then to immediately provide a correction or reward. Amazing skills can be acquired using these techniques. Richard Herrnstein and his colleagues taught pigeons to identify different species of trees, such as maples from pines, by systematically providing a reward if a pigeon pecked at the "correct" picture (Herrnstein 1979).

Terrace's apparent goal was to make use of these techniques to refute Chomsky's claim that innate knowledge of syntax was the basis for human language. In essence, Terrace and his coworkers treated a young chimpanzee as though he were a caged pigeon or rat. One of the primary determinants of normal adult linguistic ability is the presence of "normal" social interaction in childhood. It's been evident for decades, if not centuries, that a normal social environment is necessary to achieve human cognitive and linguistic ability. Children raised in aberrant environments—like Genie, a child locked in a room for years and immobilized in a toilet-training chair—fail to develop normal linguistic ability (Curtiss 1977). Seen in this light, what is most peculiar about the studies of Terrace and his colleagues is their claim that "a team at Columbia University (Terrace, Petitto, Sanders, and Bever) attempted to replicate the experiments of the Gardners, Fouts, and Patterson" (Seidenberg and Petitto 1979, p. 117). (Seidenberg no longer holds to these views.) In contrast to the human environment of the Gardner project, Nim the chimp's teachers were told that they should not treat him as though he were a child. Nim was subjected to long training sessions in an environmentally impoverished setting. From the age of 9 months, Nim was taken to a small 8-by-8-foot room. The room was empty except for a one-way mirror and was painted a uniform white. Terrace was not malicious; he believed that a rich environment would distract the chimpanzee in the three daily two-hour drill sessions. Moreover, instead of a stable "family" of human companions who were proficient in ASL, an ever-changing staff of volunteers who were themselves novices in sign language "cycled through Project Nim in a revolving door manner" (Terrace 1979, p. 108). In short, Terrace and his Columbia University team did not replicate the Gardners' experiments. Terrace applied strict Skinnerian theory to Nim as though he were a large rat or pigeon.

One interesting result of the primate research of Goodall and her coworkers (1986) is that chimpanzees have an extended childhood and adolescence that is quite similar to that of human beings. Chimpanzees under natural conditions are weaned at about five years, and they live with their mothers until they are eight to ten years old. Throughout this period, there is close interaction between the chimpanzee and its mother. The environment that the Gardners set up for their chimpanzees took advantage of the possibility of close interaction between the infant chimpanzee and its human caretakers. As Gardner and Gardner noted,

> chimpanzee subjects can be maintained in an environment that is very similar to that of a human child . . . their waking hours follow a schedule of meals, naps, baths, play and schooling, much like that of a young child. Living quarters are well stocked with pictures, tools, and toys of all kinds, and frequent excursions are made to other interesting places, a pond or a meadow, the home of a human or chimpanzee friend. Whenever a subject

> is awake, one or more human companions are present. The companions use ASL to communicate with the subjects, and with each other so that linguistic training is an integral part of daily life, and not an activity restricted to special training sessions. These human companions are to see that the environment is as stimulating as possible and as much associated with Ameslan [ASL] as can be. They demonstrate the uses and extol the virtues of the many interesting objects around them. They anticipate the routine activities of the day and describe these with appropriate signs. They invent games, introduce novel objects, show pictures in books and magazines, and make scrapbooks of favorite pictures, to demonstrate the use of ASL.
>
> The model for this laboratory is the normal linguistic interaction of human parents with their children. We sign to the chimpanzees about ongoing events, or about objects that have come to their attention as mothers do in the case of the child. We ask questions, back and forth, to probe the effectiveness of communication. We modify our signing to become an especially simple and repetitious register of Ameslan, which makes the form of signs and the structure of phrases particularly clear. And we use devices to capture attention, such as signing on the chimpanzee's body, which are also used by parents of deaf children. (1980, p. 335)

It is difficult to conceive of a protocol further removed from the Gardners' than Terrace's method.

Terrace and his colleagues, moreover, made a series of claims concerning what the Gardner chimpanzees failed to do. None had any merit. According to Terrace, the apes studied by the Gardners were simply imitating ASL signs produced by their human observers. Terrace presented ostensibly objective evidence in the form of a frame-by-frame analysis of some of the Gardner motion pictures of chimpanzees' signing. The "analysis" of Washoe supposedly showed that the chimpanzee was imitating the ASL signs that her human companion was using in the "conversation." The exercise was a charade; Terrace's criterion for imitation was absurd. *Any* sign produced by Washoe was counted as an imitation if her human companion had previously signed *anything*. Thus, following Terrace's criterion, Washoe's response of "watch" to the human signed question "What is this?" would be counted as an imitation. If we were to follow Terrace's lead, virtually all human conversations would be classified as imitations. Terrace also claimed, among other things, that chimpanzees repeat signs but human children never repeat words—a claim that is inconsistent with the published results of every study of the early utterances of children.

It is difficult to account for the fact that Terrace's objections were accepted by otherwise rational scholars. A claque of Chomsky's supporters parroted Terrace—by changing sides in the "war" between Chomsky and Skinner, Terrace had become a valuable ally. But the degree to which Terrace's demonstrably false claims have become part of the accepted body of "evidence" concerning the species specificity of human language is nonetheless surprising. Terrace's claim that the Gardner chimpanzees were simply imitating the last ASL sign that their human companion had made was nonsensical. It is, for example, inherently impossible to account for the correct reply to a Wh- question such as "What are you eating?" by invoking imitation. Washoe and the other Gardner chimpanzees were repeatedly observed answering such questions. Expert and unbiased ASL specialists such as William Stokoe observed Washoe and engaged in conversations with her. Stokoe was convinced that ASL was being used productively.

Other evidence shows that the Gardners' cross-fostered chimpanzees used language spontaneously and productively. As the Gardners (1980) noted, the chimpanzees "initiate[d] most of the interchanges by themselves, with their own questions, requests and comments" (Gardner and Gardner 1980). It was clear that Washoe was using words spontaneously before the publication of the Terrace group's 1979 paper. In a paper published in 1974, the Gardners addressed this issue. They noted that, as is the case for human children, Washoe talked to herself:

> [She] often signed to herself in play, particularly in places that afforded her privacy, i.e., when she was high in the tree or alone in her bedroom before going to sleep. While we sat quietly in the next room waiting for Washoe to fall asleep, we frequently saw her practicing signs. . . Washoe also signed to herself when leafing through magazines and picture books, and she resented our attempts to join in this activity. . . Washoe not only named pictures to herself in this situation, but she also corrected herself. Washoe also signed to herself about her own ongoing or impending actions. We have often seen Washoe moving stealthily to a forbidden part of the yard, signing *quiet* to herself. (Gardner and Gardner 1974, p. 20)

As I noted earlier, one critical, productive attribute of words is that they have elastic referents that can stretch to encompass related concepts or even extend to very different referents. For example, "dog" can refer to both a member of the domesticated species and a device that seals a ship's watertight bulkhead. It is clear that the referents of the ASL signs that chimpanzees acquire when they are raised in a humanlike environment are not restricted to single items. As the Gardner and Gardner (1984) study demonstrated, the referent of the chimpanzee sign for *car* is the class of vehicles that are cars; the sign for *tree* was used when chimpanzees identified a wide variety of trees, and so on. Moreover, the chimpanzees formed the complex referents of words without instruction. The sign for *meat,* for example, was used to identify raw meat wrapped in supermarket packaging, though meat was never presented and identified to the chimpanzees in that form. The chimpanzees through their visits to the kitchen in the Gardner household formed the association themselves. The signing chimpanzees, like children, coined new words by using signs in combination to describe objects that were not represented by a sign in their vocabulary. Washoe, for example, referred to a swan as a *water bird.* Terrace and his colleagues (1979) focused on this example and claimed that there is no evidence that chimpanzees ever create "new" words. In the *water bird* sequence, they claimed that "Washoe may have simply been identifying a body of water and a bird, in that order" (Terrace et al. 1979, p. 895).

However, Terrace overlooked many other documented instances in which chimpanzees coined new words through sign sequences to describe new stimuli. For example, Gardner and Gardner (1980) report the following examples: *listen drink* for references by chimpanzee Moja to Alka-Seltzer in a glass and *metal hot* for a cigarette lighter. Fouts (1975) reports *smell fruits* for chimpanzee Lucy identifying citrus fruits, a radish as a *cry fruit* after she tasted it, and a watermelon as a *drink fruit* or *candy fruit*.

Perhaps the most striking and debated example of a chimpanzee extending the semantic field (the meaning of a word) was Washoe's spontaneous use of the sign *dirty*. As Fouts noted:

> I was about to teach Washoe the sign monkey and while I was preparing the data sheet she turned around and began to interact with a particularly obnoxious macaque in a holding cage behind us. They threatened each other in the typical chimpanzee and macaque manner. After I had prepared the data sheet I stopped the aggressive interaction and turned her around so that she was facing two siamangs. I asked her what they were in ASL. She did not respond. After I molded her hands into the *monkey* sign three times she began to refer to the siamangs with the monkey sign. I interspersed questions referring to various objects that she had signs for in her vocabulary. Next, I turned her toward the adjacent cage holding some squirrel monkeys and she transferred the monkey sign immediately to them. After she called the squirrel monkeys *monkey* several times I turned her around and asked her what her previous adversary was, the macaque, and she consistently referred to him as a *dirty monkey*. Up until this time she had used the *dirty* sign to refer to feces and soiled items. She had changed the usage from a noun to an adjective. In essence, it could be said that she had generated an insult. Since that time she has similarly used the dirty sign to refer to me as *dirty Roger*, once when I signed to her that I couldn't

> grant her request to be taken off the chimpanzee island, and another time she asked for some fruit and I signed, I don't have any fruit. Lucy has used the *dirty* sign in a similar manner. Once she referred to a strange cat she had been interacting with aggressively as a *dirty cat,* and she has also referred to a leash (which she dislikes) as a *dirty leash.* (1975, p. 387)

Terrace dismissed the *dirty* example out of hand. In a public lecture at Brown University in 1979, he claimed that Washoe signed *dirty* in order to distract Fouts, by indicating that she had to defecate. That clearly was not the case. In that same public lecture, Terrace went on to claim that filmed records of Washoe spontaneously signing were fabricated, as was the Public Broadcasting System (PBS) documentary film of Washoe spontaneously identifying an image of a monkey in the bottom of the child's cup from which she was drinking, with the ASL sign for *monkey*—a reckless claim on Terrace's part.

Independent Replications of the Gardner Findings

The findings of the Gardner and Gardner projects were replicated in independent ape-language projects. A subsequent study by Miles (1983) replicated many of the Gardner's findings with a male orangutan using ASL. Miles (1976) also reported "common" chimpanzees *(Pan troglodytes)* understanding simple spoken English sentences, such as "bring some Coca-Cola." Savage-Rumbaugh and her colleagues (1993) showed that pygmy chimpanzees *(bonobos)* could produce approximately 150 words using manually operated keyboards that triggered synthesized words. These pygmy chimpanzees were exposed to spoken English sentences from birth onwards and understood English sentences that had relatively simple syntax. Fouts, Hirsch, and Fouts (1982) report the results of a project in which Loulis, an infant chimpanzee raised from birth in the company of adult chimpanzees communicating by means of ASL (the former Gardner chimpanzees), acquired fifty ASL words in the first two years of life, without any human instruction, intervention, or ASL communications directed at him by humans. Kanzi, the bonobo studied by Savage-Rumbaugh, acquired his manually produced vocabulary and ability to comprehend simple sentences without formal instruction. The study was directed towards his mother Matata, who never learned to communicate. The infant Kanzi acquired his linguistic ability by observation and imitation. Curiously, the claim that chimpanzees have to be taught these aspects of language, whereas young human children acquire language, persists in the Hauser, Chomsky, and Fitch (2002) proposal for comparative studies of the evolution of language.

The vehemence with which some linguists (e.g., Seidenberg and Petitto 1979; Piatelli-Palmarini 1989) deny that animals can understand words and that some animals can communicate using words is surprising. As I previously noted, domestic dogs commonly understand many words. Baru (1975) used synthesized speech stimuli to show that dogs could discriminate between the vowels [i] and [a] (the vowels of the words "see" and "ma") by means of vowel quality (their formant frequency patterns; cf. Chapter 3) rather than the tone or pitch of a speaker's voice. Field studies of monkeys show that they develop predator-specific calls on their own (Zuberbuhler 2002). Baboons learn to respond to conspecific barks that have different referents but have subtle acoustic distinctions (Fischer, Cheney, and Seyfarth 2000; Fischer et al. 2002). Other nonhuman primates vocally communicate subtle information relevant to their social interaction (cf. Slocombe and Zuberbuhler, 2005b, for evidence of referential vocal communication in chimpanzees). Dolphins learn complex songs to maintain identity and to facilitate social interactions (Sayigh et al. 1990). Parrots have been taught to communicate using "spoken" words (Pepperberg 1981, 2002). In fact, as I noted previously, animals as "simple" as pigeons can be taught to recognize species of trees (Herrnstein 1979) or tropical fish (Herrnstein and de Villiers 1980) as general classes and signal their decisions by pecking. Therefore, it isn't surprising that chimpanzees learn the meaning of spoken English words and can, under some circumstances, communicate using manual signs or by means of manually operated keyboards. The ability to perceive speech appears to be a primitive characteristic of language; limits on language acquisition by apes and many other species do not appear to hinge on speech perception.

Cueing—"Clever Hans"

One of the questions raised concerning the Washoe project and subsequent ape-language research projects was whether the animals' responses simply mirrored "cues" unconsciously provided by the human experimenters–the "clever Hans" phenomenon. At the beginning of the twentieth century, a horse named Hans appeared to be able to count numbers. Hans's trainer, Wilhelm von Osten, would call out two numbers and Hans would stamp one of his hoofs, stopping at the correct sum. Hans indeed was very clever. In 1907 it became apparent that Hans had observed that von Osten unknowingly tensed his facial muscles when Hans's hoof taps approached the correct sum (Pfungst 1907). Hans had discovered that he would get a carrot when he stopped at that moment. Inadvertent cueing effects still plague psychological and linguistic research. The linguist Thomas Sebeok even organized a conference in 1986 to show that every chimpanzee language experiment reported to that date was invalid owing to the clever Hans phenomenon. However, that simply is not the case. The Gardner and Gardner (1984) chimpanzee vocabulary experiment should have laid Sebeok's concerns to rest. The research reported by Savage-Rumbaugh and her colleagues and by Miles also took account of cueing effects.

Gestures and Speech

Chimpanzee studies often lead to the inference that human language in its first form made use of gestures (for more discussion on this point of view, see Chapter 6). Human children also start to communicate using a mixture of gestures and speech that gradually shifts to speech; manual gestures and facial expressions play a part in providing the pragmatic context for words in the early stages of speech communication (Greenfield 1991). Pointing gestures provide pragmatic cues for children, and adults will both point at relevant objects and use "baby talk" (Fernald and Kuhl 1987; Grieser and Kuhl 1988; Fernald et al. 1989) to engage a child in the "joint attention" that, for example, allows an infant to learn that the word "shoe" refers to that which is on his or her foot. Similar techniques work with apes (Tomasello and Farrar 1986). Although Tomasello (2004a) now appears to suggest that joint attention is a human characteristic, careful observations show that chimpanzee mothers interacting with their infants also make use of joint attention (Matsuzawa 2004). Manual gestures never really depart from human linguistic communication; certain manual gestures provide cues for syntax (McNeill 1985). Adults also offer subtle corrections to children when they mispronounce words or in later stages garble syntax.

Referential and Emotional Information

One aspect of human speech that is often overlooked is that it blends emotional and referential information. Speech is the medium of human language. Written systems are a recent invention. Although the inked words on a piece of paper may be viewed as conveying strictly referential information, spoken words always convey the emotional stance of the speaker. And human beings rarely rely on "primal" sounds to convey their emotional state—the "bad" words replaced by bleeps on radio broadcasts generally play a role in conveying negative affect.

Moreover, as we shall see in the chapters that follow, virtually all of the acoustic elements that specify the sounds of speech can in some circumstances convey the emotional state (e.g., Lieberman 1961; Protopappas and Lieberman 1997; Chung 2000) or gender (Sachs et al. 1972) of a speaker. The acoustic parameters that differentiate vowels and most consonants also serve to convey physical attributes such as a speaker's height and weight to a listener (Fitch 1993, 1997). The state of a speaker's health also can be assessed in some instances by his or her speech (Lieberman 1963a; Lieberman et al. 1994, 2005). Intoxication produces distinctive alterations in a person's speech (Pisoni and Martin 1989). A sharp distinction, a sort of watertight dam, is commonly assumed to differentiate the graded emotional vocal communications of other species from the referential "linguistic" communications of human beings. The supposed discreteness of the speech signal carries over into some theories concerning the evolution of language, which hold to the premise that no evolutionary process links the vocal communication of other species and human language (e.g., Burling 1993; Studdert-Kennedy 2000). However, that premise rests on the assumption that words are essentially composed of phonemes that have the properties of the movable type of printed text.

The research of the past fifty years discussed in the next chapter shows that the acoustic parameters that constitute words bear little resemblance to the letters of the alphabet, or to phonemes that can be freely permuted. Many of these acoustic parameters signal both graded emotional and referential information in human speech. Present data show that they transmit emotional information and, to a lesser degree, referential information in the vocal communications of other species. Thus, no absolute disjuncture exists between the vocal communications of nonhumans and humans. This does not mean that other species have the biologic capacity for fully human language—the differences are most likely the keys to the evolution of human capacities for language and cognition.

Animal Syntax

Many contemporary theoretical linguists claim that the communications systems of other species have no syntax whatsoever. In many cases theoretical linguists go further: they claim that no other aspect of human or animal behavior approaches the complexity of the syntactic operations of human language. These claims generally are based on an argument that entails two suppositions: (1) linguistic research has identified a set of innate syntactic rules, properties, and complex constraints that are "language universal" (cf. Chomsky 1976, 1986; Jackendoff 1994); and (2) no other aspect of human or animal behavior entails this level of complexity and has similar rules, properties, or constraints. These hypothetical language universals are said to be lacking in animal communication.

If theoretical linguists actually were able to specify the syntax of any human language by means of the serial algorithms, the rules and processes that they employ, these claims might merit serious consideration. In actuality, no present linguistic theory can describe more than a small fragment of the syntax of any human language, including English, the most intensively studied language on earth. Jackendoff, an advocate of "transformational" grammars (many have been proposed since Chomsky's (1957) first theory, each failing to achieve this goal), for example, notes that "thousands of linguists throughout the world have been trying for decades to figure out the principles behind the grammatical patterns of various languages. . . But any linguist will tell you that we are nowhere near a complete account of the mental grammar for any language" (1994, p. 26).

Jackendoff's response to this unsolved problem is curious. He concludes that the failure of the linguistic enterprise derives from the details of grammar being prewired into our genetic code—a deciphered code in DNA that is beyond the reach of present-day science. The difficulties faced by practitioners following Chomsky's linguistic theories seems to other linguists (e.g., Gross 1979; Croft 1991) and philosophers of science (e.g., Bunge 1984, 1985; Churchland 1995) to instead derive from its hermetic nature and its self-imposed mathematical and scientific limitations. As Mario Bunge (1984, 1985) points out, a true science tests the predictions of theories against data. Theories are discarded or revised in the light of observations or experiments. In contrast, theoretical linguists of the Chomskian school test their theories against *theories of data,* not actual data.

Claims for language universals sometimes rely on amusing contortions even when they attempt to place German, a language closely related to English, into a theoretical framework based on English. Studies of many languages show that commonly accepted universals based on English—the "universal" language—are false. For example, English speakers find it natural that only verbs accept tense markers that indicate whether an act occurred in the past or present. However, nouns as well as verbs are marked for tense in the aboriginal language Kayardid, which is spoken on an island off the north coast of Australia. The distinction between nouns, verbs, and adjectives—a key element in the theory of universal grammar—is virtually absent in Riau, a dialect of Indonesian. Other violations of supposed syntactic universals occur in "esoteric" languages such as Yup'ik, which 20,000 people speak in Alaska (Wuethrich 2000), as well as in languages such as Turkish, which is spoken by tens of millions of people (Croft 1991).

Noam Chomsky turned the world upside down with the claim that the properties of syntax are universal and innate. Languages that are closely related in time and place, such as English and German, have very different syntax. The common Indo-European origin of English and Nepali, separated by the Eurasian land mass and thousands of years of divergence, is evident in their vocabularies, not in their syntax. Comparative studies show that related languages can have very different systems of syntax; the similarities are most apparent in their vocabularies (Ruhlen 1994). Any claim to the effect that an animal's communications lack syntax because some hypothetical innate syntactic universal of human language is absent must first demonstrate that the putative universal exists.

The Syntactic Abilities of Chimpanzees

The question of whether chimpanzees have the biologic capacity to acquire some aspects of syntax can only be answered by studying chimpanzee "language." The functional role of syntax is to convey information; different languages make use of different mechanisms to this end, sometimes modifying words individually (morphophonemics), sometimes modifying words or word order at the "syntactic" level across the span of a sentence, and sometimes using both processes. ASL, which the cross-fostered chimpanzees in the

Gardner project were using, uses morphophonemic processes. ASL is highly "inflected," and many distinctions conveyed by prepositions or word order in English are fused in the gestures that constitute ASL signed words. The Gardner chimpanzees productively used the inflected forms of their ASL signs, though in a manner similar to that of young deaf children. The ASL sign for *quiet* is, for example, formed on the signer's lips; movement toward some person indicates the object of the verb. The Gardners' cross-fostered chimpanzees would instead place the hand gesture that forms the sign on the lips of the person or chimpanzee who was supposed to be quiet. An analysis of a portion of the chimpanzee corpus shows consistent use of inflected forms conveying syntax (Gardner and Gardner 1994). In other words, the ASL chimpanzees make use of simple syntax. What hasn't been determined is the limit of their syntactic ability.

Although the chimpanzees in the Savage-Rumbaugh project formed "utterances" by pushing buttons on synthesizer boards with their fingers, they heard spoken English throughout their lives and commonly responded to speech directed at them. The pygmy chimpanzee Kanzi was formally tested in sessions designed to avoid the clever Hans effect. Kanzi could see the face of the human talking to him in these formal tests, which show that he comprehended sentences that conformed to the canonical subject-verb-object form of English. Kanzi correctly responded to commands in which the implicit subject "you" had been deleted, such as "put the pine needles on the ball." Kanzi comprehended commands in which the subject was implied. Kanzi thus met the criterion proposed by Bickerton (1990, 1998) for "true" language. Bickerton distinguishes "protolanguage," which consists of strings of words that lack syntactic structure, from "true" language. Bickerton's litmus test for a mind that can command syntax and true language is the ability to recover deleted words like the implied "you" in the commands that Kanzi understood. Hence, we must conclude, though Bickerton (1998) demurs, that apes possess some degree of syntactic ability. Bickerton argues that "these commands don't contain any subjects." We can only hope that Bickerton would be able to comprehend the subject of a sentence like, "Keep away from the edge of the cliff."

Speech—Its Absence in Apes

One of the recurrent themes of the myths that human beings have crafted to explain their creation is how people obtained the gift of speech. In the Mayan story, the *Popol Vuh,* the translation reads:

> Having created all the birds and animals, the creators said to them: Talk and scream according to your kind, pronounce and praise our name, say that we are your Fathers and Mothers, as we are indeed. Speak, praise, invoke us. But even though this was commanded of them, they could not speak as humans, they only screamed, cackled and hissed. They tried to put words together and hail the creator, they were punished and since then their meat has been eaten by man.

The Gospel according to John is explicit:

> In the beginning was the Word, and the Word was with God, and the Word was God.

It is clear that only human beings can talk. Speech production is the derived, unique property of human language and qualitatively different from the communicative ability of even closely related species. Parrots and other passerine birds can produce acoustic signals that mimic some of the acoustic parameters that convey the sounds of human speech (Klatt and Stefanski 1974), but they cannot actually talk; their vocalizations are readily differentiated from human speech (Remez et al. 1981).

The productivity of human speech and language is enhanced by altering the sequence in which a limited number of speech sounds occur. The words "see" and "me" contain the same vowel; the initial consonants signify different concepts. Changing vowels also signifies different words—"sue, ma, sit, mat." Human speakers are able to voluntarily alter the sequence of articulatory gestures that generate the meaningful speech sounds, the phonemes that convey the words of their language. Different languages have particular constraints, but any neurologically intact child raised in a "normal" environment learns to speak his or her native language or languages. It has become apparent that apes lack the neural capacity to freely alter the sequence of muscle commands that generate phonemes. Apes lack the reiterative ability, the narrow language faculty (FLN), that supposedly is the key to human syntactic ability (Hauser, Chomsky, and Fitch 2002)—but this deficiency is most evident in their speech, not their syntax. Anatomic limitations, which are discussed in following chapters, limit the range of phonetic forms that apes can produce. However, acoustic analysis of the vocal signals that they produce in a state of nature shows that, in itself, this does not prevent them from talking. Chimpanzees and other primates produce many of the segmental phonetic elements that could be used to form spoken words. Chimpanzee and monkey vocalizations, for example, include formant frequency transitions and patterns that could convey the consonants [m], [b], and [p] and the vowel of the word "but" (Lieberman 1968; Hauser 1996; Riede et al. 2005). Computer modeling studies (Lieberman, Klatt, and Wilson 1969; Lieberman, Crelin, and Klatt 1972) also show that the speech anatomy of chimpanzees and other primates can produce the sounds [n], [d], and [t] and most vowels other than [i], [u], and [a] (the vowels of "tea, too, and ma"); but chimpanzees are unable to voluntarily recombine these individual sounds into the different sequences necessary to form English words.

Studies of chimpanzees now suggest that their vocal signals convey complex referential information (Slocombe and Zuberbuhler 2005b). This is not surprising because monkey species also transmit referential information by means of vocal signals. These field observations, however, suggest that their repertoire is limited; chimpanzees even have great difficulty suppressing their calls in situations in which that would appear to be warranted (Goodall 1986). But it is obvious that it is a mistake to think that only speech differentiates human beings from apes. The vocal signals of apes are only one aspect of their "bound" behavior; their cognitive and creative abilities are limited (Lieberman 1994b).

One avenue of research discussed in detail in Chapter 6 appears to reflect our genetic endowment. Orofacial motor control, speech production, syntax, and cognitive behavior are adversely affected by the anomalous expression of a gene (FOXP2) that regulates the development of neuroanatomic structures of the basal ganglia, cerebellum, and other neural structures in a large extended family (Vargha-Khadem et al. 1995; Lai et al. 2001; Watkins et al. 2002). The most apparent deficit of the afflicted members of this family is difficulty in sequencing orofacial gestures. The human FOXP2 gene differs from similar genes in present day apes and appears to have evolved to its human form within the last 100,000 years (Enard et al. 2002; Vargha-Khadem et al. 2005), the approximate time frame for the appearance of anatomically modern *Homo sapiens* (Ingman et al. 2000).

As I hope to demonstrate, the neural mechanisms that yield our species-specific ability to talk are gradually becoming apparent. The neural substrate that allows us to form a limitless number of words also appears to be implicated in human syntactic ability, allowing us to form a potentially limitless number of sentences. And the same neural mechanisms that grant linguistic creativity confer cognitive flexibility and other seemingly unrelated creative attributes. The question that must be answered by those who would claim that human linguistic ability is totally disjoint from other aspects of behavior is, "Has anyone ever seen an ape dancing?"

CHAPTER 3

The Singularity of Speech

As the previous chapters have noted, speech is one of the derived attributes that differentiates *Homo sapiens* from other living species. The complex process by which we produce and discern words through speech enables us to transcend the limits that would otherwise be imposed by our commonplace mammalian auditory system. Human speech allows us to transmit information vocally at rates that are ten times as fast as we could achieve with other sounds. Inner speech—silent, internal modeling of speech using the neural mechanisms that generate overt speech—keeps words active in our brain's short-term verbal working memory as we comprehend the meaning of a sentence. If you did not have the gift of speech, and you were listening to some arbitrary sequence of sounds, you most likely would forget the beginning of this sentence before you came to its end.

This chapter covers several closely related topics, beginning with the anatomy and physiology involved in producing speech, which entails an explanation of the role of the airway above the larynx, the supralaryngeal vocal tract, and the nature of "formant frequencies" in a manner that most readers hopefully will be able to follow. I will then explain how the formant frequency encoding-decoding process yields the high data transmission rate of speech. Next we tackle another set of closely related, and as yet not fully understood, issues that concern the manner in which we perceive formant frequencies as well as the process by which we take into account the length of a person's airway, which affects the formant frequency patterns that differentiate vowels and consonants.

We subconsciously take account of the fact that the formant frequency pattern that conveys a particular vowel or consonant is different for a child, an adult, or, for that matter, for an adult who is taller or shorter than another person. This leads the discussion into the properties of the supervowel [i] (the vowel of the word "see") and the species-specific anatomy of the human airway that allows us to produce that sound. The evolution of human, species-specific, speech-producing anatomy is discussed in Chapter 6, after some explanation of the neural bases of speech and language. One of the messages of this chapter is that, although human speech has singular properties, the continuity of evolution is evident when we examine the aspects of vocal communication that are common to humans and other species.

The Basic Anatomy and Physiology of Speech

Johannes Muller (1848), one of the founders of physiology and psychology, discerned the three basic systems—the lungs, larynx, and supralaryngeal vocal tract—that generate the vocal communications of most terrestrial animals, excluding some birds. Figure 3.1 blocks out these components.

Lungs

The lungs provide the air stream that powers speech. The primary, life-sustaining, "vegetative" role of the lungs is to oxygenate the bloodstream. The inspiratory muscles of the chest, abdomen, and diaphragm must first expand the internal volume of the lungs, lowering the alveolar air pressure within the lungs below the ambient atmospheric pressure so air can flow into the lungs, inflating them. The lungs are elastic and energy is stored as they expand. A useful

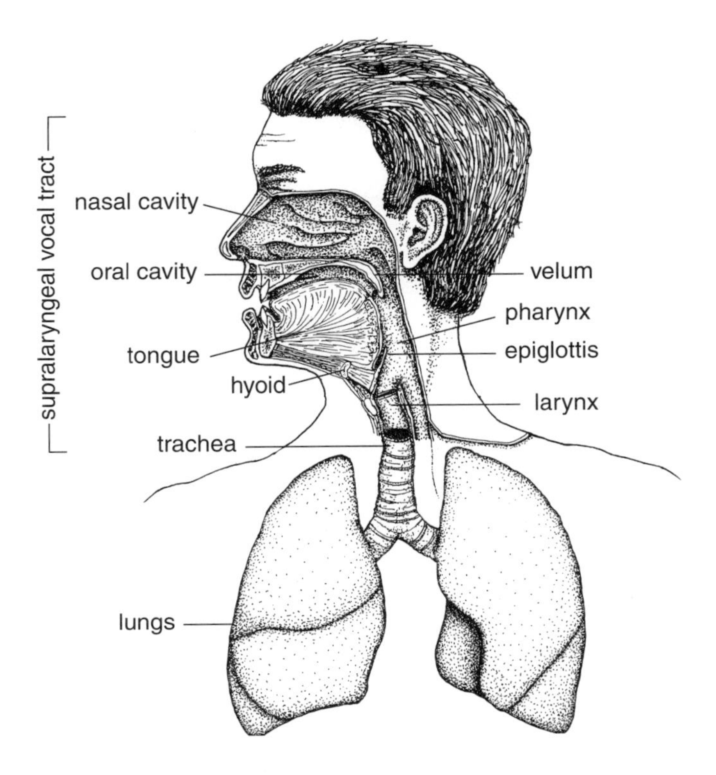

FIGURE 3.1 *The anatomic bases of speech production—the lungs, larynx, and the supralaryngeal vocal tract.*

analogy is the energy stored in a rubber balloon as it is inflated. Expiration occurs when the inspiratory muscles release their hold; the elastic force exerted by the lungs is the primary motive power forcing air out of the lungs during expiration. Expiratory muscles acting on the rib cage can also act to force air out of the lungs.

The anatomy and physiology of the lungs make "sense" in the light of evolution. The starting point for the evolution of the anatomy and basic operation of human lungs may have been in the briny deep. Fish extract air dissolved in water through their gills. Phylogenetically advanced fish achieved the ability to transfer air from their gills into two swim bladders, which are elastic sacks that change size as they are inflated or deflated. By using their swim bladders to displace a volume of water that has a weight equal to its own weight at a given depth, an "advanced" fish can effortlessly float. The basic system survives in humans though we cannot float in the depths of the sea. Although, as noted earlier, some controversy exists as to whether swim bladders evolved from lungs, with lungs then reappearing later (though that scenario seems dubious because aquatic life preceded terrestrial life), that does not change the physiology of our respiratory system. As we breathe in, inflating our lungs, our body volume increases. You can see that this is the case if you mark the water level as you lie immersed in a bathtub. As you breathe in, the level will rise slightly because the volume of water you are displacing has also risen slightly.

The primary force for expiration in humans is the elastic recoil of the two lung sacks. Chapter 7 explores how the anatomy and physiology of the lungs play a part in structuring the acoustic signals that constitute the melody, or intonation, of speech.

The Larynx and Phonation

The larynges of virtually all mammals—rats, sheep, dogs, baboons, chimpanzees, and so on—work in much the same manner as the human larynx (Negus 1928, pp. 142, 143; Van den Berg 1958). This again follows from the fact that the larynges of all terrestrial animals have a common origin, and those of mammals are very similar (Negus 1949). The larynx evolved from a valve that protected the lung in archaic fish similar to the modern lung fish (*Protopterus, Neoceratodus,* and *Lepidosiren*). The larynx in these fish is a longitudinal slit in the floor of the pharynx, with muscle fibers lying on either side of the slit that close it in a sphincteric manner (Negus 1949, pp. 1–7).

The function of this primitive larynx, which looks like a small set of human lips (another sphincteric organ), is simply to protect the fish's lungs from the intrusion of water. Victor Negus, in his comparative studies of the phylogenetic development of the larynx, *The Mechanism of the Larynx* (1928) and *The Comparative Anatomy and Physiology of the Larynx* (1949), showed the series of adaptations from the simple larynx of the lung fish to the mammalian larynx. Chapter 7 provides some of the details. In brief, a series of adaptations added cartilages that initially served to facilitate the closing of the laryngeal opening, the glottis. Further adaptations facilitated breathing or phonation.

The vocal cords of the larynx are not really cords. The term derives from a theory proposed by Ferrein (1741), who thought that the larynx worked like a stringed instrument in which strings (hence *cordes,* in French) vibrate to produce sound. This is not the case; Muller (1848) showed that the larynx instead works more like the reeds of a woodwind instrument, in which reeds flap to interrupt the flow of air through the instrument. The actual situation is more complex, but the analogy suffices. The vocal cords, often termed "folds," consist of a complex set of cartilages, muscles, and soft tissue that can be set into motion by the airflow from the lungs.[1]

Muller modeled laryngeal activity in a straightforward manner. He placed a larynx that had been excised from a cadaver over a hole on a board. A tube was placed under the hole so that air from a bellows could be blown through the excised larynx. Cartilage is a stiff, bonelike material that resists deformation; ligaments are similar to nylon strings. For a detailed description of the larynx, look to works like Negus's *Comparative Anatomy and Physiology of the Larynx* (1949). The sketch in Figure 3.2 illustrates the larynx and its supporting structures.

Though Muller obviously did not have tape recorders, computers, electronic instruments, or high-speed photography or video to monitor the output of the larynx, he obtained quantitative data that were consistent with his theory of phonation. Muller used an ingenious mechanical system of strings, weights, and pulleys to apply different forces to the thyroid cartilage while he blew air through the larynx. The thyroid cartilage is hinged on the cricoid cartilage and the complex set of muscles and cartilage that constitute the vocal cords bridge these thyroid and cricoid cartilages. The cricothyroid muscle connects the two cartilages and in life changes the tension by tilting the thyroid cartilage with respect to the cricoid. Muller, therefore, was able to simulate the effects of changing

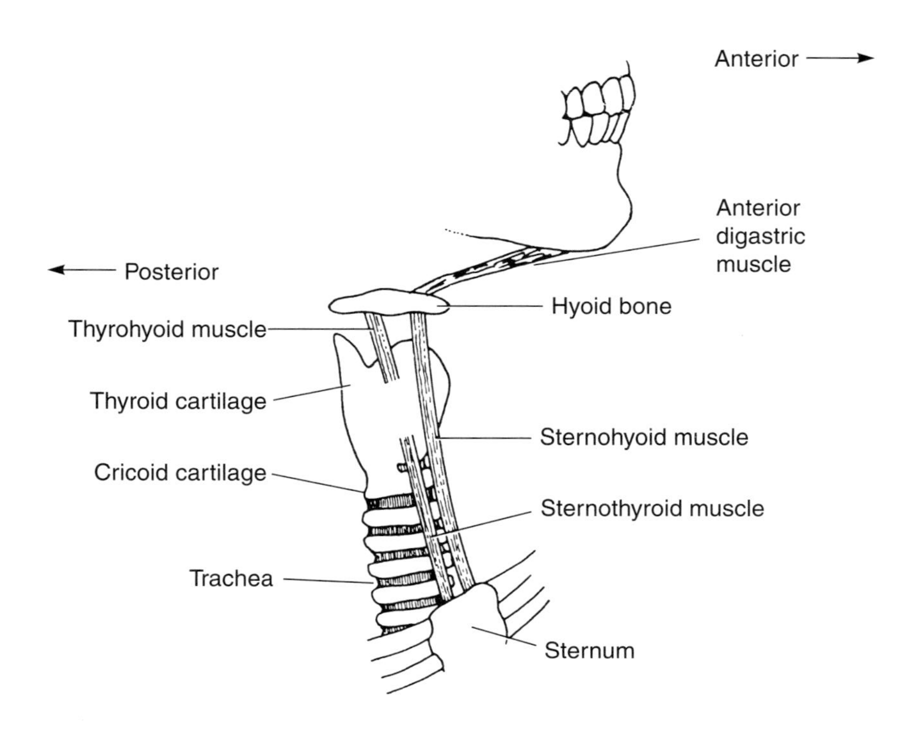

FIGURE 3.2 *The thyroid and cricoid cartilages of the larynx support the vocal cords. The larynx is suspended by ligaments and muscles from the hyoid bone. The hyoid, in turn, is suspended from the mandible and the styled process of the skull. The trachea (windpipe) connects the larynx to the lungs. The larynx can be pulled downward toward the sternum or upward. The human larynx is pulled upward and forward toward the mandible as we swallow to minimize the chance of food lodging in it.*

the tension of the cricothyroid muscle during phonation. As he added weights to the strings connected to the thyroid cartilage, tilting it, increased tension was placed on the vocal cords, simulating the real-life situation. Muller listened to the pitch of the laryngeal output as he varied the simulated muscle tension. He found that increasing the tension placed on the vocal cords by the cricothyroid muscle raised the fundamental frequency of phonation. Muller also discovered that the fundamental frequency of phonation would increase if he blew air through the vocal cords with greater force. These findings have since been replicated in experiments with excised larynges (the plural of larynx) (Van den Berg 1958) and in experiments in which the tension of the laryngeal muscles in living human beings is monitored during actual phonation by means of electromyographic techniques (Ohala 1970; Atkinson 1973). These and other experiments have monitored the force of the air against the vocal cords (the subglottal air pressure) during actual phonation (e.g., Lieberman 1967).

Figure 3.3A shows a sequence of photographs of the larynx closing for phonation. The photographs are frames from a high-speed motion picture sequence that was exposed at a rate of 10,000 frames per second (Lieberman 1967). The closing gesture took about 0.6 second. The sequence of photographs in Figure 3.3B is from the same high-speed movie during phonation. The interaction of the forces developed by the air stream from the lungs and the tension of the tissue of the vocal cords caused the vocal cords to open and close rapidly. Phonation cannot take place without a person's blowing air through the closed or partially closed vocal cords. For phonation, a speaker must first move the vocal cords inward from the open position that they hold for inspiration. Phonation then takes place when the force generated by the pressure of the air impinging against the closed or partially closed vocal cords results in a flow of air through the larynx. This generates a sequence of puffs of air. The fundamental frequency of phonation, F0, is the rate at which the puffs of air from the larynx occur. Our perceptual response to this fundamental frequency is the pitch of a person's voice. The faster the rate at which the puffs occur, the higher the pitch that we "hear."

The fundamental frequency of phonation of the larynx and the amplitude of the laryngeal output are sensitive to variations in the air pressure that drives the vocal cords. Hence the subglottal air pressure—that is, the air pressure developed by the lungs (essentially equal to the alveolar air pressure in the lungs), which impinges on the vocal cords of the larynx—must be regulated during the production of speech. Vocal cords that have a greater mass also will accelerate more slowly given the same driving forces. (Newton's second law

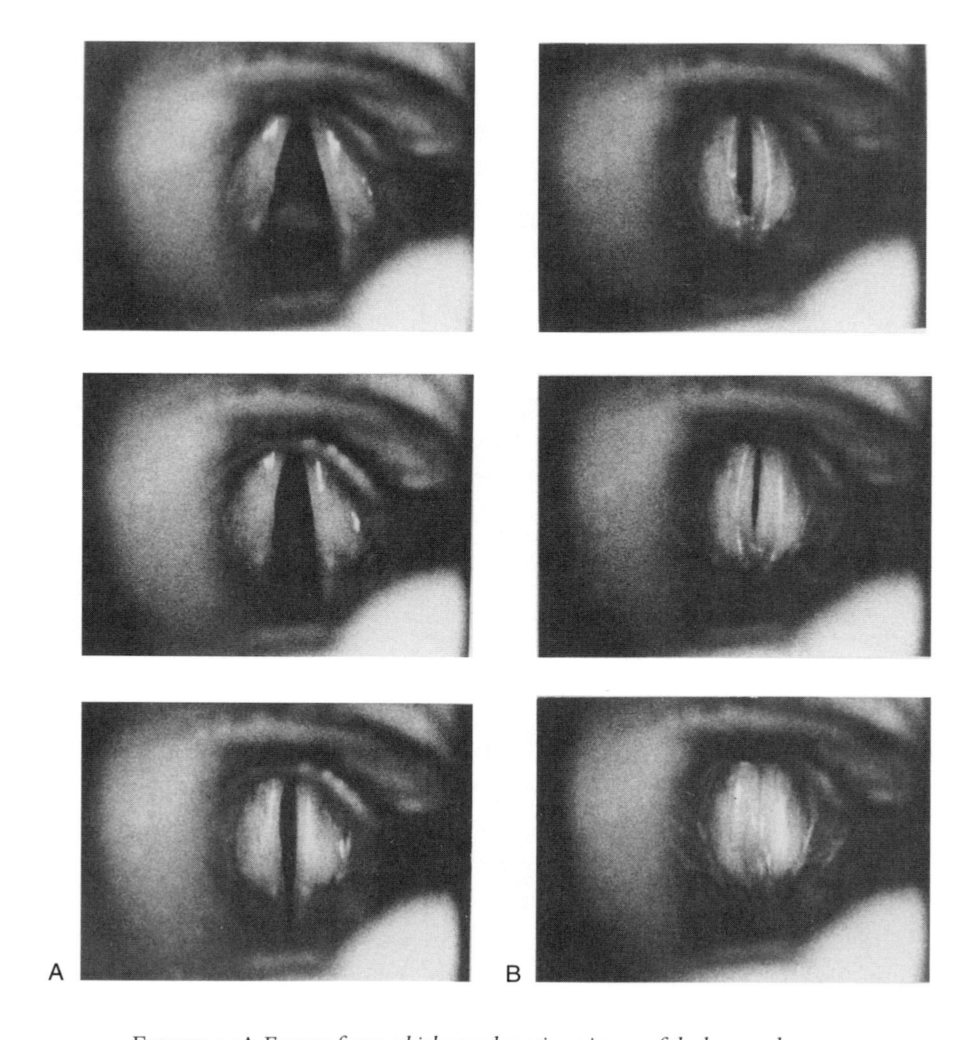

FIGURE 3.3A *Frames from a high-speed motion picture of the human larynx. The camera is looking downward on the vocal cords. The posterior part of the vocal cords is oriented toward the bottom of each frame. The uppermost frame shows the vocal cords as they start to adduct or close to initiate phonation. The lower frames were exposed 50 and 100 msec later.*

FIGURE 3.3B *Three frames from the same high-speed movie during phonation. The uppermost frame shows the maximum opening of the vocal cords. The bottom frame, which was exposed 7 msec later, shows the vocal cords closed. Note the rapidity of vocal cord motion during phonation after the vocal cords have closed. Note the rapidity of vocal cord motion during phonation compared with the closing gesture of Figure 3.3A. (After Lieberman 1967)* of motion, F=MA, or force equals mass multiplied by acceleration, applies.) Therefore, adult males who have a larger larynx with bigger vocal cords generally have lower F0s. In general, adult males have lower F0s and hence have voices with a lower pitch; women and children's larynges are usually smaller and produce higher F0s and hence higher-pitched voices.

The larynges of all primates can generate sound on the inspiratory as well as the expiratory phase of respiration. The cries of newborn infants, for example, often are phonated during part of the inspiratory phase (Truby, Bosma, and Lind 1965; Lieberman et al. 1972). However, because phonation cannot take place until the glottis closes, it cannot be sustained throughout inspiration because the glottis must then open to admit air into the lungs. Thus the basic vegetative function of the lungs and larynx—breathing—explains why phonation is generally produced on the outward expiratory flow of air from the lungs.

The primary role of the larynx in vocal communication, both for humans and for many other species, is to efficiently generate acoustic energy (sound) at frequencies that can be heard. The fundamental frequency (F0) of a periodic, repetitive event is equal to the inverse of the rate at which it occurs, 1/T, where T equals the duration of the periodic event. The mathematical tools developed by Fourier and other mathematicians in the eighteenth century show that the energy present in a periodic event can occur at F0 and its harmonics—the integral multiples of F0, that is, 2F0, 3F0, 4F0 . . . nF0. The human auditory system "hears" periodic air pressure variations that occur between approximately 20 to 20,000 Hz. The rate at which people breathe rarely exceeds 20 cycles during the course of a minute, yielding an F0 of 0.3 Hz (frequencies are generally measured by the unit Hz, the number of periodic events that occur in a second). Therefore, the fundamental frequency of the air moving outwards from a person's lungs cannot be heard because it is occurring below the lower range of the auditory systems of humans and other terrestrial animals (Heffner and Heffner 1980).

We can hear a person breathing as we listen to the noise-like acoustic energy generated by turbulent air flow, but we cannot sense the low fundamental frequency of the air flowing in and out of the lungs. The larynx converts the outward flow of air from the lungs to audible sound through phonation, generating a series of quasi-periodic puffs of air as the vocal cords rapidly open and close. Acoustic energy is present at the harmonics of F0, which in humans ranges from about 60 Hz in adult males to 1.5 kHz (1,500 Hz) in young children (Keating and Buhr 1978). Thus the glottal energy source has acoustic energy within the range of frequencies that the human auditory system perceives as sound (the term "glottal" refers to the opening between the vocal cords). Phonation occurs in amphibians, such as frogs; their larynges also serve as acoustic transducers, producing useful acoustic energy that can be heard by these creatures. The energy source for whispered speech is "noise" produced by the turbulent flow of air through the larynx, but whispers cannot be heard at any reasonable distance and they use up a great deal of air (Klatt, Stevens, and Mead 1968), limiting the duration of a sentence.

As we talk, we control fundamental frequency so as to bind together and delineate the words that constitute a sentence. The F0 contour, which is the time course of F0 over the sentence, serves as a form of vocal punctuation that can be used to signal the end of a sentence or major clause (Armstrong and Ward 1926; Jones 1932; Lieberman 1967). In the languages spoken by most people on earth, local changes in the F0 contour, limited to syllables, can differentiate words. For example, in Mandarin Chinese the syllable [ma] produced with different F0 contours can mean "hemp," "mother," or "horse." The linguistic function of laryngeal activity is evident in many species. As Chapter 2 notes, vervet monkeys signal the presence of different predators by means of alarm calls that are differentiated by different F0 contours (Cheney and Seyfarth 1990). Chapter 7 examines the linguistic uses of vocal signals, including intonation.

The glottal source's energy content also is modulated when people talk and sing. The classic registers of phonation used by singers, such as falsetto or chest, are the result of laryngeal maneuvers that yield more or less energy at high frequencies (Van den Berg 1958; Bouhuys 1974). In some languages, words are differentiated by phonation in different registers. Moreover, emotional information also is conveyed by the F0 contour and register changes, both for human beings and most mammals (c.f. Hauser 1998). The screams of an injured dog are unmistakable. Terror and tension in aircraft accidents can be assessed through measurements of F0 and the energy content of the glottal source. Both F0 and amplitude rise and the energy content of the glottal source shifts to higher frequencies (Protopappas and Lieberman 1997; National Transportation Safety Board 2002). The particular characteristics of the glottal source and F0 also play a part in identifying a person's voice; observations made by telephone engineers dating back to the 1920s noted that this is the case (Lieberman 1967).

The Supralaryngeal Vocal Tract (SVT)

Human speech involves much more than the larynx. If a larynx (sometimes termed the "voicebox") was all we had, we would only be able to change the fundamental frequency of phonation and other characteristics of the glottal source. We would not be able to produce most of the distinctions in vowel and consonant quality that typify human speech—these phenomena and the high data transmission rate that distinguishes human speech from other vocal signals depend on the filtering effects of the airway above the larynx, formed by the pharynx, mouth, and nose. Together, these constitute the human supralaryngeal vocal tract.

Johannes Muller (1848) noted the basic role of the supralaryngeal vocal tract (SVT) in his studies with excised larynges. Although Muller's focus was on laryngeal physiology, his most significant finding was what the larynx did *not* do. Muller observed that the sound that came directly from the excised larynx did not sound like speech. Speech-like quality was achieved when he placed a supralaryngeal tube with a length and shape roughly equal to that of the airway that intervenes between the vocal cords of the larynx and a person's lips.

Muller most likely was not surprised to find that the sound produced by the larynx had to be passed through a tube in order to sound natural. Kratzenstein (1780) had won a prize offered by the Academy of Sciences of St. Petersburg to the first person who could artificially produce the vowel sounds of Russian, [a], [i], and [u] (the vowels of the words "ma," "me," and "moo"), by using tubes that had different shapes and lengths, which he excited by vibrating reeds. When phoneticians place a symbol between brackets, such as [i], they are using the phonetic alphabet to specify particular speech sounds (International Phonetic Association 1949). The acoustic properties of Kratzenstein's tubes determined the synthesized vowel. The tubes acted as acoustic filters and the reeds as sources of acoustic energy. Von Kempelen (1791) produced a talking machine by the same means before he went on to his notorious chess-playing machine (a midget in a box).

Muller found that the larynx bore the same relation to the human supralaryngeal vocal tract as the reeds bore to the tubes of Kratzenstein's mechanical model. The larynx provides the source of acoustic energy for phonated vowels, while the SVT acts as an acoustic filter that determines the phonetic quality of the vowel sound. A given SVT shape lets more or less acoustic energy through at different frequencies. The frequencies at which maximum energy can pass through the SVT are termed *formant frequencies.* Hermann (1894) introduced the term. The vowels and many consonants of English, for example, are differentiated by their formant frequency patterns. The vowels [i] and [u] (the vowels of the words "see" and "sue") can be produced with identical F0s; the different formant frequencies specify these vowels. The orthographic symbol [h] signifies a vowel that is produced by a noisy, turbulent air flow through the larynx without any phonation. Human beings have an SVT that can change its shape and length, allowing maximum acoustic energy to pass through it at frequencies that continually vary.

The Source-Filter Theory of Speech Production

The relationship between the SVT, formant frequencies, and the laryngeal source is to a degree similar to the way a stained glass window works. The daylight that illuminates the window is the source of light energy. Unlike the color produced by a neon sign, which generates light energy at a frequency that our visual system perceives as red, the color that we see in the stained glass is the result of how the glass attenuates—that is, reduces—the light energy at different electromagnetic frequencies. We perceive the light frequencies that pass through the stained glass with least attenuation as different colors. For example, increased light energy passing through a portion of the stained glass at high frequencies is perceived as a blue color; light passing through at low frequencies is perceived as a red color. The electromagnetic frequency peaks that we perceive as colors are analogous to formant frequencies.

Another, perhaps more commonplace, analogy to the relationship between formants, the laryngeal source, and speech signals may be apparent when you select sunglasses. The balance of frequencies of light energy from the sun that passes through the sunglasses determines whether everything looks blue or pink. The tinted glass in the sunglasses achieves a particular effect by attenuating the amount of light energy at various frequencies. The light energy frequencies that are least attenuated determine the color. The source of light remains the same—sunlight—but the filtering by different sunglasses creates a blue or pink world. Think of formant frequencies as the acoustic frequencies that the SVT allows to pass through it with minimum attenuation. What we perceive are different speech sounds.

It would be perfectly feasible to synthesize speech by mechanical systems similar to Muller's. Magnetic resonance imaging (MRI) equipment can accurately determine the shape of the SVT for different vowels. We could make a brass tube that had the right shape for a particular vowel. If we then excited this tube with a source of acoustic energy that approximated phonation—an artificial larynx—

we would have energy at the correct formant frequencies and we would hear the particular vowel sound. However, that approach isn't very convenient and many different types of speech-synthesizing devices have already been developed in the course of modern research on the source-filter theory of speech production (Fant 1960; Henke 1966; Rabiner and Shafter 1979; Maeda 1989; Carre, Lindblom, and MacNeilage 1995; Story and Titze 1998). Contemporary speech synthesizers are instantiated in software run on digital computers.

The midsaggital view in Figure 3.4 sketches the human supralaryngeal vocal tract, the airway consisting of the oral cavity (mouth), tongue, pharynx, and nasal cavity (nose). The figure also shows the entrance to the esophagus, which leads to the stomach. A midsagittal view corresponds to what you would see if a head were sectioned in half, front to back. The primary vegetative function of the human tongue, as is the case for the tongues of other species, is to ingest food. When food is introduced into the mouth, the tongue pushes food along the roof of the mouth (the hard palate) to the point at which it can be propelled down the pharynx into the esophagus. The most extreme motions of the human tongue occur while we eat; the tongue blade starts to propel food back along the palate. The tongue body then continues to push food backward, moving upward to the roof of the mouth and thrusting backward along the palate (Kramer 1991; Palmer et al. 1992; Hiiemae et al. 2002).

In contrast, during speech production the tongue contour conforms to its rest position, hardly changing its shape for any vowel (Russell 1928; Carmody 1937; Ladefoged et al. 1972; Nearey 1979; Baer et al. 1991; Story, Titze, and Hoffman 1996; Hiiemae et al. 2002). Figure 3.5 shows the tongue's contour in the midsagittal plane during different vowel sounds; note that it has a similar contour, resembling a semicircle that delimits the lower surface of the oral cavity and anterior surface of the pharynx. This is not surprising because the human tongue and those of virtually all mammals is a hydrostat (Stone and Lundberg 1996). That means it is a muscle

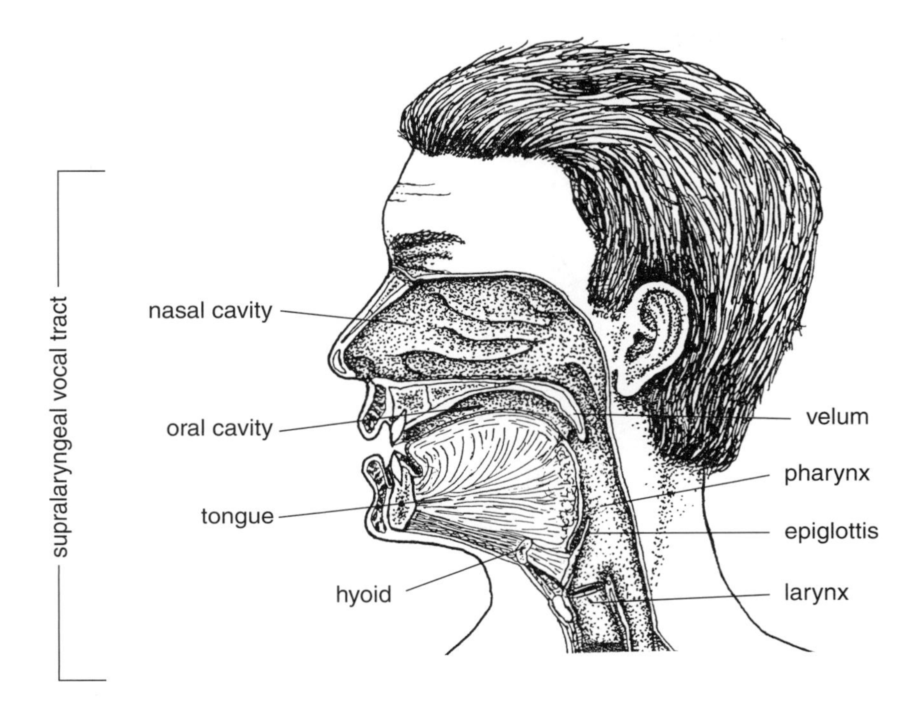

FIGURE 3.4 *Midsaggital sketch of the human supralaryngeal vocal tract (SVT). The nasal cavity can be sealed off by raising and tensioning the velum. The tongue is positioned in the oral cavity and pharynx. The position of the larynx relative to the roof of the oral cavity and skull base is low, owing to half of the tongue being positioned in the pharynx. The esophagus, which leads to the stomach, is positioned behind the laryngeal opening.*

that cannot be squeezed into a smaller volume and, when we produce different vowels, the extrinsic muscles of the tongue propel it up or down, forwards or backwards, without materially changing its shape. The changes in tongue shape seen when we swallow involve more extreme maneuvers (Hiiemae et al. 2002). Speech production appears to be an energy-efficient activity, which perhaps explains why we can talk on and on.

The Antiquity of Vocal Communication and Formant-Like Signals

All mammals can be traced back to ancestral species similar to present-day frogs. Therefore, it is not surprising to find that frequency

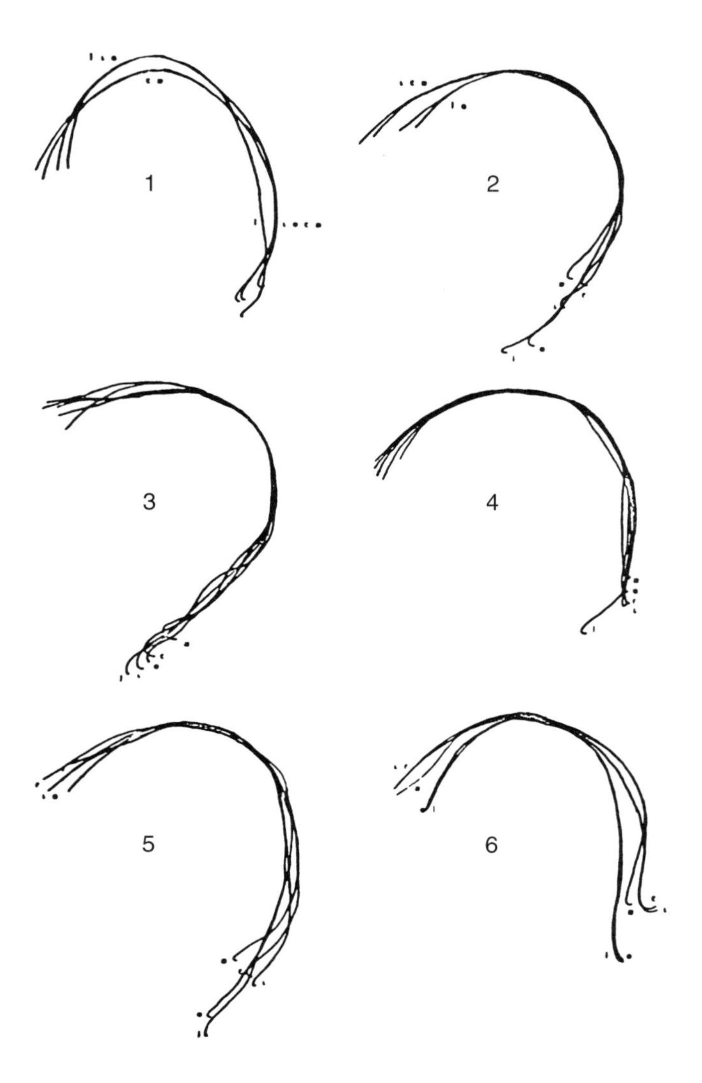

FIGURE 3.5 *Tongue contours used to produce vowels for six normal adult speakers. These tongue contours were derived from a cross-linguistic cineradiographic study of vowel production (Ladefoged et al. 1972). The contours for the vowels traditionally classified as "front" vowels (see Chapter 7; the vowels of the English words beet, bit, bet, bat) have been superimposed. Note that the semicircular tongue contour is virtually identical for all of these vowels.*

peaks similar to formants and the fundamental frequency of phonation both play a part in conveying the vocal calls of frogs. A series of experiments in the 1960s showed that frogs possess perceptual systems "tuned" to the specific acoustic properties of their mating calls. The graph in Figure 3.6 shows a frequency analysis of the acoustic output of a synthesizer generating a bullfrog's mating call. The synthesizer was designed for generating human vowel sounds, but it served equally well for some bullfrog sounds. The speech synthesizer, a POVO system (Stevens, Bastide, and Smith 1955), generated a sustained vowel-like sound that has a fundamental frequency of phonation of 100 Hz. In Figure 3.6 the fundamental frequency of phonation shows up in the lines that occur at 100 Hz intervals.

Bullfrogs (*Rana catesbeiana*) vocalize in much the same manner as

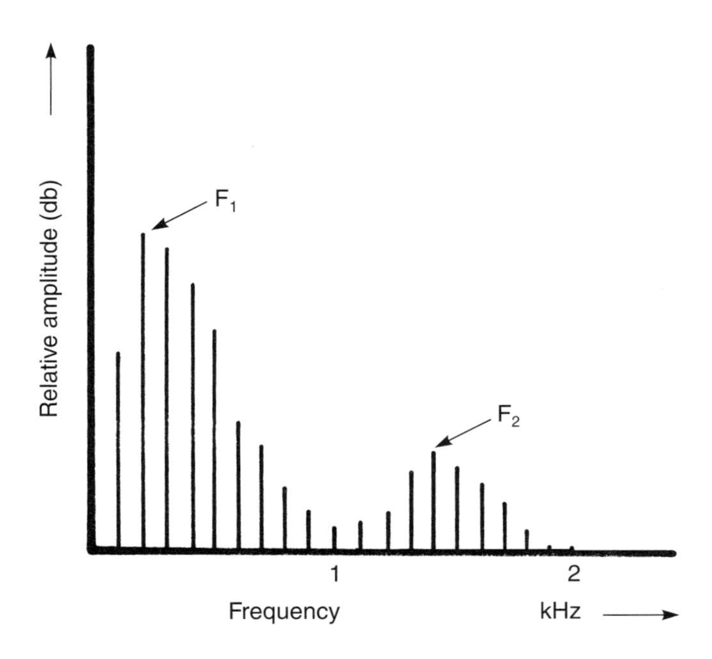

FIGURE 3.6 *Spectrum of bullfrog mating call produced on a speech synthesizer. F1 denotes the first amplitude peak at a frequency of 0.2 kHz (or 200 Hz); F2 denotes the second amplitude peak at a frequency of 1.4 kHz (or 1400 Hz).*

we do when we produce vowels. The vocal cords of the frog larynx open and close rapidly, emitting puffs of air. As is the case for all periodic sources of energy, when the bullfrogs' vocal cords open and close at a rate of 100 Hz (100 times a second), they generate an acoustic signal that has energy present at an F0 of 100 Hz and its harmonics—integral multiples of 100 Hz. The bullfrog call thus has acoustic energy present every 0 Hz, that is, 100, 200, 300, 400, . . . , 900 Hz, and so on. However, the energy content of the spectrum is modulated, producing energy peaks similar to formant frequencies. These peaks are apparent in Figure 3.6 as the relative peaks in the graph, which occur at 200 Hz and at 1.4 kHz.

Frogs can make a number of different calls (Bogert 1960), including mating calls, territorial calls that serve as warnings to intruding frogs, rain calls, distress calls, and warning calls. The different calls have distinct acoustic properties, and there are some obvious differences in the manner in which frogs produce some calls. For example, the distress call is made with the frog's mouth wide open, whereas all other calls are made with the mouth closed. The articulatory and acoustic distinctions that underlie other calls are not as obvious. Capranica (1965), however, analyzed the acoustic properties of the bullfrog mating call in detail. In behavioral experiments Capranica used a POVO speech synthesizer to synthesize the mating call and presented the synthesized stimuli to bullfrogs. The bullfrogs responded by joining in a mating call chorus, as long as the synthesized mating call had acoustic energy concentrations at either 200 Hz or 1.4 kHz or both. The presence of energy concentrations at other frequencies inhibited the bullfrogs' responses.

Frog calls shed light on human speech in that they involve a neural system for interpreting these species-specific signals that is "matched" to the frog's vocalic capabilities. Studies of the neuropsychology of human beings are always limited by the indirect techniques that have to be used. Direct techniques that record the electrical activity of the brain at the single-cell level involve using electrodes that ultimately destroy parts of the animal's brain. These direct techniques, however, can selectively record the electrical responses of cells to stimuli that are presented to an anesthetized animal. Frishkopf and Goldstein (1963), in their electrophysiologic study of the bullfrog's auditory system, found two types of auditory units in the anesthetized animal. They found units of the eighth cranial auditory nerve whose cells had maximum sensitivity to frequencies between 1.0 and 2.0 kHz and other units whose cells had maximum sensitivity to frequencies between 0.2 and 0.7 kHz. Maximum electrophysiologic responses occurred when the frogs were presented with acoustic signals that had energy concentrations at or near the two formant-like frequencies of bullfrog mating calls, at F0s between 50 and 100 Hz, the rates at which the frog larynges phonate for this call. The auditory nerve units were not simply responding to noise—adding acoustic energy between the two peak frequencies at 0.5 kHz inhibited the response of the low-frequency units. The electrophysiologic, acoustic, and behavioral data concerning bullfrog mating calls thus all complement each other. In short, bullfrogs have a neural mechanism that specifically responds to the bullfrog mating call. Capranica tested his bullfrogs with the mating calls of thirty-four other species of frogs, and they ignored them all, responding only to bullfrog mating calls. Bullfrogs do not respond to just any sort of acoustic signal as though it were a mating call; they respond to acoustic signals that have energy at these specific energy peaks. The best stimuli, furthermore, must have the appropriate fundamental frequency of the natural bullfrog mating call. The bullfrog's neural perceptual apparatus is demonstrably matched to the acoustic characteristics of its call.

Although electrophysiologic data similar in kind to that derived from frogs is obviously not available for human beings, we shall see that human beings have neural systems for speech perception that are matched to the productive limits of our speech-producing anatomy.[2] There are, of course, many differences between people and frogs, but there are also some striking similarities. The basic acoustic parameters that describe the speech signals of human beings and frog vocalizations, the fundamental frequency of phonation, and the spectral energy peaks are similar.

Speed—Fast Talking and Formant Frequency Patterns

Although we can state with certainty that frogs have neural mechanisms that are "tuned" to respond to their vocal signals, that premise isn't immediately obvious for human beings. It, in fact, would seem very unlikely if you turned the dial on a short-wave radio and listened to a broadcast in an unknown language. It is generally impossible to even identify the individual words. We often cannot even differentiate or imitate the sounds of an unfamiliar language. Phoneticians have to train for years in order to make transcriptions of uncommon languages and obscure dialects. Indeed, the obvious rejoinder to any claim that people, like bullfrogs, have innate neural mechanisms tuned to the sounds that they can produce is that all bullfrogs make the same sounds, but all people do not. Emily Lattimore in her account of travels in Central Asia more than a half century ago tells how her husband, Owen, was regarded as a sage by Siberian Tatars. In one remote encampment:

> other guests had arrived, two Tatars in Russian blouses and two Chanto [tribesmen] in skull caps and loose white coats. One of the latter delighted us particularly, a fat fellow who looked like a perspiring egg with a little black skull cap on one side of his small end and his clothes hanging on him like Humpty Dumpty and always coming unbuttoned. He was avid for information of the world and spent the day mopping his face and plying Owen with questions. In fact they all sat around all day on gay rugs and asked Owen questions as if he were an oracle, one of the best of which was, "If sheep in one part of the world make the same kind of noises as sheep in any other part of the world, why is it that men don't talk the same all over the world?" (Lattimore 1934, 116–117)

The burden of proof thus rests on anyone who claims that humans possess neural mechanisms that play a part in perceiving the sounds of speech, or, for that matter, any claim that there is anything special about the sounds of human speech. Would not a more plausible theory be that infants and young children just learn to string more or less arbitrary sounds together to form words and sentences? After all, many people learn the letters of the alphabet a few years later, and then again learn to string these letters together so they form words and sentences. These are not merely rhetorical questions that I am inventing to change the pace of this exposition. They were at one time or another implicit or explicit hypotheses that led to attempts to make complex "talking machines."

In fact, many linguists still believe that the phonetic aspects of human language are trivial. In the four international conferences on the evolution of language that have taken place in the last decade, 90 percent of the papers focused on syntax. The putative insignificance of speech is not recent; Simpson, for example, in reviewing attempts to trace the evolution of language, notes that "audible signals capable of expressing language do not require any particular phonetic apparatus, but only the ability to produce sound, any sound at all" (1966, p. 473).

If Simpson's claim were true, any sequence of sounds that people found tolerable could serve as a vehicle for linguistic communication. We could, for example, communicate with each other by using the sounds that spoons and forks make when tapped on a wall. The system is limited in its scope, but people have in fact used it. The codes that prisoners have devised to "talk" to each other through walls are examples. Morse code is a system that makes use of sequences of dots and dashes transmitted by short and long beeps. However, Morse code is slow; the fastest speed that a Morse code operator can achieve is about 50 words per minute, which translates to about 200 to 250 letters per minute. Moreover, to achieve that rate the Morse code operator would have to devote full attention to transcribing the message. Morse code operators usually do not remember what they transcribe and must rest after an hour or two to recover from the fatigue engendered by the intense concentration. In contrast, if we remain awake, we can easily follow a long lecture delivered at a rate of 150 words per minute.

The fastest rate at which people can produce and perceive speech sounds is about 20 to 30 phonemes per second. The letters of the alphabet reasonably approximate "phonemes," the sounds that differentiate words (Chomsky and Halle 1968). The speech rate (1200 to 1800 phonemes per minute) is approximately ten times faster than that achieved with other sounds. The research that I review demonstrates that speech has special properties that make possible the high rate at which we can communicate. Speech involves neural mechanisms that may be part of our genetic endowment and some of these neural mechanisms are present in other species, demonstrating the continuity of evolution.

Vocoders and Speech Synthesis

The research program that demonstrated the special status of the sounds of speech started in the years before World War II. Before the breakup of telephone companies in the United States, Bell Telephone Laboratories was arguably the world's premier center for the study of human speech. Telephones are used to transmit human speech, and telephone systems that are better, more reliable, or less expensive can be devised if one knows more about the nature of speech. By the mid-1930s the Bell Telephone research group under the direction of Homer Dudley knew how to make what engineers call a real-time analysis-synthesis speech system. The system was cumbersome and expensive. But the device they invented, the Vocoder, was to become one of the components of top-secret speech cryptographic transmission systems during World War II. The Vocoder later was a key element in a project directed at developing a reading machine for blind people. In the course of that engineering project, researchers found that the sounds of human speech are not discrete entities similar to the letters of the alphabet. The particular acoustic characteristics of the sounds of speech, which follow from the basic physiology of speech production, confer the speed of speech.

However, the Vocoder was developed by Bell Telephone Laboratories for none of the above reasons. It was developed to make money for the telephone company. I hope that the connection between engineering projects and advances in theoretical science will become evident as we follow this story. The divide between engineering and science becomes meaningless when engineers find that the theories they were supposed to be implementing are incorrect.

Speech production entails regulating a source of acoustic energy and a supralaryngeal filter. As we talk, we must control the source of acoustic energy—for example, the larynx—for a "voiced" sound like the vowel [i]. We also must control the positions of our tongue, lips, and velum (which closes our nose from our mouth) to set up the supralaryngeal "filter" that specifies this vowel. (Recall that the SVT produces formant frequencies by attenuation, i.e., by filtering out energy at other frequencies.) The Vocoder operates by first electronically tearing the speech signal apart into (a) electrical signals that specify some of the characteristics of the source and (b) electrical signals that specify the ensuing output of the filter. These two sets of electrical signals, (a) and (b), are then transmitted to a synthesizer that puts them together, reconstituting the speech signal. Vocoder systems thus consist of an analyzer and a synthesizer. The analyzer generally represents the character of the source using three parameters, or signals, that must be extracted and transmitted. The filter function is represented by at least sixteen parameters that also must be derived from the incoming speech signal and transmitted. The system was quite complicated; the 1936 version filled a small room with a mass of vacuum tubes.

The immediate objective for this mass of electronic equipment was reducing the frequency range necessary to preserve speech intelligibility in telephone calls from about 5000 to 500 Hz. The range of frequencies that can be transmitted on any particular electronic channel is inherently limited. A particular circuit, or channel—for example, the telephone cable between the United States and England—has a given capacity of, say, 50,000 Hz. By using electronic techniques, it is possible to squeeze about eight separate conversations that need a channel of 5000 Hz each into the 50,000 Hz available. (Some frequency "space" has to be left between the 5000 Hz channels necessary for each individual conversation.) Vocoders would have squeezed ten times as many conversations into the same cross-Atlantic telephone cable. Such a cable is very expensive with a limited lifespan. The cables deteriorate as salt water leaks into them. If you can squeeze ten times as many messages onto a cable and avoid laying nine additional cables, you have saved tens of millions of dollars.

Vocoders never were placed into regular service on commercial cross-Atlantic telephone cables because the Bell executives judged the voice quality to be too poor and the intelligibility of voices too low (it was often impossible to identify a speaker's voice). Vocoders, however, were adapted for use in secret cryptographic systems in which the parameters derived by the analyzer were digitized and scrambled. The scrambled signals were put together at the receiving end in the Vocoder synthesizer by using suitable cryptographic devices. Winston Churchill and Franklin Roosevelt apparently communicated using Vocoders. It was not possible to keep the principles on which Vocoders are designed secret during World War II, because they had been described in the open scientific literature in 1936. However, the actual equipment was classified. Nazi Germany attempted to build Vocoders but never succeeded, despite a great deal of effort. Work on Vocoder systems continued for many years after World War II. The Soviet Union had not been supplied with Vocoders; the Russian scientists imprisoned in Solzhenitsyn's (1965) account of Stalin's Gulag, *The First Circle,* apparently were working on building a Vocoder.

Another application of the Vocoder was developed at Bell Telephone Laboratories. The Vocoder synthesizer could be used to generate artificial speech sounds. Because the synthesizer started with a set of electrical signals that specified the acoustic signal, it was possible to alter the electrical signals to produce modified versions of actual speech signals or create synthetic speech signals that had never been spoken by a human speaker. Completely synthetic sounds could be generated by specifying the entire set of electrical control signals. Modified and synthetic speech sounds provided a new tool for determining the salient acoustic correlates of the linguistic distinctions of human speech. Research programs that combined speech analysis with synthesis revolutionized the study of speech.

Phonemic distinctions are defined as the sounds that differentiate the words of a language or dialect. For example, in English the difference between the vowels [i] and [I] is phonemic because this vowel distinction differentiates words like "heat" and "hit." In contrast, the difference between [i] and a nasalized [i] is not phonemic in English because English words are not differentiated by means of nasal versus nonnasal vowel oppositions. (In Portuguese nasal versus nonnasal vowel distinctions are phonemic.) Speech synthesizers allowed experimenters to determine systematically the acoustic factors that actually conveyed phonemic distinctions. An experimenter could, for example, determine whether the fundamental frequency of phonation played a part in determining whether a listener, whose native language was English, would hear a vowel as an [i] or an [e]. Vowel stimuli that had the formant frequencies of these vowels could be synthesized with different fundamental frequencies. The synthesized vowels could be copied a number of times in random sequence on a tape recording. A psychoacoustic listening test could then be conducted in which listeners would be asked to write down on an answer sheet whether they perceived each particular synthesized vowel to be an [i] or an [e]. Psychoacoustic tests with synthesized speech signals, in fact, showed that fundamental frequency had only a slight effect on vowel perception (Fujisaki and Kawashima 1968).

Encoding

Formant frequency patterns and *transitions,* which are time-varying formant frequency patterns, turned out to be the critical acoustic cues that make human speech special. The sound spectrograph had been designed at Bell Telephone Laboratories in the 1930s for the analysis of speech as a companion piece for the Vocoder. The sound spectrograph made it possible to see the dynamic pattern of formant frequencies as functions of time. Although it is possible to

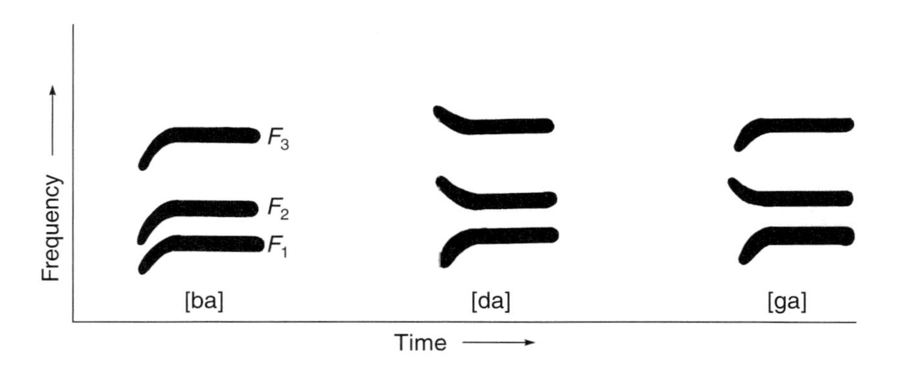

FIGURE 3.7 *Formant patterns used to control a speech synthesizer to produce the syllables [ba], [da], and [ga]. The dark bars constitute instructions to the synthesizer to produce a speech signal that has formants at these frequencies.*

produce sustained vowels in which the formant frequency pattern does not change, the formant frequencies of the speech signal continually change in the course of normal speech production. In Figure 3.7 the formant frequency patterns of the sounds [ba], [da], and [ga] are shown as dark bars. Note that the formants of the sound [ba] all start at a lower value and then abruptly rise to a rather steady level. The symbols F1, F2, and F3 in Figure 3.7 have been used to identify the formant frequencies of the [ba]. The changing part of the formant frequency pattern at the start of the [ba] is the *transition* and the steady part is the *steady state.* (Engineering jargon is transparent compared to the jargon of other professions.)

Because virtually all linguists believed that phonemes were "beads on a string," similar in nature to movable type, the question that could be answered using the technique of speech synthesis was obvious. Given that each phoneme was thought to be a discrete entity similar to a piece of movable type, a logical question was, What part of the formant frequency pattern of the syllable [ba] is the acoustic correlate of the [b]? What parts of the acoustic patterns of [da] and [ga] were the acoustic correlates of [d] and [g]? The phonetic representation of speech since the invention of the alphabet over five thousand years ago uses discrete segments that can be permuted.

It thus seemed to be an almost trivial problem to isolate the discrete segments of the acoustic signal that corresponded to consonants such as [b], [d], and [g]. It instead became evident that human speech was a highly encoded signal, in which the acoustic cues that specified a consonant were spread across an entire syllable and merged with the acoustic signal that conveyed the vowel.

The research program that led to this discovery had as its immediate goal a practical device. Haskins Laboratories, a privately endowed laboratory, was established in New York City shortly after the end of World War II in an old factory building. The laboratory spread over three floors served by a freight elevator. The speech research program was directed towards developing a machine that would read books aloud to blind people. A reading machine must solve three different problems. First, the machine must have a print reader, a device that can recognize printed alphabetic characters and punctuation. The second necessary element is an orthographic-to-phonetic converter, in effect a device that could regularize the spelling of English. The third major problem was making a speech synthesizer, a device that could convert the string of phonetic symbols to speech.

The key problem was the speech synthesizer. Although optical print readers were not then available, the machine could "read" new books. During the production process, a book's text was typeset. Thus there was a possibility of starting with a machine representation of the printed text, using a special typesetting machine that provided a control tape for the reading machine. The process of converting the orthographic text to a phonetic text was not trivial, but at worst, it would entail a large dictionary in which all the words of English would appear with their orthographic and phonetic spellings. At the time, the speech synthesizer did not seem to present a problem. The individual phonetic symbols would have to be sequentially converted to acoustic signals. As already noted, the prevailing view was that the sounds of speech were individual, discrete beads on a string. That is what phonetic transcriptions implied. The word "cat," for example, in a phonetic transcription is segmented into the initial stop consonant [k], plus the vowel [ae], plus the final stop consonant [t]. (A stop consonant is one that involves abruptly closing, or stopping, and then abruptly opening the supralaryngeal airway.) The stop consonant [k] in the word "cat," furthermore, was thought to be identical to the initial [k] consonant in the word "kite" or the final consonant of the word "back." In a phonetic alphabet a given symbol has the same sound quality wherever it appears. If the phonetic units did not have the same sound quality, you would not be able to have a phonetic transcription. Theories still are minted that claim that the sounds of speech can be freely permuted. Studdert-Kennedy's (2000) "particulate" theory for the evolution of language proposes that the sounds of speech have the status of the letters of the alphabet that can be freely permuted.

The first attempt at speech synthesis made use of the linear properties of tape recording in which a magnetic image of the speech signal was imprinted on a tape. Suppose that you recorded a sequence of sounds on a tape recorder. If you recorded, for example, a person clapping his hands, followed by a person sneezing, followed by a person singing, you would be able to unreel the tape and isolate each event by cutting out that piece of the tape. With the tape recording, you could segment the sequence of acoustic events. You would be able to take the short sections of magnetic tape and splice them together in a different sequence, so that the sneeze came after the singing or the singing before the clapping. This technique seemed to be an appropriate tool for isolating the phonetic segments of English. If you started by recording a speaker reading a selected list of words, you could record the total phonetic inventory of English. The tape recording could then be copied and sliced up into sections of tape that each corresponded to a phonetic element (Harris 1953; Peterson, Wang, and Sivertsen 1958). If, for example, you had a tape recording of the word "cat," you should be able to slice off the section of tape that corresponded to the initial [k] and then slice off the final [t], leaving the third remaining [ae] section of the tape recording. If you went through the same process with a tape recording of the word "tea," you would be able to segment out a [t] and an [i]. The segmentation of the word "ma" would yield [m] and [a]. If a phonetic transcription were simply a linear sequence of individual invariant segments, you then would be able to combine the individual sounds recorded on the sections of tape into new words. Even with the small set of words that I used as examples, you would be able to form the English words "mat," "me," and "tack."

An apparatus was constructed that could link the excised tape segments together to form words. However, when new words were put together out of the phonetic segments that had been segmented, the speech signal was scarcely intelligible. It was impossible to even segment out sounds like stop consonants without hearing the vowels that followed or preceded them in the words where they had occurred. It was, for example, impossible to cut out a pure [k] from the tape recording of the word "cat." No matter how short a segment of tape was cut, you could hear the vowel sound [k] plus the [ae] vowel that followed it.

The Haskins Laboratories research program showed that it was inherently impossible to isolate a stop consonant without also hearing the adjacent vowel. The difficulty was not that the speech-segmenting technique lacked precision. The speech signal is inherently encoded at the acoustic level (Liberman et al. 1967). The formant transitions, which are generated by the human supralaryngeal vocal tract, meld the consonants and vowels of the speech signal. The acoustic cues of the stop consonants and vowels in sounds such as [ba], [da], or [tu] are fused together. This, in part, arises from the inertial properties of the anatomy involved in speech production. A human speaker, when producing a sound like [tu], starts with the vocal tract (lips, tongue, larynx height, soft palate) in the positions that are necessary to produce a [t]. The speaker initially raises his tongue tip (blade) to produce the stop, but then lowers it abruptly while moving his tongue toward the shape necessary to produce the vowel [u]. However, the tongue cannot move instantly; there is an interval in which it is moving toward the steady-state position of the [u]. During this interval the shape of the SVT is changing; the net effect is that formant transitions are produced. The formant transitions represent the acoustic signal that results as the speaker goes from the articulatory configuration necessary to produce a [t], to that for [u]. A person listening to these transitions hears the syllable [tu].

However, speech encoding does not simply derive from the inertial properties of the organs of speech. If you look at your lips in a mirror as you prepare to say the syllables [ti] and [tu] (the words "tea" and "too"), you will see your lips protruded and constricted (the linguistic term is "rounded") at the very start of "too" so you can produce the vowel [u]. In contrast, you slightly retract your lips to produce the [i] of "tea." The entire syllable is planned in advance and the formant frequencies and resulting transitions differ from the very start of the two words. The timing, moreover, differs for speakers of different languages. For native speakers of English, the lips begin to round 100 msec (0.1 seconds) before the start of vowel phonation. For native speakers of Swedish, rounding can begin much earlier in words like "stew" (Lubker and Gay 1982). Children apparently learn the different timing sequences as they acquire their native language (Sereno et al. 1987). Planning ahead occurs over the span of a syllable; Ohman (1966) showed that the second vowel of a bisyllabic word affects the initial vowel's formant frequencies.

In short, no matter how fine a slice of the transition one cuts, if it is interpreted as speech, it will be heard as the consonant plus the vowel. When listeners are asked to identify the consonant and vowels for synthesized stimuli limited to 5, 10, 20, or 40 msec from the onset of a syllable such as [bi], [tu], or [du], they can identify the stop consonant and the following vowel 90 percent of the time when they are presented with only the first 5 msec of the stimuli (Stevens and Blumstein 1978). The acoustic information present at the onset of the syllable allows listeners to identify the vowel in the absence of a complete specification of the formant frequency transitions or the steady-state vowel. The effect occurs even when we present 15-msec onset signals computer-edited from the first words of 70-week-old children (Chapin, Tseng, and Lieberman 1982). The speech signal is encoded into a unit whose minimum span is a consonant followed by a vowel.

The speech-encoding process is not limited to consonant-vowel syllables. In producing a word like "bat" in fluent discourse, a speaker does not necessarily attain the articulatory configuration or the formant frequency pattern of the steady-state vowel after producing the [b]. The speaker instead starts toward the articulatory configuration of the [t] before reaching the steady-state values that would be characteristic of an isolated and sustained vowel. The articulatory gestures are melded together into a composite. The sound pattern that results from this encoding process is itself an indivisible composite. Just as there is no way of separating the [b] gestures from the vowel gestures (you cannot tell where the [b] ends and the vowel begins), there is no way of separating the acoustic cues that are generated by these articulatory maneuvers. The result is that the acoustic cues that characterize the initial and final consonants are transmitted in the interval of time that would have been necessary to transmit a single, isolated vowel.

Encoding and the "Motor" Theory of Speech Perception

The encoding of speech has a critical property: it yields a high rate of signal transmission. A simple experiment that you can perform without any complicated equipment will illustrate the difference between the transmission rate of encoded human speech and that of segmental nonspeech sounds. The experiment requires only two persons, a pencil, and a watch. The object of the experiment is to determine the fastest rate at which you can count the number of pencil taps that occur in five seconds. One person taps the pencil while the other counts and looks at the watch. If you perform this experiment, you will discover that most people can, with a little practice, tap a pencil much faster than the rate at which you can count the taps. The fastest rate possible is about seven to nine taps per second, that is, thirty-five to forty-five taps in five seconds. In contrast, the rate at which we identify the individual sounds of speech ranges from twenty sounds per second for normal rates of speech, to about thirty sounds per second (100 to 150 "taps" in five seconds). The Haskins Laboratories research team directed by Alvin M. Liberman and Franklin S. Cooper systematically tried to use various types of nonspeech sounds—Morse code, musical tones, beeps, and combinations of sounds and vibrators attached to a person's hands and arms—to convey words in their reading machine project. They found that it was impossible to attain the speech rate with any sequence of segmented nonspeech sounds (Liberman et al. 1967). At rates that exceed fifteen sounds per second, the individual sounds merge into a buzz. *We cannot even resolve the individual sounds, much less identify them.*

The process that allows human speech to attain this high rate of transmission involves encoding phonetic events into a speech signal that consists of units whose minimum length is a syllable. Human listeners appear to decode the speech signal by using a neural system that inherently takes account of the constraints of speech production. The listener's perceptual system "knows" the set of encoded signals that reflect speakers' execution of motor acts that correspond to a series of individual phonetic segments. Motor theories of speech perception are not new; Zinkin (1968), for example, presents a comprehensive review of Russian research that dates back to the nineteenth century. However, the Haskins experiments of the 1950s and 1960s provided the first convincing evidence that speech perception was structured in terms of the constraints of speech production. The Haskins experiments relied on careful, controlled psychoacoustic experiments in which listeners were asked to respond to synthetic speech stimuli. Synthesized speech allowed experiments to be run using stimuli that differed along known acoustic parameters.

For example, Figure 3.8 shows stylized spectrograms of the formant frequency patterns that were used in a typical Haskins experiment. The dark bars of the spectrogram plot only the first and second formant frequencies. These two formant patterns, however, when transformed to speech signals by the synthesizer, are heard as the sounds [di] and [du] (the words "dee" and "due" are reasonable

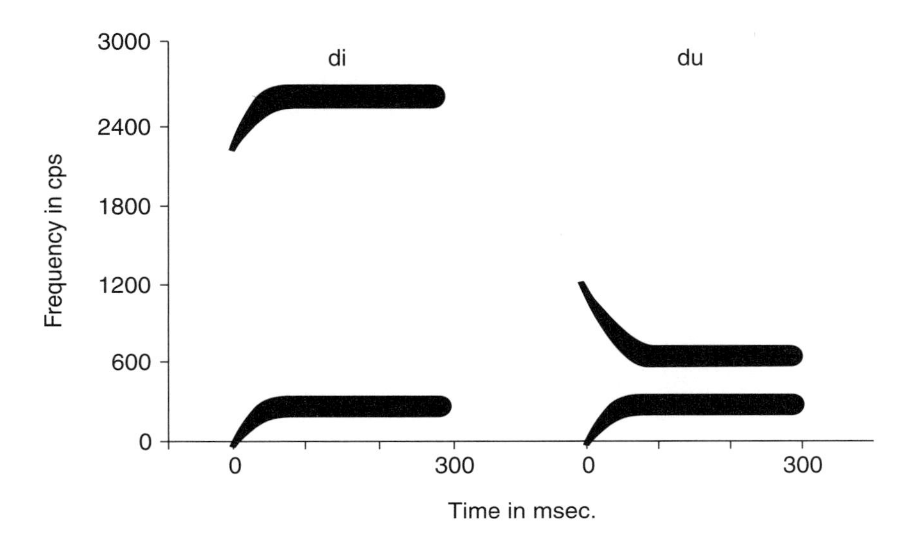

FIGURE 3.8 *Formant frequency patterns for F1 and F2 that will produce the sounds [di] and [du] when they are used to control a speech synthesizer. Note that the F2 formant transitions are very different, yet they convey the "same" consonant [d].*

approximations). Note that the second formant transition is different for the two synthesized syllables [di] and [du]. The transition is falling in the [du], rising in the [di]. However, listeners hear an identical [d] consonant at the start of these two synthesized sounds. If the identical first formant frequency track is removed, the synthesized sound no longer resembles speech and the F2 tracks again sound different. The identical [d] consonant is only heard when the two formants are synthesized, producing a speech-like signal. The motor theory of speech perception developed at Haskins Laboratories proposed that these two different acoustic patterns are heard as the same [d] because the listener is decoding the signal in a "speech mode," using an internal neural process that implicitly "knows" that these different acoustic patterns will occur when a speaker puts his or her tongue into the configuration that is used to produce a [d] before either [i] or [u]. In other words, we hear the same [d] because, when we talk, a [d] articulation produces different formant frequency transitions in the encoded speech output before either an [i] or a [u]. The Haskins theory does not claim that we interpret speech by moving our lips and generating silent speech signals that we use to compare with the incoming speech signal. Older, traditional versions of motor theories of speech perception claimed that we, in effect, talk to ourselves quietly in order to perceive the sounds of speech.

The early Haskins experiments have been replicated many times with improved computer-implemented synthesizers. Current techniques that observe neural activity while a person listens to words or interprets the meaning of a sentence show that similar processes occur at this "higher" stage of language comprehension. The structures of the human brain that are involved in producing speech are activated when we listen to words (Wilson et al. 2004). As we shall see, although some aspects of speech perception make use of neural systems that also respond to other auditory signals, we seem to have neural devices that respond in a special manner to human speech. In other words, our speech perception system may be regarded as an exceedingly elaborate version of the system that can be discerned in frogs.

Perceiving Formant Frequencies

The bullfrog system appears to involve "mechanisms" that are located in the peripheral auditory system. Frishkopf and Goldstein (1963) monitored the electrical output of the frogs' eighth cranial nerve, which in essence is the channel from the basilar membrane of the ear.[3] The findings of decades of research on the perception of human speech suggest that a similar, though exceedingly more complex and flexible, system exists in human beings. The pattern of formant frequency variation arguably is the most important factor that yields the high data transmission rate of human speech.

Some form of formant frequency "extraction" must exist in other species. As I noted in Chapter 2, dogs, apes, and most likely other species can clearly identify human speech distinctions that depend on formant frequency patterns. Similar formant frequency variations mark the calls of chimpanzees and other apes (Lieberman 1968, Hauser 1986; Riede et al. 2005) that appear to be meaningful (Goodall 1986). As we shall see, many other species pay attention to formant frequencies to estimate the size of the animal who is vocalizing. The ability to derive formant frequency patterns is one of the basic steps in the perception of human speech, and it is almost a certainty that similar processes exist in other species.

The process by which formant frequencies are derived—in other words, "extracted" or "calculated" from the acoustic signal—is complex. It is striking that formant frequencies are not realized directly in the acoustic speech signal. Formant frequencies are instead properties of the supralaryngeal airways that filter the acoustic source or sources involved in speech production (Hermann 1894; Fant 1960). The formant patterns plotted in Figures 3.7 and 3.8 were schematic diagrams of the control pattern of a speech synthesizer. The actual acoustic signals that were presented to human listeners had nothing that could be directly identified as a formant frequency. Formant frequencies can be calculated with certainty only if one takes into account the anatomic and physiologic constraints of human speech production, most of which are similar in principle (as Negus 1949, showed) to those of the vocal anatomy of most terrestrial animals.

Figure 3.9 illustrates a simple fact: formant frequencies are not always directly present in the acoustic speech signal. The top graph shows the filter function of the supralaryngeal airway that would yield the vowel [i]. The three relative maxima—F1, F2, and F3—are the first three formant frequencies. Note that these three local peaks necessarily form part of a total, continuous filter function that extends across the audible frequency spectrum. The filter function passes acoustic energy through at all the frequencies that are plotted in accordance with the values shown. Relative energy maxima occur at the formant frequencies, but energy passes through the filter at other frequencies. The formant frequencies thus are

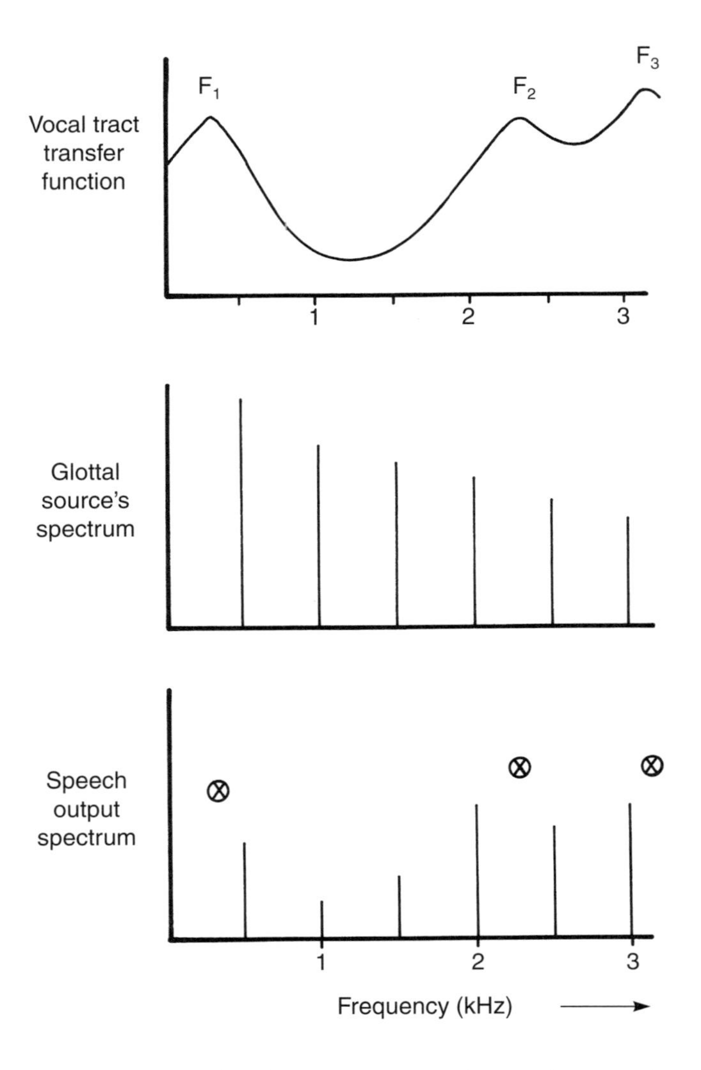

FIGURE 3.9 *Above: Filter function of the supralaryngeal airway for the vowel [i]. Middle: Spectrum of glottal source for a fundamental frequency of 500 Hz. Below: Spectrum of the speech signal that would result if the filter function shown in the top graph were excited by the glottal source noted in the middle graph. Note that there is no acoustic energy present at the formant frequencies, which are noted by the circled Xs.*

the frequencies at which maximum energy passes through the supralaryngeal airways relative to nearby frequencies. The formant frequencies are thus local properties of the supralaryngeal filter.

The middle diagram shows the spectrum of the periodic energy that could be generated by the larynx. The fundamental frequency of phonation is 500 Hz, which is represented by the position of the first bar of this graph. This vertical line or bar represents the amplitude of the fundamental component of the glottal excitation. The bars that occur at integral multiples of the fundamental frequency 500 Hz, 1.0 kHz, 1.5 kHz, and so on represent the amplitudes of the harmonics of the fundamental. The empty spaces between the bars indicate that there is no acoustic energy between the harmonics. Together the ensemble of the fundamental frequencies and its harmonics represent the power spectrum of the glottal sound source in the frequency domain. A human listener would never have access to this acoustic signal because it is filtered by the supralaryngeal airway that is interposed between the larynx and the speaker's lips. If a speaker's supralaryngeal airway assumed the configuration that is appropriate to generate the vowel [i], a listener would hear an acoustic signal whose spectrum would be the product of the spectrum of the glottal excitation and the filter function. The result would be the spectrum sketched in the bottom graph of Figure 3.9.

Note that the listener also does not have direct access to the filter function sketched in the top graph. The signal that the listener hears instead has energy at the fundamental frequency and the harmonics, which only provide clues to the formant frequency locations. Note that in this particular instance, no energy is present at the formant frequencies. The absence of any acoustic energy at the formant frequencies in the speech signal is a consequence of there being no energy in the periodic laryngeal source, except at the fundamental frequency and its harmonics. The fundamental frequency of phonation and its harmonics are controlled by the activity of the laryngeal muscles and the air pressure generated by the lungs. The control of the fundamental frequency of phonation and its harmonics is independent of the control of the supralaryngeal airway's filter function. In general, the harmonics of the fundamental frequency do not fall on the formant frequencies. The bottom graph sketched in Figure 3.9 does not represent some extreme, odd condition. The fundamental frequency of phonation of many women and some men can exceed 400 Hz during connected discourse. The fundamental frequency of children under the age of 6 years typically exceeds 500 Hz. The upper limit of fundamental frequency for two- to three-year-old children is at least 1.5 kHz (Keating and Buhr 1978).

The spectra of speech sounds that tend to be reproduced in introductory texts on speech analysis have low fundamental frequencies, but most human beings (women, children and many adult men) speak with relatively high fundamental frequencies. If we were to limit our examples to speech produced with low fundamental frequencies, we would be misled regarding the acoustical reality of formant frequencies. In Figure 3.10 the fundamental frequency is low—100 Hz—and it is easy to see the location of the formant frequencies. The supralaryngeal filter function has, in effect, been sampled by the closely spaced harmonics of the low fundamental frequency of phonation. The 100-Hz fundamental in the figure samples the filter function 35 times between 100 Hz and 3.5 kHz. On viewing the resulting spectral analysis in Figure 3.10, we might be led to the belief that the formant frequencies are present in the speech signal. A 500-Hz fundamental, in contrast, samples the filter function only seven times in this same frequency interval. The increased uncertainty of formant measurements from the bottom graph in Figure 3.9 follows from the fact that we have to reconstruct the shape of the continuous filter function from the six points sampled by the discrete harmonics of the fundamental. Only if we can reconstruct the filter function with sufficient detail to determine the local maxima, can we then determine the formant frequencies. In other words, formant frequency extraction is a two-stage process. In effect, we first have to connect the sampled points to reconstruct the supralaryngeal filter function. The second stage involves finding the local peaks of the reconstructed filter function.

It is notoriously difficult to calculate the formant frequencies of

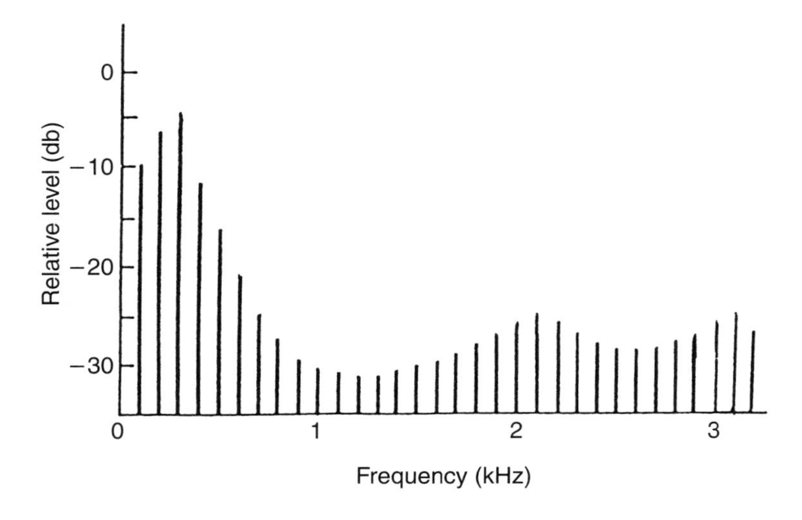

FIGURE 3.10 *Spectrum of the speech signal filtered by the vocal tract transfer function sketched in the top graph of Figure 3.9, excited by a glottal source having a fundamental frequency of 100 Hz.*

sounds that have high fundamental frequencies by using sound spectrograms or other computational procedures involving simple interpolation between the sampled points of the spectrum. There simply are not enough available data points that would readily reveal the formant frequencies in speech signals produced with high fundamental frequencies. Psychoacoustic experiments show that human listeners also have somewhat more difficulty in determining the formant frequencies of vowels as the fundamental frequency of phonation rises. In an experiment with computer-synthesized vowel stimuli that differed only in their formant patterns, the error rate for their identification increased slightly for higher fundamental frequencies (Ryalls and Lieberman 1982). However, human listeners are exceedingly better formant extractors than are human viewers using sound spectrograms or other methods limited to reconstructing the supralaryngeal filter function from only the data provided by the harmonics of the fundamental frequency. Computer-implemented systems using techniques such as linear predictive coding (LPC) that take into account the way in which the supralaryngeal filter function is produced, achieve better accuracy by bringing more information to bear on reconstructing the supralaryngeal filter function (Bell et al. 1958; Atal and Hanauer 1971). The computer system has a store of all the possible formant frequency patterns that a human SVT can produce and a typical glottal source spectrum. The computer mindlessly generates every possible formant frequency pattern and looks for the one that best matches the incoming speech signal.

However, these primitive computer systems are poor compared to human listeners and, as comparative studies suggest, birds, dogs, monkeys, apes, and animals as far removed from humans as alligators.

The Psychological Reality of Formant Frequency Patterns

Because formant frequency patterns are not directly present in the speech signal, one reasonable question is whether human listeners actually derive formant frequencies at any stage of the perceptual interpretation of speech. Do formant frequency patterns, in other words, have a psychological reality? This question was one of the issues in the nineteenth-century controversy concerning the "harmonic" and "inharmonic" theories of vowels. Helmholtz in 1863 derived the harmonic structure of voiced vowels by means of acoustic analysis. Hermann (1894), in contrast, followed the path taken by Hellwag in 1781 and took note of the inharmonic formant frequencies of the supralaryngeal vocal tract that differentiated vowels. (We'll return to Hellwag shortly.) Because the harmonic structure of a vowel is the direct physical manifestation of a vowel and is physically present in the acoustic signal, it was reasonable to suppose that speech perception in human beings might not involve any stage of processing entailing a representation of the signal in terms of formant frequencies. This controversy, as we shall see, bears on the probable evolution of the peculiar, species-specific anatomy of the human supralaryngeal vocal tract.

The hypothesis that we perceive linguistic distinctions directly from the spectrum of the acoustic signal instead of extracting formant frequencies would be reasonable only if we were to limit speech perception to ideal acoustic environments, such as sound-treated rooms, anechoic chambers, or open fields on quiet days. In the real world we would run into trouble because the spectrum of the acoustic signal that a listener hears is constantly subject to distortion from environmental sources. The spectrum of an acoustic signal that corresponds to the same sound, for example, can vary in different rooms. The acoustic spectrum that corresponds to a given sound can, in fact, vary as you turn your head (Beranek 1949). Speech recognition schemes that have to identify words from measurements of the "raw" acoustic spectrum have to take account of the effects of the external environment. Many of the software packages used for computer-recognition of speech require the user to speak into a directional microphone.

It is easy to demonstrate that changing the overall shape of the frequency spectrum that you hear has little effect on speech intelligibility. Most music systems have tone controls that can effect changes in the overall acoustic spectrum. The bass or low-frequency boost controls can tilt the acoustic spectrum up toward the low-frequency part of the spectrum, changing the overall energy balance of the spectrum. However, this does not change the identification of speech sounds, even when isolated short-term onset spectra are presented to listeners in psychoacoustic tests (Blumstein, Isaacs, and Mertus 1982). Moreover, human listeners can perceive sinusoidal replicas of formant frequency patterns as speech (Remez et al. 1981). The formant frequency pattern then is represented by a set of varying sinusoids, that is, "pure" tones, whose amplitudes are adjusted to follow the amplitudes of the formant frequencies derived from the speech signal. Remez and his colleagues "primed" their subjects. When the subjects were told that they were listening to "science fiction sounds," they heard computer bleeps. When the subjects were instead told that the signals were speech, they heard these sounds as speech and were able to identify the original sentences. The only property of these sinusoidal signals that bears any relation to the original speech signal is that sinusoids tracked the formant frequencies of the unaltered sentences. The hypothesis that appears to explain the behavior of the listeners is that there is a psychologically "real" stage of formant frequency speech processing at which formant frequency patterns are represented and that the listeners were able, when primed, to relate the time-varying sinusoidal pattern to this stage of speech perception.

Talking Birds

The data of Remez and his colleagues (1981) are consistent with earlier, more limited data (Bailey, Summerfield, and Dorman 1977). They also explain why many "talking" birds appear to talk. Talking birds such as mynahs produce their calls by means of two syringes. One syrinx is located in each of the bifurcations of the trachea that lead to the bird's lungs. The larynx, which is positioned at the top of the trachea, plays a minor part in the production of sound in these birds (Greenewalt 1968). Each syrinx can produce a different sinusoidal, pure tone. Mynah birds mimic human speech by producing acoustic signals in which a sinusoidal tone is present at the formant frequencies of the original human speech sound mimicked by the bird (Greenewalt 1967; Klatt and Stefanski 1974). We perceive these bird calls as speech because they have energy at the formant frequencies.

Computer-Implemented Formant Frequency Extraction

Many theories for the perception of formant frequencies follow the general framework of the Haskins motor theory, albeit without claiming that we actually produce invariant motor gestures. The processes implemented in computer programs for formant "extraction" involve making use of knowledge concerning the physiologic constraints of speech production. In one of the earliest proposals, "analysis by synthesis" (Bell et al. 1961), the formant extraction algorithm makes use of the fact that there is a fixed relationship among the frequency of a formant, its amplitude, and its effect on the total form of the overall filter function. The process is based on two crucial facts noted by Fant (1956). First, the relative amplitudes of the formants of an unnasalized vowel sound are a function of the values of the formant frequencies. Second, each formant frequency determines a partial specification of the total filter function. We can derive the total filter function by adding up the combined effects of a set of individual formant frequencies. Conversely, it turns out that a supralaryngeal filter function that is made up solely of formants can be resolved into the sum of a set of partial filter functions, each of which corresponds to an isolated formant frequency. We thus can determine the formant frequencies of a vowel even if there is no energy present at the exact formant frequencies in the acoustic signal, so long as we have enough acoustic information to establish the overall shape of the filter function. The procedure of linear predictive coding (LPC) devised by Atal and Hanauer (1971), which is still in general use, essentially is a modified version of this procedure; the computer algorithm matches the general shape of the spectrum with a set of individual filter functions, each of which has the correct overall shape of a single formant.

Some of the perceptual effects that have been noted in psychoacoustic experiments with human listeners can be explained if we hypothesize that human beings extract formant frequencies by using the entire spectrum and by having some implicit knowledge of the physiology of speech production—that is, an analysis by synthesis in which we internally interpret the input speech signal in terms of how it could have been produced. The analysis-by-synthesis procedure, for example, takes into account the low-frequency part of the spectrum below the first formant frequency. This is consistent with Holmes's (1979) observation that high-quality speech synthesis requires a careful spectral match below the first formant frequency. Analysis-by-synthesis methods are also sensitive to the fundamental frequency of phonation. They become less accurate as F0 increases, but that also is the case with human listeners (Ryalls and Lieberman 1982).

Avoidance of Nasal Speech

The presence of antiformants, which are frequencies at which the SVT absorbs acoustic energy, degrades the perception of speech sounds. Antiformants occur when nasalized sounds are produced and introduce errors in LPC analysis (Atal and Hanauer 1998) and all other analysis-by-synthesis programs of the type first developed by Bell and colleagues (1959). Nasalization involves adding air passages through the nose to the SVT filter, which yields antiformants. Human listeners also have more difficulty in identifying these nasalized speech signals. Bond (1976), in an experiment in which vowels were excerpted from nonnasal and nasal contexts, showed that the error rates of listeners who had to identify these stimuli increased 30 percent for the nasalized vowels. The Bond study employed English-speaking listeners who normally are not exposed to phonemic variations that differentiate words by means of nasalized versus nonnasalized vowels. However, the effect is not due to unfamiliarity with nasal vowels because similar results occur when the responses are analyzed of native speakers of Portuguese, a language that makes use of a full set of nasalized vowels at the phonemic level. Surveys of the speech patterns of the world's languages show that they avoid distinctions that depend on having many nasalized vowels (Greenberg 1963; Maddieson 1984), most likely because of the higher perceptual error rates that nasalization introduces. One interesting aspect of the evolution of the human SVT (discussed in Chapter 6), is that it facilitates producing vowels that are not nasalized.

Adjusting for a Person's Vocal Tract Length—SVT Normalization

If formant frequency patterns are once removed from the physical attributes of the speech signal, their linguistic interpretation is at least one step more "abstract." Because formant frequency patterns are properties of the supralaryngeal vocal tract, the absolute values of the formants that convey a particular vowel or consonant always depends on the length of a speaker's SVT (Chiba and Kajiyama 1941; Fant 1960; Nearey 1979; Baer et al. 1991; Story, Titze, and Hoffman 1996). For example, the formant frequencies of an [i] produced by a two-year-old child are almost twice as high as those produced by the child's mother, because the child's SVT is approximately half the length of the parent's (Fitch and Giedd 1999; Lieberman and McCarthy 1999; Vorperian et al. 2005). People come in all sizes, and the length of their SVTs varies. The absolute values of the formant frequencies of the "same" sound produced by different adults can vary (Peterson and Barney 1952; Hillenbrand et al. 1995). Therefore, a neural process that "normalizes" SVT lengths to take account of differences in SVT length is a critical element of speech perception.

One of the first experiments that showed that SVT normalization occurs when human listeners perceive speech was published by Ladefoged and Broadbent in 1957. They made a tape recording of a synthesized English word and presented it to panels of listeners. Their experiment showed that the synthesized word's vowel was heard as an [I], an [ε], or an [æ] by listeners, depending on the listeners' estimate of the length of the SVT that had produced the signal. A speech synthesizer was used to generate three "carrier" sentences that introduced a "test" word with the incomplete sentence, "The word you will hear is . . ." One carrier sentence had low formant frequencies, another somewhat higher formant frequencies, and a third still higher formant frequencies. The word "bet" synthesized with formant frequencies appropriate for the mid-frequency carrier phrase was heard as "bet" when it was presented to listeners immediately after that carrier phrase. The same tape-recorded word was identified as "bit" after the low-frequency carrier phrase and as "bat" after the highest-frequency value carrier phrase. In short, the listeners perceived the vowel of the same, identical tape-recorded stimulus as an [i], [ε], or [æ] depending on the average formant frequencies of the carrier phrase.

The basis for the listeners' decisions is apparent if we consider

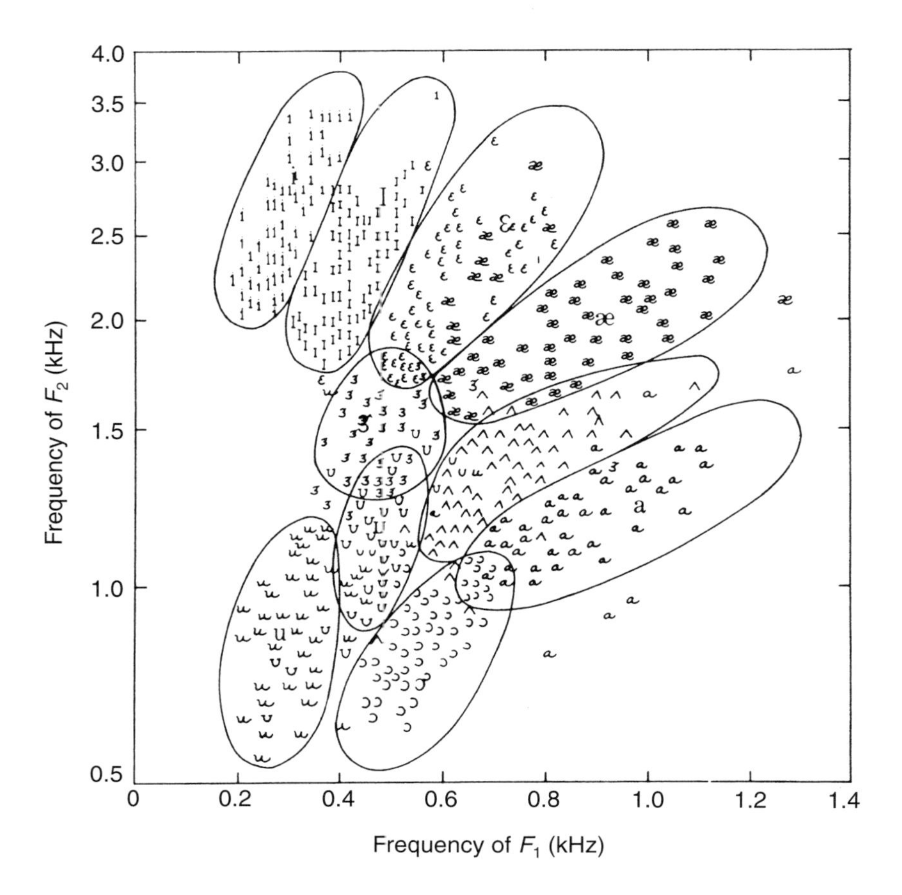

FIGURE 3.11 *Plot of first and second formant frequencies derived by Peterson and Barney (1952) for the English vowels produced by seventy-six different speakers. The frequency of F2 is plotted with respect to the ordinate for each vowel token, the frequency of F1 with respect to the abscissa. The labeled loops enclose 90 percent of the tokens produced for each of the vowel categories of English by the speakers.*

how speakers who have different-length supralaryngeal vocal tracts will produce different formant frequencies for the same vowel. Figure 3.11 shows the Peterson and Barney (1952) plot of seventy-six adult male, adult female, and adolescent male and female speakers. The vowel symbols are plotted with respect to the absolute values of their first and second formant frequencies. Each phonetic symbol corresponds to a measured vowel; the value of its F1 (the first, lowest-frequency formant frequency) is plotted on the abscissa, F2 (the second, higher formant) on the ordinate.

The vowels' formant frequencies were measured from spectrograms of each speaker reading a list of English words with the form "heed," "hid," "head," and so on. (The words had vowels that differed; the word-initial [h] sound produced noise excitation of the vowels that facilitated formant frequency measurement.) The words were then identified by panels of listeners who had to identify each token without previously listening to a long stretch of speech produced by the same speaker. This was achieved by presenting a set of all the words produced by ten speakers in random order to the listeners. The listeners thus did not know whose voice or what word was coming next. The listeners sat in an auditorium to listen to the recorded words on a loudspeaker system, and they checked off each successive stimulus against a scoring sheet on which the intended spoken words were listed. A vowel symbol that fell into a loop marked with the same phonetic symbol signified a token that was heard as the intended vowel target. The loops on the plot in Figure 3.11 enclose the vowel tokens that made 90 percent of the vowels that the speakers intended to convey. Note that the loops overlap even though they do not include 10 percent of the stimuli that fell into a nearby vowel class. The data show, for example, that a sound intended by many speakers to signify the vowel [e] had the same formant frequencies as other speakers' [I]'s.

Figure 3.12, which is from Nearey (1979), perhaps makes the phenomenon clearer. Nearey has plotted the averages of formant frequency for F1 and F2 that Peterson and Barney (1952) derived for their adult male, adult female, and adolescent speakers. Nearey has converted the plots of formant frequencies to a logarithmic scale. The logarithmic plot shows that the vowel "space" is similar if we take into account the frequency spread that occurs with shorter supralaryngeal vocal tracts (see Lieberman and Blumstein 1988, for

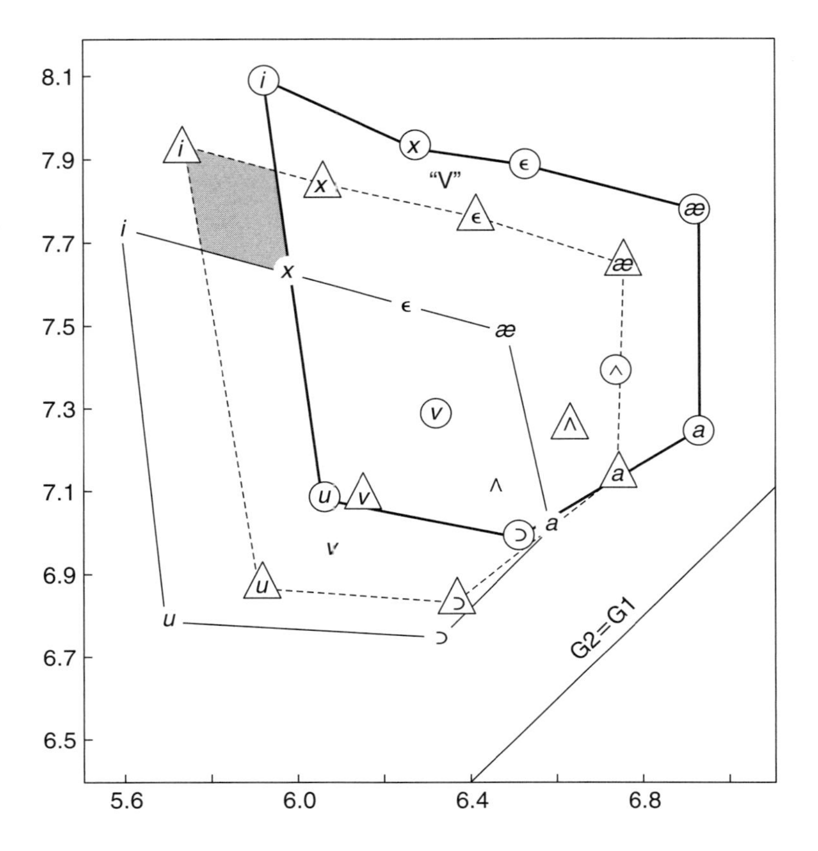

FIGURE 3.12 *Average values of formant frequencies measured by Peterson and Barney (1952) for adult male speakers, adult female speakers (vowel points enclosed by triangles), and adolescents (vowel symbols enclosed by circles). The frequency values have been plotted on logarithmic scales. The vowel points for each class of speakers are connected and form relative vowel spaces. Note that the vowel spaces for the adult females and the adolescents are transposed upward in frequency from the vowel space of the adult males. (After Nearey 1979)*

a detailed discussion of the physiology of speech production). I have entered the stimulus marked "V" in Nearey's diagram. If a listener heard this stimulus and thought that it was being produced by an adolescent child, it would fall into the range of child [I]'s. If the listener instead thought that the stimulus V had been produced by an adult female, it would fall into the class of female [e]'s.

The Supervowel [i]

Vowel normalization and the linguistic status of formant frequencies were systematically studied by Nearey (1979). Nearey pointed out the unique status of the vowel [i] in specifying the probable length of a speaker's SVT. It is obvious that human listeners do not always have the opportunity to hear a complete sentence before they estimate the probable length of a speaker's vocal tract. Thus if vocal tract normalization were a psychologically "real" stage in the perception of speech, there would have to be immediate acoustic normalization cues in the speech signal. Nearey realized that the perceptual data of the Peterson and Barney (1952) experiment involving isolated words furnished a clue. That speech perception experiment showed that virtually all the [i] vowels produced by the speakers were correctly identified as [i]'s by the listeners. There were only 2 [i] errors out of over 10,000 trials. The [u]'s were subject to slight confusion, whereas other vowels showed high error rates; for example, [e] and [I] were confused hundreds of times, yielding error rates in excess of 5 percent. Nearey reasoned that [i] might be a "supervowel" and that [u] might serve as a perceptual anchor point for vocal tract normalization. If so, it should be possible to build an automaton to identify tokens of vowels, that is, to assign acoustic signals to the phonetic categories intended by their speakers, in terms of acoustic cues derived from the vowels [i] and [u].[4]

The data plotted in Figure 3.12 point out the status of [i]. The formant frequency overlaps that can be seen in this figure are consistent with the perceptual confusions that listeners made in the Peterson and Barney (1952) vowel study; for instance, [a] and [ɔ] confusions are quite common. Note that the formant frequency patterns of vowels in the interior of the vowel space overlap in the plot of the adult male, adult female, and adolescent averages. Vowels on the lower-right boundary of the vowel space also can overlap; note that the adult males' and adolescents' [a] and [ɔ] overlap. The only vowel that has a formant pattern that inherently cannot be confused with a token of some other vowel is [i]. This is apparent in the topography of Figure 3.12, where the [i] vowels of different speakers form the extreme upper-left margin of the vowel space plotted for different-sized SVTs. Nearey (1979) thus predicted that a token of a formant frequency pattern that specifies an [i] will always be heard as an [i] produced by a supralaryngeal vocal tract that has a particular effective length. If he is correct, a listener would immediately "know" the supralaryngeal vocal tract length of the speaker to whom he or she is listening.

Nearey (1979, pp. 98–149) tested this hypothesis in a psychoacoustic experiment in which a listener first heard a "calibrating" [i] followed by a test vowel, followed by the same calibrating [i]. Nearey selected two different calibrating [i] vowels, one with a formant frequency pattern that corresponded to an [i] produced by an adult male's relatively long SVT, and one to an [i] produced by an adolescent child's shorter SVT. Juxtaposed with the calibrating [i]'s were formant patterns that ranged over almost the total possible range of vowels for adult speakers and adolescent children. These stimuli were presented to listeners, who were asked to identify the vowel that they heard. The listeners were also asked to rate the naturalness of each [i-V-i] combination (where V = the vowel corresponding to the varying formant frequency patterns). There were four categories of naturalness judgment, from "OK" to "very bad."

The listeners' responses showed that they were "normalizing" SVTs using the single token of an [i]. Figure 3.13 shows the overall shift in the listeners' vowel identification responses for the two supralaryngeal vocal tract length conditions. The phonetic symbols connected by lines represent the boundaries at which listeners identified formant frequency patterns as particular vowels. The boundaries for the tokens presented between the [i]'s of a longer vocal tract are connected with a solid line, and those for the [i]'s from a shorter vocal tract with a dashed line. Note that there is a general shift throughout the vowel space. The listeners categorized the vowel formant frequency patterns that they heard in terms of the presumed length of the supralaryngeal vocal tract corresponding to

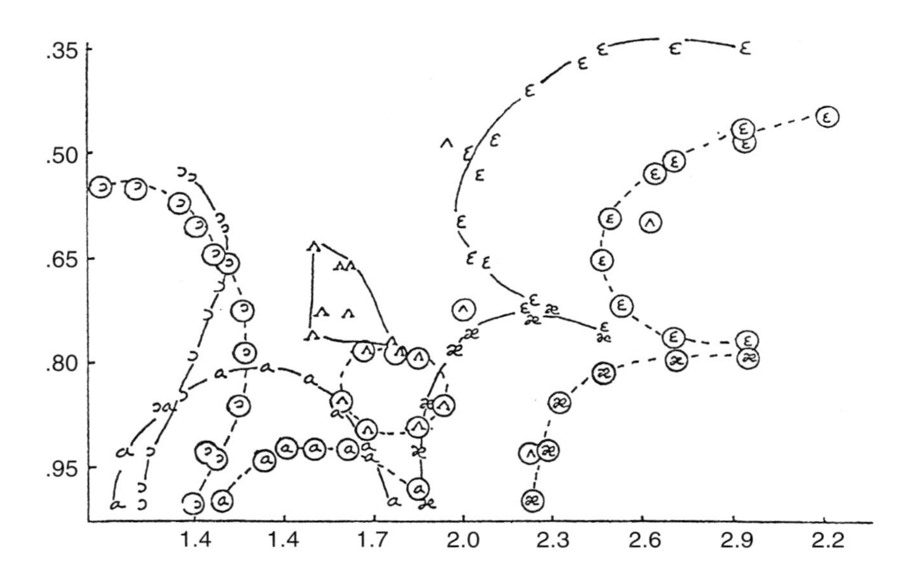

FIGURE 3.13 *Shift in listeners' categorizations of synthesized vowel stimuli when they heard the stimuli after an [i] vowel that could be produced by either a long or a short supralaryngeal vocal tract (circled symbols). Note that the identification boundaries shift upward in frequency for the short vocal tract condition marked by the circled symbols. (After Nearey 1979)*

the "carrier" [i]. The boundary for the vowel [e], for example, shifts from F2s of 2 kHz to F2s of 2.5 kHz.

The listeners' "naturalness" responses, moreover, demonstrate that they interpreted these synthesized speech stimuli using a mental procedure that "knew" the range of formant frequencies that can be produced by the SVT length signaled by the calibrating [i]. In other words, these signals were perceived in a "speech-mode," by means of neural processing that took account of the speech-producing capabilities of the human SVT. For example, formant frequency patterns such as that indicated by the "X" in Figure 3.14 were judged to be extremely good speech signals with a short SVT [i]. A short SVT would be able to produce the formant pattern at X.

In contrast, when the listeners heard the formant frequency pattern with the values at point "X" and with a long SVT [i], they

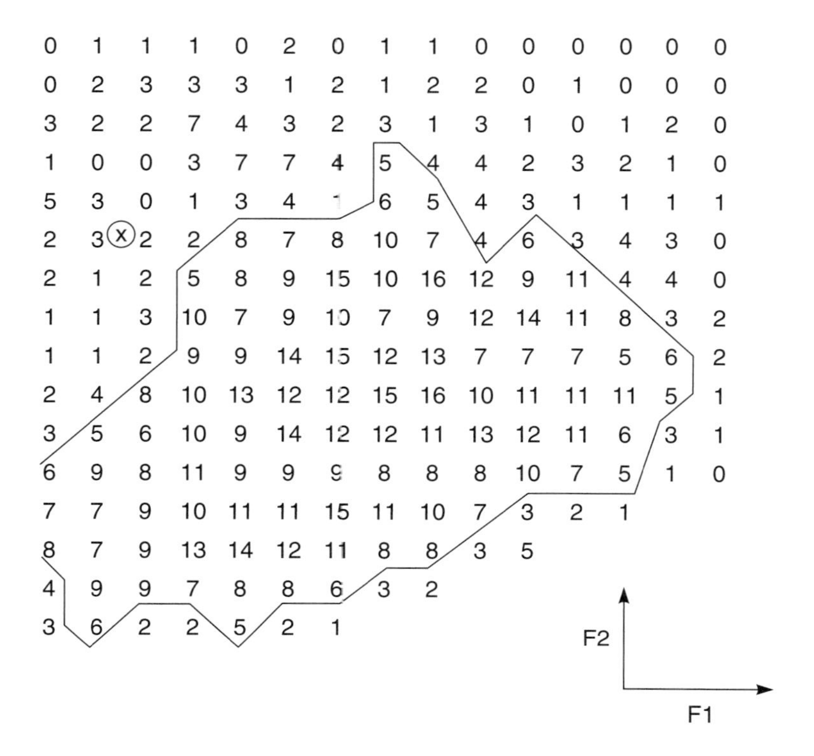

FIGURE 3.14 *The F2-F1 vowel formant frequency combinations synthesized by Nearey (1979) and the "naturalness" judgments made by the listeners who heard these vowels preceded and followed by an [i] having F2-F1 values that corresponded to a long male supralaryngeal vocal tract (SVT). Higher numerical ratings signify vowels that the listeners judged more natural; 10 was a perfect score. The F2-F1 combination at point "X" was judged to be extremely "unnatural"; it fell outside the possible range of a long SVT. This same F2-F1 pattern was within the range of a short SVT and, when the same listeners heard the vowel specified by this F2-F1 pattern preceded and followed by an [i] corresponding to a short SVT, it was judged to be extremely natural. (After Nearey 1979)*

judged it to be a terrible nonspeech signal. The listeners unconsciously "knew" that formant frequency pattern "X" could not have been produced by a long SVT. Human listeners, therefore, clearly interpret vowel sounds by means of a perceptual process that involves knowledge of the physiologic constraints of the SVT, including its length.[5]

Replicating the Peterson and Barney (1952) Study

James Hillenbrand and his colleagues (1995) replicated the Peterson and Barney experiment, recording a larger sample of children and controlling for dialect. Peterson and Barney (1952) deliberately recorded the speech and measured the formant frequencies of subjects who spoke different dialects of English, as well as non-native speakers of English, because they were working on a telephone system that would enable a person to dial by speaking the telephone number. It was therefore necessary to find acoustic determinants of spoken numbers that would hold over different dialects. A telephone that required a caller to punch in a code that indicated whether the person was a native of Brooklyn or Memphis would not be very useful. Hillenbrand and his colleagues may have thought that dialect variations were responsible for many of the speech perception errors in the 1952 study. The vowel formant frequencies were measured using state-of-the-art computer-implemented LPC spectra (Atal and Hanauer 1968) and checked using smoothed spectra. In the Hillenbrandt experiment, subjects were individually tested as they listened to the recorded words played through a loudspeaker positioned 70 cm from the listener's head to eliminate artifacts from room acoustics. In all, forty-five men, forty-five women, and forty-six ten- to twelve-year-old children (twenty-seven boys and nineteen girls) were recorded. The majority of the speakers (87 percent) were raised in Michigan's lower peninsula. The remainder were from nearby areas of the Midwestern United States. All of the speakers were screened to be certain that they differentiated between the vowels [a] and [ɔ]. Hillenbrand and his colleagues wanted to be certain that the high rate of confusion between these vowels in the Peterson and Barney (1952) study was not the result of some speakers' lacking this distinction in their dialect: The [a]–[ɔ] distinction is not maintained by many speakers of American English. The speech signals were digitally recorded; F0 and vowel durations were also measured. Figure 3.15 shows the formant frequencies of the vowels measured by these techniques.

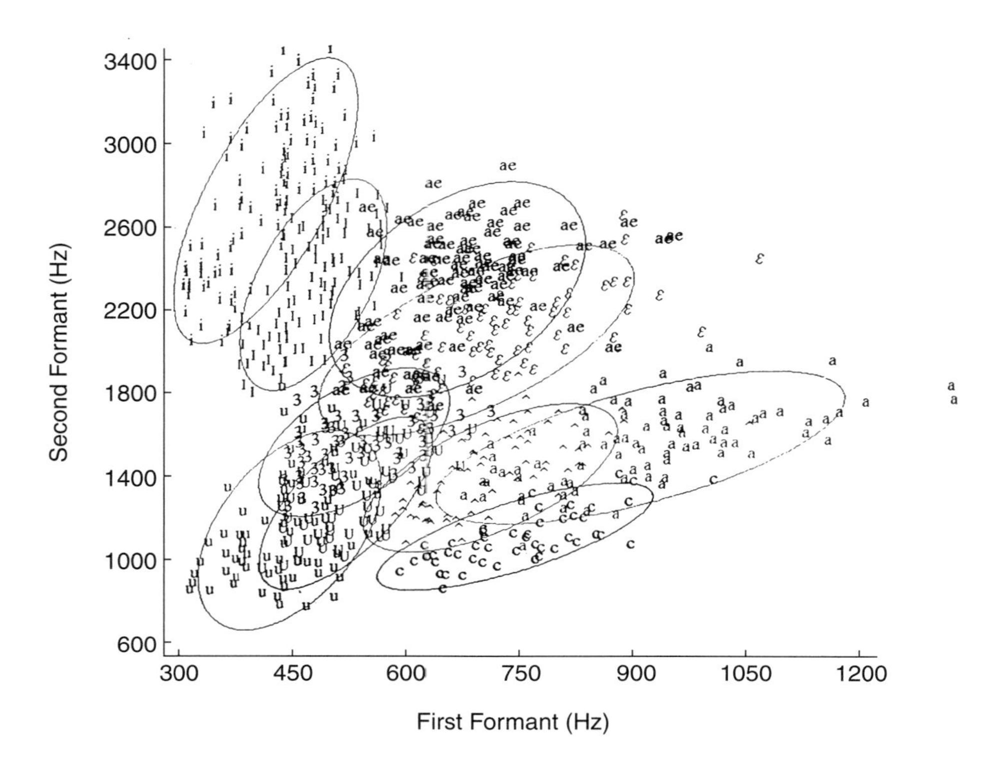

FIGURE 3.15 *Formant frequencies for the vowels of 136 men, women, and ten- to twelve-year-old children speaking the same dialect of American English. The formant frequency overlap is even greater than that found by the Peterson and Barney (1952) study, eliminating the possibility of overlap deriving from different dialects. (After Hillenbrand et al. 1995)*

The net result was that the vowel overlaps were greater than those noted by Peterson and Barney (1952); as Hillenbrand et al. (1995, p. 3103) note, "The degree of crowding among adjacent vowel categories appears much greater than in the PB data." The absolute values of the formant frequencies for each vowel differ from those identified in the 1952 study, perhaps due to the dialect studied by Hillenbrand and his colleagues or the greater precision afforded by computer-implemented formant frequency measurements. Because dialect was strictly controlled by Hillenbrand et al. (1995), the formant frequency overlaps noted in both studies cannot be ascribed to the fact that dialect was not controlled by Peterson and Barney.

The listening tests in the 1995 study yielded an error pattern similar to that found by Peterson and Barney. The vowel [a] was confused with [ɔ] in both studies, as one might expect from the formant frequency overlap. The vowel [i] was most resistant to confusion and [u] somewhat less so.

The vowel [o] was also resistant to confusion. As Hillenbrand et al. (1995) note, human listeners also make use of duration and dynamic formant frequency shifts to identify vowels. English, for example, has "long" and "short" vowels, often termed "tense" and "lax" (Chomsky and Halle 1968). The long vowels gradually shifted their formant frequencies and became diphthongs in the course of the historic English vowel shift. Long-short durational distinctions separate many of the neighboring vowels in the F1-F2 space. As measured by Hillenbrand et al. (1995), the long vowel [i] has an average duration of 243 msec. The average duration of [I], the adjacent short vowel, is 192 msec. The average duration of [e], the next long vowel, is 267 msec; the average duration of [ɛ] is 189 msec, the average duration of the long vowel [æ] is 278 msec. Perceptual effects such as these appear to be useful in explaining the phonetic inventories used in human vocal communication. Chapter 7 suggests other phenomena that theoretical linguists might consider.

Innateness and SVT Normalization

One of the major debates in linguistic theory centers about the question of "innateness." In contrast to the uncertainty noted earlier concerning the details of syntax, vocal tract normalization clearly seems to be an innate, genetically transmitted attribute of *Homo sapiens.* Some infants and mothers establish a pattern of interactive vowel imitation when the infant is as young as three months (Lieberman, Ryalls, and Rabson 1982). Figure 3.16 shows the frequencies of F1 and F2, the first two formants of the sustained vowels, and the endpoints of the diphthongs that were produced by a Japanese-speaking mother and her three-month-old son while they were imitating each other. The speech sample was recorded in the infant's home in the course of a study investigating the early development of speech. A recording crew visited the infant's home at

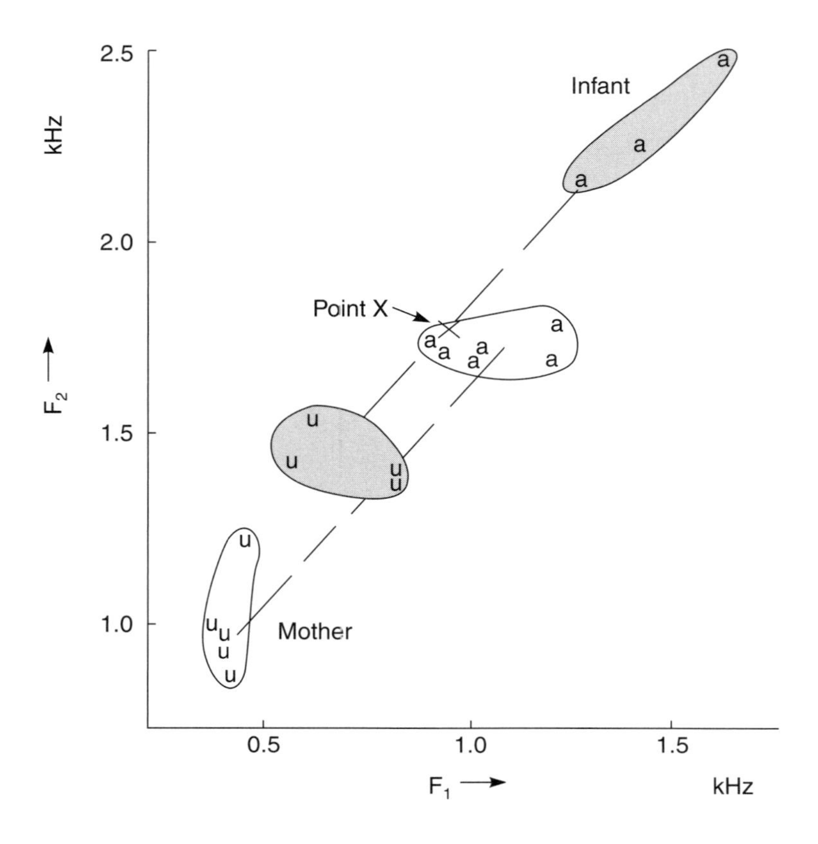

FIGURE 3.16 *Formant frequencies of vowels produced by a mother and her three-month-old infant who was imitating her utterances. The frequency of F2, the second formant, is plotted with respect to the ordinate; the frequency of F1, the first formant, is plotted with respect to the abscissa. Note that the infant's imitations are scaled in frequency with respect to the mother's utterances.*

two-week intervals and recorded an hour-long sample of spontaneous discourse between the infant and the mother. Because the recording crew was part of the child's normal environment, it was possible to derive data that minimized the effects of strangers intruding into the home. Directional microphones and attention to tape-recording techniques yielded signals that could be analyzed to determine formant frequencies.

The vowel imitations whose formant frequencies are plotted in

Figure 3.16 were produced after an interlude in which the infant was spontaneously producing sustained [æ]-like and [u]-like vowels as well as [æ]-to-[u] diphthongs. The duration of these utterances ranged from 950 to 1120 msec. The mother initiated the "conversation" when she placed her head near her son and imitated one of his vowels that had a phonetic quality that was intermediate between [æ] and [a]. After a short pause, she produced a sustained 870-msec-long diphthong in which she started with [a] and ended with [u]. The infant responded even as she was phonating, starting with an [a]-like vowel and ending with a [u]-like vowel. The duration of the infant's imitation was 700 msec. The mother then responded by producing a second diphthong starting with a slightly centralized [a] and ending with [u] in a 966-msec-long diphthong, which the infant again imitated. The infant's imitation was 790-msec long. The mother then produced a third [a]-to-[u] diphthong, which the infant again imitated. The tempo of the conversation then changed as it came to an end. The mother produced a series of shorter, 300-msec [æ]'s and [a]'s, which the infant at first imitated. The conversation ended with the infant's producing a [u]-like vowel in response to his mother's [a].

The tape recording indicates a pattern of conversational turn-taking developing in which the infant and mother respond to each other's utterance. The data in Figure 3.16 show that the child did not attempt to mimic the absolute values of the formant frequencies of his mother's vowels. The infant's formant frequencies were always higher than his mother's. He imitated the phonetic class into which her vowels fell in her acoustic vowel space; that is, he imitated her [a]-class vowels with [a]'s that had higher formant frequencies that were appropriate to his acoustic vowel space. The infant was able to produce a closer match to the absolute values of the formant frequencies of his mother's [a]'s. Point "X" lies within the infant's vowel space. He could have imitated his mother's [a]'s with a sound having these formant frequencies, which would have approximated the absolute values of her formant frequencies for this vowel, but he did not do so. The infant's [u] imitations also have higher absolute formant frequencies appropriate to his acoustic vowel space. The infant acted as though he were equipped, at age three months, with an innately determined neural vocal tract normalization device.

Behavioral studies of six-month-old infants have replicated this finding. The procedure involved monitoring infants turning their heads (Kuhl and Meltzoff 1996). Infants at this age will direct their attention toward interesting events. Infant subjects first learned to associate an animated toy (for example, a bear playing a drum) with a synthesized [i] vowel that could have been produced by a short supralaryngeal vocal tract. An [a] vowel that could have been produced with the same supralaryngeal vocal tract was associated with another animated toy. The infants soon learned to turn to look at the appropriate animated toy when they heard [i]'s, even though the [i]'s were produced by SVTs of different lengths. The infants consistently responded correctly to vowels and speech sounds with formant frequencies produced by supralaryngeal vocal tracts that differed in length from those on which they were trained.

How Big Are You? The Evolutionary Basis for Vocal Tract Normalization

There is a tendency to assume that any aspect of linguistic behavior that is manifested by human infants is necessarily innate and species-specific. Pinker (1986), for example, notes infants' perception of the stop consonants [b], [d], and [g], as documented by Eimas et al. (1971), and concludes that their perception is both innate and species-specific. Pinker fails to note that chinchillas, monkeys, and birds also perceive these sound contrasts (Kuhl 1978, 1981, 1988; Kluender, Diehl, and Killeen 1987). Computer-simulated neural nets can also "acquire" this behavior after exposure to about 100 stop-consonant formant frequency patterns (Seebach et al. 1994). Therefore, the early perception of these speech sounds is neither demonstrably innate nor specifically human.

The two claims—innateness and specificity to humans—are separable. A better case for the innate nature of some aspect of human behavior can be made if we can show that a similar innate mechanism exists in other species. This clearly is the case for supralaryngeal vocal tract normalization. Human SVT normalization appears to have a long evolutionary history, dating back at least to animals ancestral to present-day frogs. Fitch has demonstrated that many species use formant frequency patterns to gauge body size. Fitch (1994) first showed that formant frequencies provide a clue to human listeners concerning the height of a speaker. Listeners were asked to estimate the height of a person speaking an isolated vowel; they were best able to estimate the speaker's height when they heard an [i]. The fundamental frequency of phonation plays no part in this process.

Rhesus macaque monkeys also can estimate the size and weight of another monkey from the formant frequencies of its pant-threat vocalizations. Fitch (1997) developed a robust method of estimating formant frequencies based on the dispersion between formants (a shorter supralaryngeal vocal tract length yields higher formant frequencies spaced farther apart). He simply subtracted the value of F3 (the third formant frequency) from that of F1, the first and lowest formant frequency. The correlation between formant frequencies and oral vocal tract length was −0.918 for the monkeys' pant-threat vocalizations (negative because a longer vocal tract yields lower formant frequencies). Oral supralaryngeal vocal tract length was correlated with both body weight and length ($r = 0.9$). Formant frequency information thus is a reliable predictor of body size to monkeys. A monkey could determine the size of another monkey by listening to its threat call. Primates competing for food and mates need to know when they may be confronted by a larger, stronger competitor. Hence vocal signals that provide advance information are useful in the Darwinian struggle for existence. Similar correlations suggest that dogs, deer, and even animals as far removed phylogenetically from humans as alligators also make use of formant frequencies to estimate height and weight.

Playback experiments in which whooping cranes *(Grus americana)* listened to both natural and synthesized crane calls show that they perceive and attend to changes in the formant frequencies of their own species-specific vocalizations (Fitch and Kelley 2000). Correlations between skull dimensions and body size in a variety of nonhuman mammals suggest that formant frequencies play a role in animal communication by providing immediate information about the body size of a conspecific (Fitch 2000). Several conclusions follow from this evidence. First, human beings possess an innate neural mechanism that allows them to estimate supralaryngeal vocal tract length. Second, similar innate mechanisms exist in many other species and probably in all other mammals, including the hominid species that were the ancestors of present-day human beings. Third, this mechanism has been put to a "new" use in human speech, allowing us to decode the formant frequency patterns that yield the high data transmission rate of human speech. And fourth, as Chapter 6 suggests, the unique morphology and evolution of the human supralaryngeal vocal tract was driven in part through selection that enhanced the process of SVT length estimation.

In short, although innate supralaryngeal vocal tract normalization is a key element in the perception of human speech, it appears to involve redirection of a neural mechanism present in many other species. Therefore, research directed at explicating the precise nature of the neural mechanism that carries out this process in other species will bear on our human capacity to carry out supralaryngeal vocal tract normalization.

What We Don't Know about the Vocal Communications of Other Species

The vocal communications of frogs, cats, and virtually all mammals (Hauser 1996) make use of variations in the fundamental frequency of phonation; some theorists therefore have supposed that formant frequencies have no role in animal communication. However, as we have seen, this clearly is not the case in terms of conveying size information and mating. It is probable, as preliminary studies suggest, that formant frequencies convey referential information in nonhuman species. The acoustic analysis of rhesus macaque and chimpanzee calls of (Lieberman 1968) showed formant transitions. The studies noted in Hauser (1996) and Fitch (2000a) as well as the data of Riede et al. (2005) replicate this finding. The early modeling study of Lieberman, Crelin, and Klatt (1972) also shows that nonhuman primates have the anatomic capability to produce a wide range of vowels, excepting the quantal vowels [i], [u], and [a] (Stevens 1972), which are discussed in Chapters 5 and 6. Riede et al. (2005) show that the calls of at least one species of monkey approach [a]'s—though, contrary to the claim of their study, the formant frequencies are not those of an [a].[6] As I noted in Chapter 2, chimpanzees raised in an English-speaking environment clearly understand speech distinctions based on formant frequencies. The bonobos (pygmy chimpanzees) studied by Savage-Rumbaugh and Rumbaugh even responded to computer-generated signals (Savage-Rumbaugh et al. 1986; Savage-Rumbaugh and Rumbaugh 1993). The Gardner and Gardner chimpanzees learned to respond to spoken English (Fouts, Hirsch, and Fouts 1982; Fouts, Fouts, and Van Cantfort 1989). Dogs commonly learn to respond to spoken words. Baru (1975) trained dogs to respond to synthesized [i] and [a] vowels that differed only with respect to their formant frequencies. Many birds can learn to respond to the acoustic parameters that convey human speech (Hauser 1996). Blackbirds and pigeons can discriminate steady-state vowels that are differentiated only by their formant frequency patterns (Heinz, Sachs, and Sinnott 1981). Frogs, as we have seen, communicate using calls that differ in their formant frequencies (Capranica 1965; Frishkopf and Goldstein 1963). It thus is evident that, although no other species appears to be able to produce human speech, most of the acoustic-perceptual characteristics of human speech are neither species-specific nor of recent origin. The open question is whether other species can readily and voluntarily form "new" calls through different combinations of formant frequency, temporal and F0 patterns or have genetic constraints on their vocal communications.

However, further studies of animal vocal communication that focus on acoustic cues which also convey linguistic distinctions in human language may yield some surprises.

Take-Home Messages

- Speech plays a critical role in human language, permitting a high rate of information transfer through the encoding of formant frequency patterns into units whose minimal size is the syllable.
- The source-filter theory of speech production accounts for the physiology of human speech production as well as the vocal signals of many other species. The larynx is a "transducer" that efficiently generates audible acoustic energy powered by the flow of air from the lungs. The airway above the larynx, the supralaryngeal vocal tract (SVT), acts as a dynamic acoustic filter, allowing maximum energy through at formant frequencies. Formant frequency patterns play a major part in specifying vowels and consonants.
- Laryngeal modulation continues to play a role in language and vocal communication. Most human languages make use of tones, which are fundamental frequency (F0) patterns that can differentiate words, as well as the intonation contours that play a role in delimiting sentences. Emotion and affect are conveyed by means of F0 variations and the spectral content of the laryngeal source. Animals can perceive most of the basic acoustic parameters that play a role in conveying spoken words.
- Formant frequency patterns reflect changes in the shape and length of the SVT effected by movements of a person's tongue, lips, and soft palate as well as by the position of the larynx. Encoding automatically occurs as the movements of the tongue, lips, and larynx that generate "phonemes" (roughly, the letters of the alphabet) are melded together into segments that span a syllable or more. We also continually plan ahead, producing articulatory gestures that anticipate the speech sounds that will

occur, such as the vowel of the word "too" at the moment when we produce the gestures associated with the initial consonant.

- Objective data from radiographic and MRI studies show that the traditional measures of tongue position used in many linguistic studies to specify different vowels—tongue height and position—do not differentiate these sounds. Formant frequency patterns instead define these sounds. Except for the quantal vowels [i], [u], and [a] (the vowels of "see," "do," and "mama"), different tongue positions can be used in concert with lip protrusion, lip constriction, and laryngeal position to produce the appropriate vowel formant frequency pattern.
- Formant frequencies are not directly present in the acoustic speech signal. They appear to be perceived by means of a neural process that has implicit knowledge of the constraints imposed by the human SVT.
- The process by which a given formant frequency pattern is interpreted as signifying a particular vowel or consonant involves determining the length of the speaker's SVT. A shorter SVT yields higher formant frequencies for the same vowel or consonant than a longer SVT. Humans unconsciously take into account the probable length of the SVT of the person to whom they are listening. This process of vocal tract normalization occurs in infants and most likely is an innate neural capacity. The process of vocal tract normalization has a "primitive" evolutionary basis; many species determine the size of a conspecific by means of the formant frequencies of a vocalization.
- The vowel [i], the vowel of the word "see," is the "supervowel" of human speech, being less subject to confusion. The vowel [i] yields an optimal signal for vocal tract normalization.

CHAPTER 4

The Neural Bases of Language

AT BEST, any present account of the neural bases of human language is tentative. However, I will attempt to provide an overview of some current views on the nature of the neural bases of human language. My focus is on the cortical-striatal-cortical circuits that yield the reiterative ability that, according to Hauser, Chomsky, and Fitch (2002), confers human recursive syntax. But the studies that I review demonstrate that this creative faculty also plays a critical role in human speech and confers the flexibility of human thought processes. I present evidence that this faculty is linked and probably derives from elements of neural circuits that regulate motor control. Many aspects of these circuits are still a mystery, so I will not attempt to provide a solution to how the mind or brain "works." But reiteration, in itself, cannot be the "key" to language; without words it would be impossible to convey thoughts. Therefore, I also discuss studies that are exploring the neural bases of the brain's dictionary, as well as some aspects of the neural control of speech production, because words must be conveyed. Some related issues such as neural plasticity and the lateralization of the brain are also noted.

I start by reviewing well-established facts concerning the structure of the brain and then attempt to explain distributed neural networks, which appear to be reasonable models for associative learning and other local operations that occur in particular neural structures. The chapter also briefly reviews procedures that are commonly employed in neurophysiologic studies. I present evidence for cortical-striatal-cortical circuits that regulate motor control, syntax, and cognition, including studies of Broca's syndrome, Parkinson's disease, and hypoxia. The findings of these studies contradict traditional theory by showing that Broca's and Wernicke's cortical areas are not the brain's language organs. Converging evidence from studies of experiments-in-nature—i.e., the effects of trauma or disease that damages the human brain, studies of neurologically intact human subjects, and studies of the neural bases of motor control and other aspects of behavior in humans and other species—have provided new insights on the neural bases of observable behaviors. As is the case for other complex behaviors that can be studied in both humans and other species, human language is regulated by cortical-striatal-cortical "circuits" that constitute systems linking processes in different parts of the brain.

Many different regions of the cortex, including Broca's and Wernicke's areas and subcortical structures, play a part in the circuits that confer both human linguistic and cognitive ability. The basal ganglia and other subcortical structures that traditionally were associated with motor control are key elements in these neural circuits. The findings of independent studies spanning almost twenty years show that the permanent loss of language due to damage to these cortical areas, aphasia, derives from damage to subcortical structures implicated in motor control, language, and cognition. Similar problems associated with the so-called language gene also derive from impaired subcortical brain activity. Memory and the brain's dictionary likewise appear to involve activity in circuits linking frontal and posterior regions of the cortex with the subcortical hippocampus. The behavioral deficits of Parkinson's disease involve the subcortical basal ganglia. A new paradigm has emerged that can lead to better understanding of the nature and evolution of the neural bases of human language. And some useful techniques are being developed to evaluate and treat trauma and disease.

The following background information should enable most readers to understand the findings of studies demonstrating that neural structures that regulate motor control are critical components of circuits that confer human cognitive and linguistic ability. These studies also demonstrate beyond reasonable doubt that the traditional Broca-Wernicke language organ theory, though simple, is wrong. Hopefully, readers will not become stranded in a thicket of facts, unable to see the path. However, the detail is necessary to avoid asking readers to take my claims on faith.

Neuroanatomy

Most of the brain is divided in two roughly equal halves, or, cerebral hemispheres, along the front (anterior) to back (posterior) axis. Virtually all of the structures that make up the brain come in pairs, one in each half. In dogs, horses, and other quadrupeds, the term "rostral" also refers to the front of the brain (toward the beak in Latin), while the term "caudal" refers to the back (toward the tail in Latin). In upright humans, the direction toward the back is "dorsal" (the Latin word for "back"), while the direction toward the stomach (the front in upright humans) is "ventral." Structures closer to the midline of the brain are "medial." Structures farther out from the midline are "lateral."

The internal structure of the brain is often viewed in sections that traditionally were obtained by slicing a brain after death. Sectional views of living brains can now be obtained by the computerized imaging techniques (CT and MRI scans) discussed in this chapter. The lateral section that is formed by dividing the brain in two along the anterior-posterior midline is the sagittal plane. Sections parallel to the midsagittal plane are called "parasagittal." Sections parallel to the ground are called "horizontal." Sections perpendicular to the midsagittal and horizontal planes are called "coronal" or "frontal." The surface of the brain is shown in the lateral view in Figure 4.1.

The neocortex, the outer layer of the cerebrum, consists of right and left hemispheres. The part of the cortex lying under the forehead is the frontal lobe. The sylvian fissure, a deep groove, separates the frontal region from the temporal lobe, while the shallower central sulcus marks the border between the frontal and parietal lobes. Neocortex, which occurs only in mammals, has a characteristic structure, with neurons arranged in six layers. The paleocortex is located within the posterior part of the frontal lobe. Basal ganglia structures, which are located within the cerebrum, are bilateral. Thalamus, hippocampus, cerebellum, and other subcortical structures in the basal ganglia and other subcortical structures are represented right and left. Pathways project from many of these neuroanatomic structures for channeling sensory information to

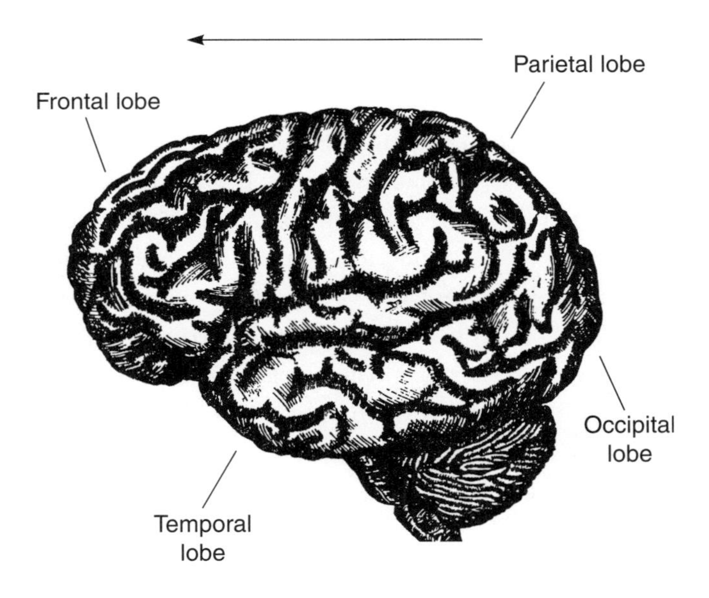

FIGURE 4.1 *Lateral view of the left hemisphere of the human brain, showing the major regions of the neocortex. The arrow points to the anterior, frontal regions.*

different cortical areas and sending signals down to midbrain structures. Pathways to the spinal cord also project from certain neocortical areas. Electrical stimulation of the motor cortex (Brodmann's area (BA) 4; see Figure 4.5 later in this chapter), for example, elicits the contraction of muscles because of direct pathways from this cortical area to the spinal cord.

Neurons

The basic computing elements that make up the nervous systems of animals are neurons. I cannot delve into the details concerning the structure of neurons or how they function here, except to note that neurons connect to each other and transmit information by means of dendrites and axons. A cluster of dendrites (from Latin, referring to the arbor, or tree-like image formed by the dendrites branching out from each cell body) is associated with each neuron. A cortical neuron typically receives inputs on these dendrites from a few thousand other neurons. Each neuron has an output axon, which again typically is arborized and transmits information to a few thousand neurons. Incoming axons from other neurons transmit information into the dendritic tree of a neuron through synapses on the tree as well as on the cell body. Synapses are structures that determine the degree to which an incoming signal will cause the cell body to generate an electrical pulse, the action potential or spike that it transmits out without decrement on its axon. The output action potential can be visualized as an abruptly rising electrical spike and can be monitored by microelectrodes, which are exceedingly fine electrodes typically positioned in, or in the near vicinity of, a neuron.

Note that the synapses sketched in Figure 4.2 do not cross the cell membrane boundary of the neuron. Synapses can occur on the cell body or on its dendritic tree. The concept of the synapse as a connecting, modifiable element was proposed by Sherrington (1947), though the discussion and description of the neuron goes back to the end of the nineteenth century. The synapse acts as a coupling device in the transfer of a signal from a dendrite to the neuron. The all-or-nothing response of the neuron to incoming sig-

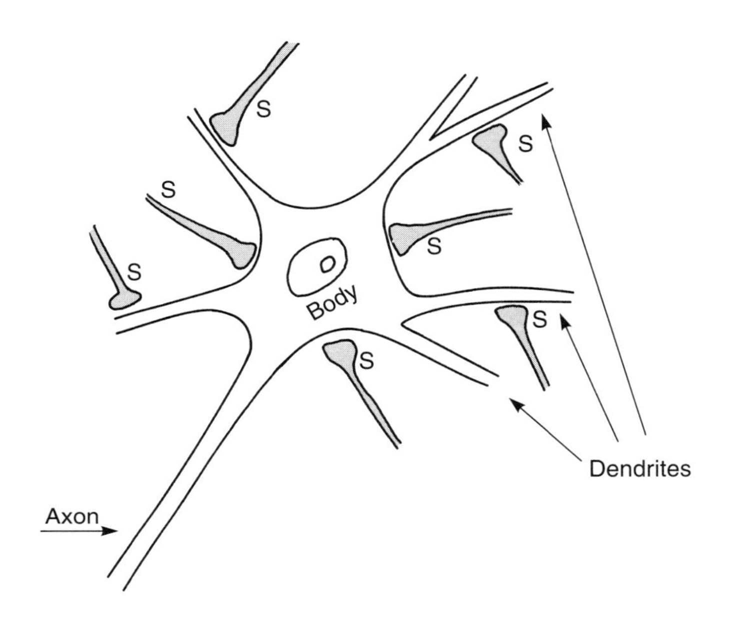

FIGURE 4.2 *Sketch of a neuron showing axon, dendrites, and some synaptic connections.*

nals from the dendrites is an action potential, an electrical discharge that, in turn, is transmitted through the axon to another neuron. Both inhibitory and excitatory synapses exist, which result in an incoming signal having less or more effect, respectively, in triggering an outgoing action potential. As a loose analogy, think of a synapse as the biologic equivalent of the volume control of an audio amplifier: the incoming electrical signal remains the same, but the volume of sound changes depending on the setting of this control. The setting of the volume control, the synaptic weight, determines the extent to which the incoming electrical signal will result in a louder or quieter sound.

The synaptic weights of a large group (a population) of neurons, in effect, constitute an adaptable distributed memory; there is some evidence that the presence of more synapses contributes to adaptability. The neuronal basis of associative learning, which was proposed by Hebb (1949), hinges on the modification of synaptic weights by the axon of one cell consistently and repeatedly firing the axon of another cell. Hebb proposed that the activity of neurons tends to become correlated as they continually respond to some stimuli that share common properties. According to Hebb, when "cell A is near enough to excite a cell B and repeatedly or persistently takes part in firing it, some growth or metabolic change takes place in one or both cells such that A's efficiency as one of the cells firing B is increased" (1949, p. 62). Hebb's theory is consistent with experimental data that show that conduction across the synapse is enhanced as animals are exposed to paired stimuli (Bear, Cooper, and Ebner 1987). Long-term potentiation—that is, changes in synaptic weights—is also affected by various processes in the dendrites themselves (Sejnowski 1997; Markram et al. 1997). Dendrites were once thought to simply transmit information without modifying it. However it is becoming evident that dendrites play a role in modifying synaptic weights (Magee and Johnston 1997). The process of Hebbian synaptic modification appears to be the key to associative learning. Massive interconnections link most cortical areas with considerable, though not complete, reciprocity (Bear, Conners, and Paradiso 1996) and extensive dendritic arborization exists everywhere, connecting various neuronal populations and possibly linking circuits.

Distributed Neural Networks

Virtually all current theories for the neural bases of associative learning and the "local" processes carried out in particular regions of the brain invoke distributed neural networks in which synaptic modification occurs. The extremely simplified diagram in Figure 4.3 indicates some of the complexity of a neural network. Close inspection of this spaghetti-like drawing reveals that every neuron in set α projects to (that is, has a synapse with) every neuron in set β. This drawing, where N = 6, grossly understates the size and connectivity of the nervous system (there are at least 10^{10} neurons in the human brain) and single neurons and single synapses have little effect on the discharge patterns of the group as a whole. The individual in-

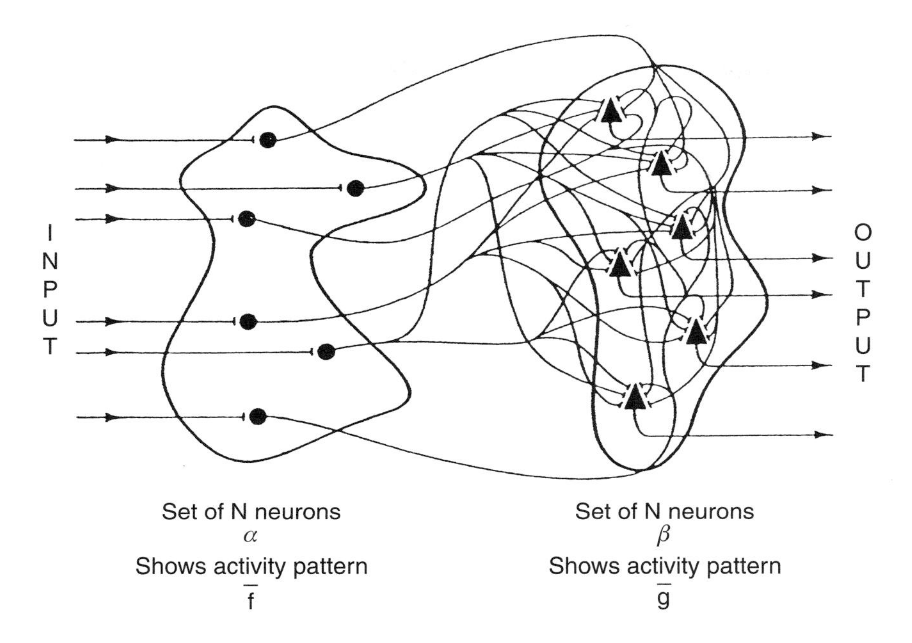

FIGURE 4.3 *An extremely simplified distributed neural network. The diagram shows the interconnections of two sets of six neurons. (After Anderson et al. 1977)*

put units are triggered by particular inputs and fire, thereby transmitting signals through modifiable connections. The connection weights in computer models simulate the modifiable synaptic connections that exist in a biologic brain.

The connection weights are modified by "learning rules" (e.g., Rumelhart et al. 1986). As the network is exposed to stimuli, pathways transmitting signals more often attain higher conduction values. The memory of the net is instantiated in the totality of connection weights that hold across the entire system. Neural nets have a number of properties that differ from those of conventional computers. Distributed neural network models are massively redundant. Representation is distributed; damage to the network reduces resolution or accuracy, but the breakdown is gradual and graded. Unlike a system in which memory is local, damage to some discrete part of a neural network designed to recognize faces will not, for example, destroy the memory of a grandmother's face (Kohonen 1984). The distributed, redundant computational processes carried out by these simulated neural networks appear to be reasonable approximations to neural computation. Different versions of distributed systems have been implemented on digital computers (for example, Kohonen 1984; Anderson 1988, 1995; Sejnowski, Koch, and Churchland 1988; Elman et al. 1997) using somewhat different synaptic modification rules.

Both rules and representations are coded in a similar manner in neural networks. Networks can, for example, code either a large list of the regular plural nouns of English (e.g., "boy" and the plural "boys") or the rules that generate the plural nouns. If we were to observe the output of competing neural models that either coded lists of regular and irregular nouns (e.g., "man" and its plural "men") or "generated" regular plurals using rules, we would not be able to detect how they worked. The behavior of human children likewise does not indicate whether their brains contain lists of different noun forms or rules that generate regular forms plus lists of irregular forms.

To followers of Noam Chomsky's theories (for example, Pinker 1994; Ullman 2004), the fact that children often overgeneralize the process of forming regular nouns and verbs is often taken as evidence for innate rule-governed linguistic process. However, the fact that a child formed a regular version of an irregular English noun (e.g., "sheeps") does not demonstrate that she was overgeneralizing an innate rule. The child's behavior could mean that she made a statistically based decision: she said "sheeps" because most of the plural nouns in her neural lexicon have that form. Moreover, in distributed neural computation, statistical regularities that can be represented as rules emerge as the network is exposed to a set of exemplars. In short, in systems that have large memory stores, either lists or rules suffice to capture the phenomena that contemporary theoretical linguists attribute to rule-governed operations. As studies of the behavior of human infants and young children demonstrate, they make use of statistically driven, associative learning to "acquire" the syllabic structure of words (Saffran et al. 2001) and syntax (Singleton and Newport 1989). Curiously, though the data of these studies argue for statistically driven learning, the authors of these studies hold to Noam Chomsky's claims for innate specification of this linguistic knowledge.

Understanding Distributed Neural Networks

Understanding how a distributed neural network stores memory traces is difficult to explain without resorting to analogies that are just as difficult to understand. The source of the difficulty lies in the fact that most of the devices with which we are familiar are discrete. For example, when we go to a library to recover the "memory" content "coded" in a particular book, we look for it on a shelf in a particular, discrete location. If the book is out on loan or shelved in the wrong place, we cannot access its contents. Likewise, if the segment of my computer's hard drive on which this book was coded had become corrupted, I would not have been able to access the text.

Distributed neural networks do not work in this way. Large sections of the network can be destroyed without losing memory traces. This may account for the seeming paradox documented in many of the studies discussed in this chapter. Broca's and Wernicke's areas are active when a person comprehends or produces a word or sentence. However, as we shall see, people do not lose their preexisting language or their native dialect when these cortical areas are destroyed. Transient loss of language may occur; but after a period of months, a person's premorbid linguistic ability returns. The linguistic knowledge that existed before a stroke or other trauma occurs clearly has a redundant, distributed representation in people's brains.

Electrical Power Grids: A Metaphor for Distributed Networks

The electrical power grid that supplies power to your home is a reasonable, though highly simplified, metaphor for a distributed system. The properties that make the electrical grid a distributed system are its interconnections and the multiplicity of generators. These systems are designed to continue to provide power even when individual generators or transmission lines are out of service. They can rapidly change connections: there are parallel, redundant lines and synapses, switching points that can rapidly direct the flow of power through alternative pathways. They have sufficient excess generating power, and the supply of electrical power can be interrupted by cuts only at the periphery of the system where single lines go to individual subscribers. In actual fact, blackouts have occurred in the United States, Norway, England, and Italy because the systems were not able to rapidly change connections and did not have excess generating power.

Let's explore our metaphor further. In the hypothetical, simplified system diagram in Figure 4.4, the electrical grid continues to supply power to Springfield, Massachusetts, even though the supply line from Pittsfield is cut. The grid's switching points are rearranged to take up the Springfield load by using the alternative interconnections that feed into Springfield. And the rearrangement of the network is not confined to the immediate vicinity of Springfield. The changes in the vicinity of Springfield trigger changes distant from that city as the system distributes the perturbations in the load. The circled nodes in the figure, for example, show the switching links that assumed new values after the failure of the Springfield-to-Pittsfield feed transmission line. The switching points are junctions at which the flow can be changed in the lines or branches that meet at that node. These switches have to send more current through the remaining lines into Springfield, given the failure of the Pittsfield line.

Changes in the values of the switching points throughout the distributed system can also follow from changes in the electrical load at a particular point. They do not occur only when a line falls. For example, on a hot New England summer night, the distributed power grid's synapses, or switching points, continually rearrange the flow of electrical power. Power from Hydro-Quebec streams south toward the hot, humid cities. Additional generators may come

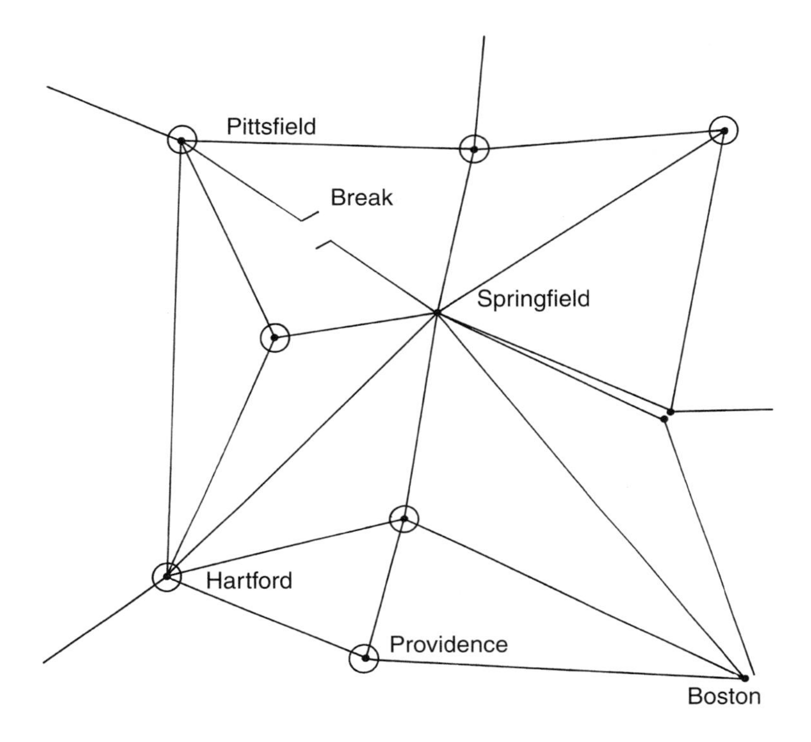

FIGURE 4.4 *A hypothetical electric power distribution network that illustrates some of the properties of distributed neural networks. The circled nodes indicate switching points that assumed new values after the direct connection between Pittsfield and Springfield was interrupted.*

onto the power grid. Then, as the night cools, the pattern changes and the distributed system takes on new synaptic conditions. The record of the distributed system's activity is the settings of all of the electrical switching points that connect the lines between the "generator-neurons."

Let me designate the array of cells represented by the loads of cities of southeastern Massachusetts and Providence, Rhode Island, as the input to the model. The input is a set of numbers that represents an observation of the "behavior"—the electrical appetite of these cities; how much electricity each is using at a given moment. The network model responds to this input by distributing the load into the power grid. In supplying this load, the synapses (the switching points throughout the entire electrical power grid) assume a particular configuration. The representation of this input in the distributed system is the set of numbers that specifies the settings of the synapses throughout the entire network. No individual cell or node or synapse represents the input; the "memory" of the event coded in the network is the entire set of synaptic values, the set of numbers that represents the settings of each and every one of the network's switches. Different input power requirements would result in a different set of values for the synapses of the entire system. Note that there is no single physical "address" for the location of the distributed system's memory trace of any particular input. That follows from the fact that the memory trace is the configuration of the entire system.

A local change in a power grid can result in a distributed change throughout the system. No single generator-neuron necessarily is active in response to a local change in electrical demand, and no single generator-neuron is essential in a properly designed network. The nodal-synapses change their values to distribute changes in load throughout the system.

The Robustness of Distributed Neural Networks

Another property of a distributed network—the robust nature of memory traces—follows from the fact that information is coded by the synaptic settings of the entire system. A fragment of the network thus bears a set of synaptic weights that reflects the settings of the entire network. Reconstructing a particular event coded in the synaptic weights of a fragment would be less reliable (noisier) than the weights of the complete network, but the event won't be completely lost. As most people who have experienced blackouts now realize, a distributed network also can be perturbed by events taking place far away from them. An isolated event that overloads a switching point can propagate through the network if switches fail to correct the problem in time. But if the network is properly designed and monitored, a single power plant failure or downed transmission line would not interrupt your electrical supply.

In an analogous manner, your brain does not stop functioning even though thousands of neurons die each day. You do not forget your cat's name after drinking a glass of wine at dinner (which can hypothetically destroy hundreds of neurons). This property of distributed networks may, in part, explain why an individual who suffers a stroke that destroys Broca's or Wernicke's area but spares subcortical structures does not have to relearn language. The memory traces that code the motor pattern generators that result in speech, the cognitive pattern generators specifying the rules of syntax, and the lexicon clearly have widely distributed cortical representations. Less redundancy appears to characterize the subcortical structures of the basal ganglia and thalamus, where damage can result in permanent losses of motor, linguistic, or cognitive capabilities (Stuss and Benson 1986; Kuoppamaki, Bhatia, and Quinn 2002).

Associative Learning

Imitation and associative learning are powerful "general" procedures for acquiring knowledge. The classic example of associative learning is Pavlov's dogs, which learned to associate the sound of a bell with food. Pavlov continually rang a bell before feeding the dogs. After a short period the dogs, anticipating the food, salivated when the bell rang; they had learned to associate the bell's sound with food. Seabiscuit, the celebrated underdog racehorse of 1938, learned to associate the clang of a bell with a fast-breaking start and, as a result, was able to defeat the iconic "perfect" racehorse of the period, War Admiral (Hillenbrand 2002).

Pavlov believed that the site of associative learning was cortical, but this type of learning can occur in very simple animals that lack a cortex. Associative learning has been demonstrated in *Aplysia californicus,* a gastropod mollusk (Carew, Walters, and Kandel 1981; Walters, Carew, and Kandel 1981). These invertebrates lack a cortex or anything that approaches the complexity of the brain of even "simple" mammals such as mice. The mollusks learned to associate a conditioning stimulus, shrimp extract, with an aversive unconditioned stimulus, an electric shock, through the classic Pavlovian paradigm. The training sessions involved first presenting the shrimp extract to the mollusks. Six seconds after the start of the presentation of the stimulus, an electric shock was applied to the head of the animal. Twenty mollusks were trained in this manner and received six to nine paired stimuli. Another twenty mollusks served as a control group and received unpaired electric shocks and shrimp extract stimuli that were presented at 90-minute intervals.

The mollusks were then tested 18 hours after the training sessions. Shrimp extract was applied to the heads of all the animals for one minute, and weak electric shock was then applied to the tail of each mollusk. The animals that had been exposed to the paired stimuli in the training sessions reacted more forcefully by escape locomotion and inking than the animals that had not been exposed to paired stimuli. The behavior of the animals was monitored both by observing the number of steps that they took to move away from the weak electric shock and by observing the electrical activity of the motor systems by means of intracellular microelectrode recording from identified motor neurons for each response.

The main point of the mollusk training was to test Hebb's (1949) synaptic modification theory for learning and memory. The correlate of learning in the mollusks' nervous system was synaptic modification. Electrophysiologic data show that the conditioning stimulus produced by stimuli that trigger defensive responses acts to enhance synaptic input to the motor neurons. What is startling is that the mollusks quickly learned to associate a benign stimulus (shrimp extract) with an unpleasant stimulus (strong electrical shock). They adapted to their new environment without the help of any neural structures that remotely resemble the mammalian cortex. Associative learning has a selective value to any animal because it allows rapid phenotypical changes in response to new environmental conditions. The mollusk experiments signify that the neural bases of "cognitive" acts can be traced back to very simple animals.

Richard Herrnstein and his colleagues used similar techniques to teach pigeons general principles of biologic classification. As briefly noted in Chapter 1, the pigeons were first trained to peck at photographs of specific trees to get a reward. The picture set was gradually expanded and generalized to many trees of different species, photographed at different angles and in different settings. In time, the pigeons were trained to distinguish different species of trees (Herrnstein 1979). Objections were raised by nativists to Herrnstein's claim that he had taught the pigeons to respond differentially to various species of trees. The nativists, who emphasized the role of innate, genetically transmitted knowledge, instead proposed that the pigeons had innate knowledge of the tree species. Because pigeons roost on trees, they suggested that "knowledge" of the characteristics of different species of trees was an innate property of pigeon brains—the result of natural selection acting to optimize biologic fitness. To counter this argument, Herrnstein trained pigeons to recognize different species of tropical fish using the same associative techniques (Herrnstein and de Villiers 1980).

Cortical Architecture—Cortical Maps

Although the processes carried out in different parts of the brain appear to involve "local" distributed neural networks, the human brain is not a big blob—a unitary distributed neural network. Particular regions of the cortex and particular subcortical structures appear to perform specific "local" operations. Studies that attempt to relate activity in different regions of the brain to behavior must use some agreed-upon "map" of these presumed functionally distinct structures. The focus of most studies of motor and cognitive behavior has been the cortex, and the map commonly used derives from a series of anatomic studies carried out in the early years of the twentieth century by the neuroanatomist Korbinian Brodmann. Brodmann showed that different parts of the cortex have different cytoarchitectonic structures—in other words, different distributions of neurons in the layers that form the cortex.

The cytoarchitectural maps of brains made by Brodmann (1908, 1909, 1912) generally were thought to reflect discrete functional distinctions. Some cortical areas were supposedly dedicated to auditory or visual perception, others were dedicated to motor control, and still other areas presumably stored memory traces. However, it is becoming clear that the situation is more complex than Brodmann thought. Certain areas of the cortex—for example, area 17 in the Figure 4.5—are, as traditional theories note, implicated in visual perception. Area 17 receives signals from part of the thalamus, which, in turn, receives signals from the eye. People who lack area 17 are blind. However, area 17 also is active is visual mental imagery, when human subjects are asked to visualize scenes (Kosslyn et al. 1999).

Many other cortical areas are active when a person views or recalls a scene (Velanova et al. 2003; Wheeler and Buckner 2003). Brodmann's area 4, the primary motor cortex, is only part of the complex assemblage of cortical and subcortical neural structures involved in motor control; and, as we shall see, it is active in perceptual and cognitive tasks when a person hears or reads a word or views a picture that has a reference to or appears to involve internal modeling of motor activity. And many details are at present unknown. The situation is similarly complex and unclear for language. Subcortical structures play a crucial part in regulating human language. Although Broca's area, which consists of Brodmann's areas 44 and 45 in the dominant, usually left hemisphere of cortex, is implicated in language, children who have had their left hemisphere surgically removed don't completely lose the ability to acquire language. If surgery occurs before the child is eight to ten years old, he often develops language within normal ranges (Elman et al. 1997; Bates, Vicari, and Trauner 1999).

Broca's and adjacent cortical areas also are implicated in manual motor control (Kimura 1979; Krams et al. 1998; Rizzolatti and Arbib 1998). As is the case for many, if not most, cortical areas and subcortical structures, it carries out different local operations and plays a part in circuits regulating different aspects of behavior. It clearly is not a "module" dedicated to language and language alone. The feed-forward and back connections that typify all corti-

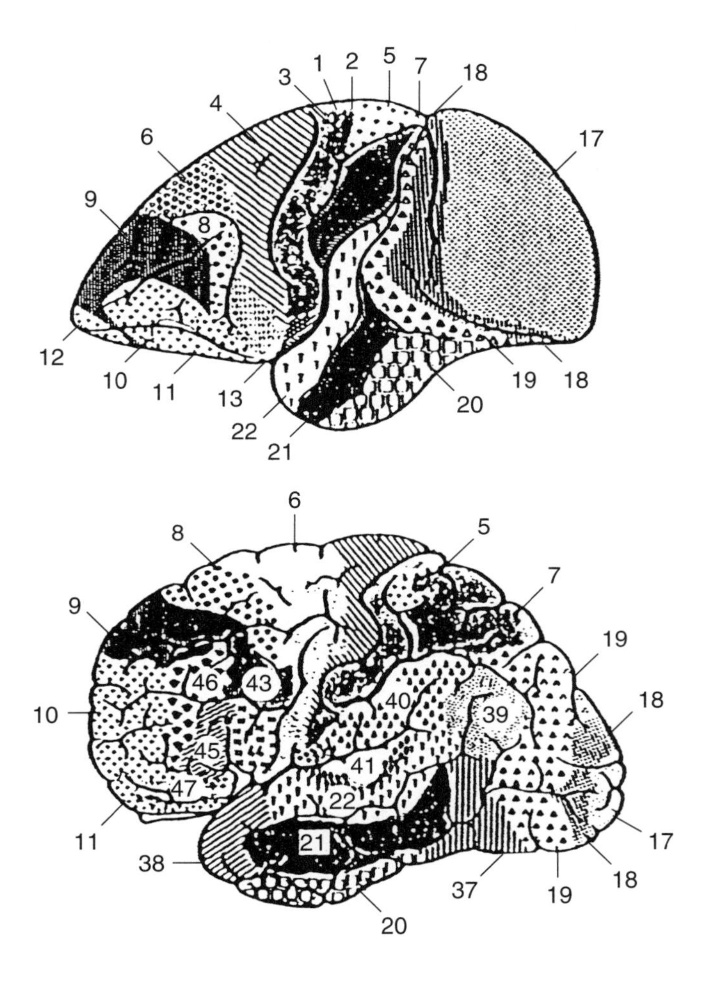

FIGURE 4.5 *Brodmann's 1909 cytoarchitectonic maps of the cortical areas of the macaque monkey (top) and human (bottom) brains. The size of the brains has been equalized here, but the monkey brain's volume is less than one-third that of the human brain. The frontal regions are to the left. Areas 44 and 45 of the human brain are the traditional sites of Broca's area.*

cal areas (Bear, Conners, and Paradiso 1996) manifest themselves in Broca's area being activated during speech perception and sentence comprehension (Just et al. 1996; Paus et al. 1996). Broca's area also is activated when subjects listen to music (Maess et al. 2001). And experimental data show that different cortical locations regulate similar aspects of language in the brains of different people (Ojemann and Mateer 1979; Ojemann et al. 1989).

Furthermore, as we shall see, many areas of cortex are malleable and can take on new functions due to damage to the brain or birth defects (Merzenich et al. 1984; Merzenich 1987a, 1987b; Sanes and Donoghue 1994, 1997; Donoghue 1995; Elman et al. 1997). For example, the visual cortex in humans who were born blind or became blind at an early age appears to be recruited to process tactile perception (Cohen et al. 1997). Sign language activates auditory cortex (Nishimura et al. 1999).

The Neurophysiologic Toolkit

Ernst Mayr, in his preface to the 1964 facsimile edition of *On the Origin of Species,* points out the paradigm that Darwin introduced to biologic research. Theories are based on preliminary data; are tested against further data; and, if they are useful, lead to refined theories that explain a greater range of phenomena. Research techniques, data, and theory are inextricably linked. We nonetheless often overlook the subtle relationships that hold between research techniques, data, and theory. A theory is necessarily formulated on the basis of initial data. Therefore, all theories are inherently structured by the technical constraints that limit experiment and observation. A theory is then subject to test by further experiment and observation, but the theory's explicit and implicit assumptions inherently determine the experiments and techniques that appear to be relevant. Subsequent data usually result in the theory being modified. The process is neither strictly deductive nor inductive. Experimental data do not merely serve to refute or confirm the predictions of a theory. Experimental techniques, the interpretation of data, and the theoretical claims deriving from these data are constrained by a common set of implicit and explicit assumptions. Advances in technology that allow the acquisition of additional data, therefore, have led to a better understanding of the neural bases of human language.

Direct Electrophysiologic Recording

Direct observation of neuronal activity in the brain of a living animal involves the placement of microelectrodes. These electrodes conduct electrical signals induced by the electrochemical communications between neurons. In principle, the techniques are similar to those employed by a wiretapper. A small antenna, the microelectrode, is placed into, or close to, the electrical circuit (the neuronal signal path) that is to be monitored. The microelectrode picks up the signal, which is then amplified and recorded. But whereas a wiretapper can surreptitiously place a miniature antenna close to the circuit connections in a telephone junction box and then remove it without disrupting the telephone system, microelectrode recording techniques are invasive.

The microelectrode first has to be placed in the brain of an animal whose skull has been opened. Elaborate precautions must be taken to achieve useful data. The microelectrode or electrodes first must be positioned in the intended neuroanatomic structure. If the experiment is monitoring neuronal activity connected with the perception or processing of visual, tactile, olfactory, or auditory signals, a representative range of appropriate stimuli must then be presented to the prepared animal. Appropriate motor tasks must be executed by the animal if the focus of the experiment is motor control. Often the interpretation of the data is skewed by the range of sensory inputs, motor activities, or the context that was *not* explored. Finally, the exact positions of the recording electrodes must be determined by sacrificing the experimental animal. The brain must then be sectioned, stained, and microscopically examined. In certain limited circumstances, direct electrophysiologic recordings can be made in human patients prior to brain surgery. George Ojemann and his colleagues have obtained many insights on brain function in that manner by electrically stimulating the surface of the brain. However, the range of stimuli, the duration of recording, and the number of locations that can be explored are necessarily limited.

Tracer Studies

One of the properties of axons is that amino acids, the building blocks of proteins, are transmitted from the cell body down the length of an axon. Research in the late 1960s showed that radioactively labeled amino acids injected into a cell body also would be transported down the axon to its terminal. The pathways from neuron to neuron can therefore be determined by mapping the radioactive axon terminals. Another tracer technique involves the enzyme horseradish peroxidase (HRP), which has an odd interaction with neurons: it is taken up at the terminal ends of axons and is transmitted retrogradely back to the cell body. Other tracer compounds as well as viruses such as herpes can be used to map out neuronal circuits. However, all traditional tracer techniques involve injecting substances into the brain of a living animal, waiting for the tracer to be transported, and then sacrificing the animal. After the animal's brain is sectioned, chemical reactions are employed to stain the pathways and visualize the HRP transport. Other means can be used to map out the transport of radioactive tags. Then, using microscopic examinations of sliced sections of the animal's brain, the pathways can be mapped. Typically populations (groups) of neurons "project" (connect) to populations of neurons in other neuroanatomic structures. The circuits usually are segregated, which means they are anatomically independent. A given neuroanatomic structure typically contains many segregated microcircuits that project to different segregated neuronal populations in other neuroanatomic structures. A recent advance in tracer techniques employs a variation of MRI technology that allows neural circuits to be traced in living human subjects (Lehericy et al. 2004).

MRI, fMRI, and PET Imaging

It is impossible to map out neuronal circuits in human beings using these invasive techniques. However, noninvasive imaging techniques make it possible to infer neuronal activity associated with various aspects of behavior in human beings. Functional magnetic resonance imaging (fMRI) is the most recent technique. fMRI is a variant on the structural magnetic resonance imaging (MRI) systems that are routinely used for imaging brain structures. The basic operating principle involves generating an intense magnetic field that perturbs the electrons of molecular compounds. When the magnetic field is suddenly released, the resetting electrons emit characteristic electromagnetic "signatures" that are mapped by complex computer algorithms. Structural MRIs can map out "slices" of the brain. Diffusion tensor fiber tracking, a technique derived from MRI scanning, can permit noninvasive determination of neural circuits in living humans by tracking the direction of fiber bundles in the white matter that connects neuronal populations; the results from this technique have been validated in postmortem animal studies (Lehericy et al. 2004).

fMRI can map out the flow and transport of deoxygenated blood (the BOLD signal) that is an indirect marker of metabolic activity, hence neural activity, in the brain (Logothetis 2001). Event-related fMRI involves close synchrony of tasks with the fMRI BOLD signal, permitting investigation of the time course of neural activity. Positron emission tomography (PET), another imaging technique, involves injecting a short-half life radioactive tracer into the bloodstream. Blood flow increases as metabolic activity increases in the parts of the brain that carry out some task. Radioactively tagged glucose also can be injected into the bloodstream. A computer system then interprets the signals picked up by sensors that monitor the level or amount of the tagged blood or glucose. The half-life of the injected radioactive compounds is short and dose levels low.

CT Scans

Computerized tomography (CT) ushered in the modern period of brain research. CT scans differ from conventional X-rays (radiographs) in that they show slices of brain anatomy reconstructed by computer processing of multiple X-ray exposures taken at different angles and planes. This made it possible to determine the site and extent of damage in a living patient. MRIs provide better images.

One problem in interpreting brain function common to all of these procedures (CT, PET, and fMRI) involves comparing activity in one person's brain with that in another person's brain. It is clear that people's brains differ as much as faces, feet, hearts, teeth, and other aspects of anatomy (Ojemann et al. 1989; Ziles et al. 1995; Fink et al. 1997). In fact, recent MRI studies by Mazziotta at UCLA of identical twins show variations even in the gyral morphology that provides the landmarks that are commonly used to locate Brodmann areas in fMRI and PET studies.

ERP

Event-related potentials (ERPs) provide a complementary technique to both PET and fMRI imaging data. PET and fMRI can determine metabolic activity in a particular part of the brain, but temporal response times are sluggish. The metabolic activity recorded by these techniques represents neural activity averaged over many seconds. ERPs can reveal transient electrical activity recorded by means of electrodes placed in contact with a person's scalp. The technique involves recording by means of the procedures commonly used to record EEG brain activity in clinical settings. The difference is that a particular stimulus—for example, a spoken word—is presented many times. The electrical signals recorded in response to the stimulus are then summed and averaged. A characteristic negative electrical trace, for example, occurs 400 msec (the N400 signal) after a person comprehends the meaning of a word (Kutas 1997; Friederici 2002). ERPs lack topographic resolution, but they complement the slow temporal resolution of PET and fMRI data.

Brain-Behavior Models

As I noted in the first chapter of this book, historically, the most complex piece of machinery of an epoch serves as a metaphor for the brain. The metaphor seems to take on a life of its own and becomes a neurophysiologic model. In the eighteenth and early years of the nineteenth centuries the most complex machines that a sedentary scholar typically encountered were clocks. A case can be made that ocean-going ships were actually the most complex "machines" of that period because they relied on interactive systems. However, mechanical clocks and chronometers were the apparent brain model. In a clock, one mechanical system, a set of parts, counted out the interval of time, a different set of parts moved the clock's hands, and so on. The discrete systems that made up a clock each carried out a particular operation in a particular location.

Phrenology

Their explicit locationist model led phrenologists to systematically identify specific places on the skull to locate the parts of the brain that were the regulators of various skills and qualities. The reasoning was valid—in a clock some discrete mechanical system carried out an operation. If those parts were damaged, the clock failed. Therefore, it was reasonable to propose a neural model that sought to find the locations of the specific parts of the brain that regulated various aspects of human behavior. The outside of the skull is a reasonable map of the surface of the neocortex, whose size comparative anatomists had discovered was the most apparent singular characteristic that differentiated human beings from other species. It, therefore, was quite reasonable to locate the sites of "higher" human capacities in the neocortex—larger sites would signify a better developed capacity, reflected in the area of the skull covering that portion of the neocortex.

Although phrenology is often portrayed as a quack science, it constituted a scientific theory subject to falsification. Gall (1809) and Spurzheim (1815) claimed that language, mathematical ability, musical ability, and various aspects of human character such as ambition, charity, and veneration were regulated in specific locations of the neocortex. Phrenologists, acting on the principle that surface regions of the skull corresponded to underlying cortical areas, partitioned the exterior of the skull into regions that were each the "seat" of a "faculty" (Figure 4.6). According to phrenological theory, the size of these seats, the areas of the protuberances and bumps of

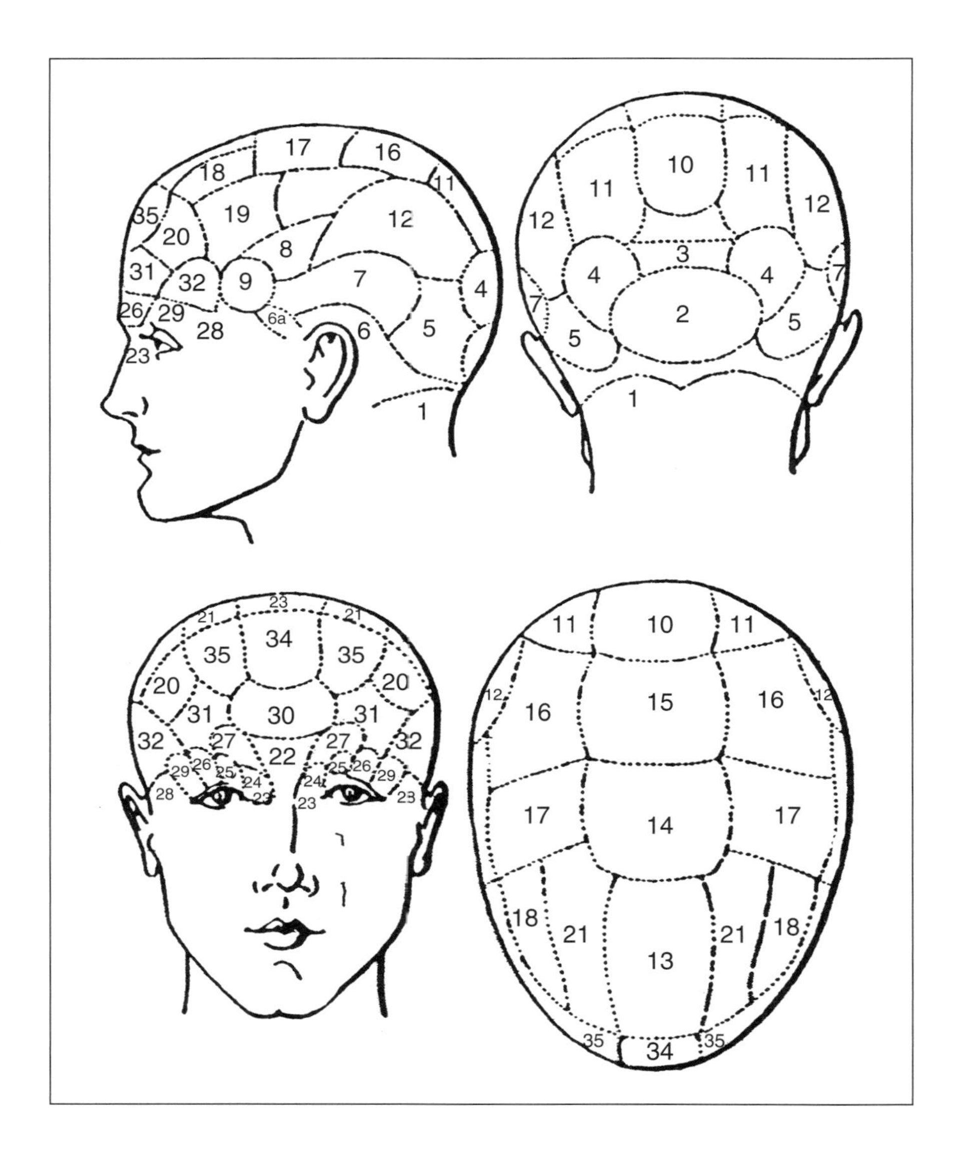

FIGURE 4.6 *Phrenological map of faculties of the human mind. (After Spurzheim 1826)*

the skull of a given person, were innately determined. The area of each region was a measure of the degree to which that complex behavior or particular aspect of character was manifested in an individual. Phrenological theory was tested through empirical studies that, for example, determined whether people whose skulls had a larger expanse in area 14, the seat of veneration, manifested this attribute to a greater degree in their daily lives than people whose skulls had a smaller area 14. Gall measured skulls in such places as prisons and lunatic asylums, correlating behavior with skull measurements. Other studies measured the skulls of clerics, professors, poets, artists, and the like. The skulls of homicidal felons often had larger areas for compassion than clerics, and distinguished mathematicians were sometimes shown to have small mathematical areas. Thus phrenology fell into disfavor. However, the underlying premise that guided phrenological research—that all aspects of a complex behavior are regulated in an anatomically discrete, separable area of the cortex—survives to the present day in the Broca-Wernicke language area theory.

Modular Theories

Confusion often arises with regard to the meaning of the term "module" in neurophysiologic and linguistic studies. The term is often used in neurophysiologic studies (for example, Graybiel 1995, 1997) to refer to the complex neural circuits that regulate observable behaviors. In contrast, theories of mind grounded in linguistics, such as Fodor (1983) and Pinker (1998, 2002), often use the term "module" to refer to localized neuroanatomic structures that hypothetically regulate specific aspects of language. Moreover, according to these modular theories, the module that regulates language, or some aspect of language such as syntax, has no anatomic or physiologic relation to other neural modules devoted to talking, walking, manual motor control, and so on. Each module is an independent structure or system, similar in nature to the systems that form a conventional digital computer. In principle, these modular theories claim that the functional organization of the human brain is similar to that of a conventional digital computer that has a discrete hard disk, a discrete electronic memory, a display, a modem, and so on.

The discrete, localized, modular structural architecture of computers is reflected in current modular mind-brain theories for language. The central processing unit of a digital computer is a discrete device; RAM memory is discrete; and hard drives are discrete, modular devices. These discrete devices translate to discrete areas of neocortex and other parts of the brain. The serial, algorithmic, computational architecture of the digital computer likewise translates into modular linguistic and psycholinguistic theories. Modular models claim that language is comprehended and produced by means of a series of independent operations. The first stage in the comprehension of spoken language hypothetically is a process whereby phonetic units or "features" (Jakobson, Fant, and Halle 1952) are derived from the acoustic signal, perhaps mediated by vocal tract modeling (Liberman et al. 1967). High-level top-down information (the semantic and pragmatic constraints conveyed by the words or word fragments that are being specified by incoming acoustic information) supposedly provides only secondary corrective information.

However, experimental data show that these claims are incorrect. Human speech generally is a sloppy, underspecified signal that deviates from textbook phonetic transcriptions. This applies to the speech of learned professors. Tape-recorded lectures are notoriously difficult to transcribe because speakers almost always underspecify the acoustic cues that convey phonetic contrasts. As I noted earlier, even "well-formed" speech recorded under ideal conditions is often completely incomprehensible unless the listener fills in missing information with expectations of what was probably intended (Lieberman 1963). Pollack and Pickett (1963) showed that listeners need to hear at least a 600-msec segment to identify the content. Samuel (1996, 1997, 2001) in a series of clever experiments has shown that a listener's expectations strongly influence perception. Segments of degraded speech signals that listeners were able to identify were excised by means of waveform editing. Semantic priming occurred with these degraded speech signals. Semantic priming is evident when a listener is asked to indicate whether a speech signal is an actual word; the listener's response is faster if he first hears a semantically related word—for example, "dog" preceding "cat." We are generally unaware of these phenomena because we continually fill in missing information, overriding acoustic disturbances that conflict with our internally generated hypotheses concerning what was *probably* said. We take into account semantic and pragmatic information derived from parallel, highly redundant processing. People "hear" what they wish to hear.

Neural Circuit Models

Neurophysiologic activity must be considered at two levels if we are to understand how the brain regulates complex behaviors, such as reaching for a pencil, walking, talking, or comprehending the meaning of this sentence. First, although complex brains contain many distinct neuroanatomic structures, these structures usually do not, in themselves, regulate an observable behavior. An individual neuroanatomic structure instead generally contains many anatomically segregated groups, or populations of neurons that carry out a *local* operation or process. Activities such as walking, talking, or moving your finger are not local operations that are regulated by a single, discrete region of your brain. They result from linked local operations carried out in different structures of the brain. The neuronal population that carries out the local operation in a given part of the brain projects to anatomically segregated neuronal populations in other neuroanatomic structures. Successive links between segregated neuronal populations in different neuroanatomic structures form a neural circuit. The linked local neural operations carried out in the circuit constitute the brain basis of a complex, observable aspect of behavior that generally has a name, such as striking the key of a piano. As the sketch in Figure 4.7 indicates, within a given neuroanatomic structure, distinct anatomically segregated neuronal populations project to neurons in different brain structures to form other circuits. Circuits linking anatomically segregated populations of neurons form neural systems carrying out processes in different parts of the brain. The systems are the neural bases of complex behaviors.

As Mesulam notes, "complex behavior is mapped at the level of

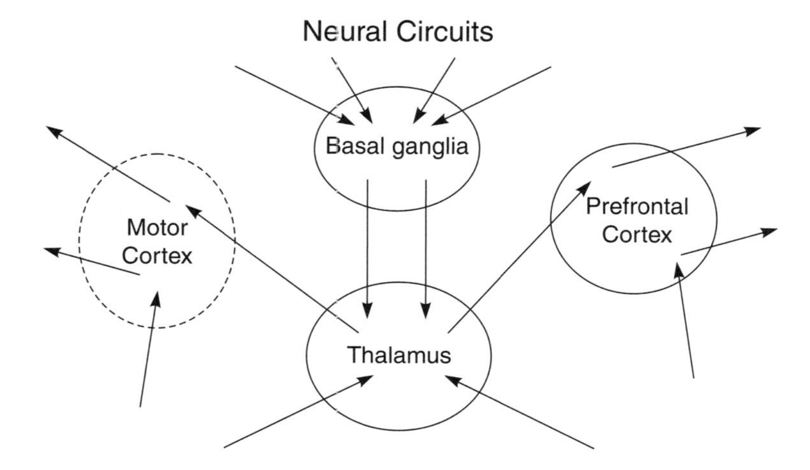

FIGURE 4.7 *Anatomically segregated populations of neurons in a particular structure or region of the brain can project to distinct anatomically segregated populations of neurons in different parts of the brain, forming "circuits" that regulate different aspects of behavior. Thus damage to a particular part of the brain can result in a "syndrome," an ensemble of seemingly unrelated behavioral deficits. Here, neuronal populations from different cortical areas project into the putamen, and from there indirectly into different regions of the cortex. The resulting circuits regulate mood, motor control, and different aspects of cognition, including the comprehension of syntax.*

multifocal neural systems rather than specific anatomic sites, giving rise to brain-behavior relationships that are both localized and distributed" (1990, p. 598). In other words, "local" neural operations occur in particular regions of the brain. However, these localized operations in themselves generally do not constitute an observable behavior such as walking or talking. Evidence from hundreds of independent studies that span three decades show that different regions of the neocortex and different subcortical structures are specialized to process particular stimuli, visual or auditory, while other regions perform specific operations that regulate aspects of motor control (such as coding the direction of a movement or its force) or holding information in short-term (working) memory (for example, Polit and Bizzi 1978; Mitchell et al. 1987; Marsden and Obeso 1994; Sanes et al. 1995; Mirenowicz and Schultz 1996; Monchi et al.

2001). However, these local operations form a set of neural computations that link together in complex neural circuits, yielding actions such as walking, pushing a button, or comprehending this sentence.

For example, within the putamen, one of the structures of the subcortical basal ganglia, anatomically segregated populations of neurons form part of a system that sequences the submovements that constitute an overt movement of a monkey's hand, a rat's grooming sequence, or a person's walking or speaking (Aldridge et al. 1993; Marsden and Obeso 1994; Cunnington et al. 1995; Lieberman 2000). But the putamen, in itself, is not the seat of the motor act. It is part of a basal ganglia complex that inhibits and releases pattern generators that code both motor gestures and cognitive acts (Marsden and Obeso 1994; Graybiel 1995, 1997, 1998). The putamen, like many other neuroanatomic structures, supports anatomically segregated neuronal populations that project to different parts of the brain, forming a number of circuits that regulate other aspects of behavior. Distinct, anatomically segregated neuronal populations in the putamen project through other subcortical structures to cortical areas implicated in motor control, higher cognition, attention, and reward-based learning (for example, Alexander et al. 1986; Parent 1986; Alexander and Crutcher 1990; Aldridge et al. 1993; Cummings 1993; Kimura, Aosaki, and Graybiel 1993; Graybiel et al. 1994; Marsden and Obeso 1994; Middleton and Strick 1994; Graybiel 1995, 1997, 1998; Lieberman 2000, 2002). In short, the neural mechanism that carries out the instruction set manifested in my pecking at my computer's keyboard is a circuit, linking neuronal populations in different neuroanatomic structures in many parts of the brain.

Experiments in Nature

The study of the neural bases of human language began long before it was possible to perform electrophysiologic, tracer, or imaging studies. Because only human beings possess language, the study of the brain bases of language relied on the permanent deficits, or aphasia, induced by experiments-in-nature when particular parts of a person's brain were destroyed by accidents, gunshots, strokes, tumors, or other pathologies. Paul Broca's (1861) observations arguably were the most influential experiment in nature. The interpretation of the behavioral deficits of aphasia that I present here is quite different from Broca's model. However, experiments-in-nature still are germane to the brain-language question, particularly when their findings are integrated with the data of noninvasive imaging studies of neurologically intact subjects as well as comparative studies of neural processing in other species.

The Traditional Broca-Wernicke Model

If one assumes that discrete localized regions of the brain, in themselves, regulate an observable aspect of behavior, it follows that removing or destroying that region should disrupt the behavior. Paul Broca (1861) studied the brain of a patient with the masked name of "Tan" who had suffered a series of strokes and then died. The strokes had caused extensive brain damage including, but not limited to, one part of the brain, "the third frontal convolution," an anterior area of neocortex. Broca followed the phrenological model and concluded that damage to this area (called "Broca's area") was the basis of the patient's speech deficit. Tan's most obvious linguistic problem was his limited speech ability; the only utterance that he was able to produce was the syllable "tan." Overlooked was the fact that Tan also had extensive subcortical damage and extensive nonlinguistic motor impairments.

Wernicke in 1874 found that patients who had suffered damage in the second temporal gyrus of the cortex in the posterior left hemisphere had difficulty comprehending speech. Again, Wernicke's conclusion was that receptive linguistic ability was localized in a neocortical area. Because making use of language involves both comprehending and producing speech or alternative phonetic systems such as writing or sign language, Lichtheim (1885) proposed a hypothetical cortical pathway linking Broca's and Wernicke's areas. Broca and his successors, Wernicke and Lichtheim, essentially translated the phrenological theories of Gall (1809) and Spurzheim (1815) to cortical areas. According to the Broca-Wernicke model proposed by Lichtheim, spoken language is perceived in Wernicke's area, a posterior temporal region associated with auditory perception. Information is then transmitted via a cortical pathway to Broca's region, which is adjacent to cortical areas implicated in motor control. Broca's region is the hypothetical neural site regulating speech production. Broca's and Wernicke's areas are sketched in Figure 4.8. Geschwind's (1970) theory, which continues to shape brain and language theories, is essentially a restatement of Lichtheim's views.

Although the Broca-Wernicke model has the virtue of simplicity, it is wrong. Clinical evidence shows that permanent loss of language does not occur without subcortical damage even when Broca's or Wernicke's areas have been destroyed. Moreover, damage to sub-

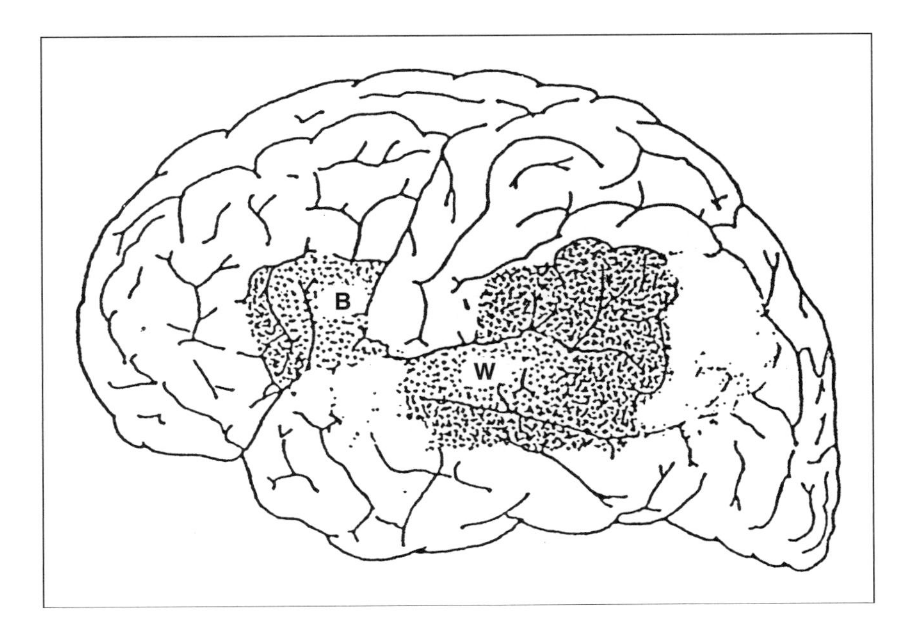

FIGURE 4.8 *The traditional mapping of Broca's and Wernicke's "cortical language" areas.*

cortical structures, sparing the cortex, can produce aphasic syndromes.

Studies of aphasia, or the permanent loss of language, were the basis for the Broca-Wernicke theory. However, subsequent studies of aphasia were among the first to note the deficiencies of this traditional model. Doubts were expressed in the early years of the twentieth century (Jackson 1915; Marie 1926). In the past two decades, computer-aided tomography (CT) scans and magnetic resonance imaging (MRI) have provided noninvasive information on the nature and extent of the brain damage that results in permanent language loss. The putative basis of Broca's syndrome in the Lichtheim (1885) model is damage to Broca's neocortical area. However, clinical studies have shown that permanent loss of the linguistic abilities associated with the syndrome does not occur unless subcortical damage is present. This was noted as early as 1986 by Stuss and Benson in their book *The Frontal Lobes,* which was written for neurologists.

The independent studies of Dronkers and her colleagues (1992) and D'Esposito and Alexander (1995) report similar results. Patients with extensive damage to Broca's area generally recover linguistic ability unless subcortical damage also occurs. Moreover, patients suffering brain damage to subcortical structures with an intact Broca's also can manifest the signs and symptoms associated with Broca's syndrome. As Stuss and Benson in their 1986 review of studies of aphasia conclude, damage to "the Broca area alone or to its immediate surroundings . . . is insufficient to produce the full syndrome of Broca's aphasia . . . The full, permanent syndrome (big Broca) invariably indicates larger dominant hemisphere destruction . . . deep into the insula and adjacent white matter and possibly including basal ganglia" (1986, p. 161). A number of independent studies show that subcortical damage that leaves Broca's area intact can result in Broca-like speech production and language deficits (for example, Naeser et al. 1982; Benson and Geschwind 1985; Alexander et al. 1987; Mega and Alexander 1994).

Alexander and his colleagues (1987), for example, reviewed 19 cases of aphasia resulting from lesions in these subcortical structures. Language impairments occurred that ranged from fairly mild disorders in the patient's ability to recall words to "global aphasia" in which the patient produced very limited nonpropositional speech. In general, the severest language deficits occurred in patients who had suffered the most extensive subcortical brain damage. Damage to the internal capsule (the nerve fibers that connect neocortex to subcortical structures) and basal ganglia structures, the putamen, and caudate nucleus resulted in impaired speech production similar to that of the classic aphasias, as well other cognitive deficits. Subsequent studies appear to rule out damage to the internal capsule as the basis for subcortically induced aphasia. Deliberate surgical lesions of the internal capsule aimed at mitigating obsessive-compulsive behavior do not induce aphasia (Greenberg, Murphy, and Rasmussen 2000). Damage to the medial cerebral artery, which snakes through the basal ganglia, often results in strokes that yield aphasia. As D'Esposito and Alexander (1995) flatly state in their study of aphasia deriving from subcortical damage, whether "a *purely* cortical lesion—even a macroscopic one—can produce Broca's or Wernicke's aphasia has never been demonstrated" (1995, p. 41).

The Basal Ganglia and Cortical-Striatal Circuits

The inherent deficiency of the Broca-Wernicke model is that it fails to take account of current knowledge concerning the computational architecture of biologic brains. In particular the role of the basal ganglia in cortical-striatal circuits is neglected. As we shall see, basal ganglia dysfunction can produce the constellation of motor, syntax, and cognitive deficits associated with Broca's syndrome. The basal ganglia are subcortical structures located deep within the brain. They are primitive neural structures that can be traced back to anurans such as frogs (Marin, Smeets, and Gonzalez 1998). The basal ganglia in humans and other primates include the caudate nucleus and the lentiform nucleus, which constitute the striatum. The lentiform nucleus itself consists of the putamen and globus pallidus (or palladium; the terms refer to the same structure). It is cradled in the internal capsule that forms a bundle snaking through the caudate and lenticular nucleus. Figure 4.9 shows the general topography.

The putamen receives sensory inputs from most parts of the brain. The globus pallidus is an output structure receiving inputs from the putamen and caudate nucleus. The caudate nucleus, putamen, and globus pallidus are interconnected and form a system with close connections to the substantia nigra, thalamus, other subcortical structures, and cortex. The thalamus, in turn, connects to different cortical areas. The connections with cortex are complex and not fully understood (DeLong, Georgopoulos, and Crutcher 1983; Alexander, DeLong, and Strick 1986; Parent 1986; Alexander and Crutcher 1990; DeLong 1993; Hoover and Strick 1993; Marsden and Obeso 1994; Lieberman 2000). Detailed discussions of the anatomy and physiology of basal ganglia and their role in regulating motor control, language, and cognition are presented in these studies.

Disruptions in behavior that are seemingly unrelated, such as obsessive-compulsive disorder (Greenberg, Murphy, and Rasmussen 2000), schizophrenia (Graybiel 1997), and Parkinson's disease (Jellinger 1990), derive from the disruption of neural circuits that link cortical areas with the basal ganglia structures of the striatum. Anomalous basal ganglia development also appears to play a major role in a genetically transmitted deficit affecting orofacial motor control, speech production, syntax, and cognition—the so-called language gene (Fisher et al. 1998; Vargha-Khadem et al. 1998, 2005; Lai et al. 2001; Watkins et al. 2002; Liegeois et al. 2003).[1] Behavioral changes once attributed to frontal lobe cortical dysfunction can be observed in patients having damage to subcortical basal ganglia (for example, DeLong 1983; Cummings and Benson 1984; Flowers and Robertson 1985; Alexander, Delong, and Strick 1986; Lange et al. 1992). Cummings' 1993 review article, which was based on clinical studies, identifies five parallel basal ganglia circuits of the human brain (three of which are illustrated in Figure 4.10). Cummings notes,

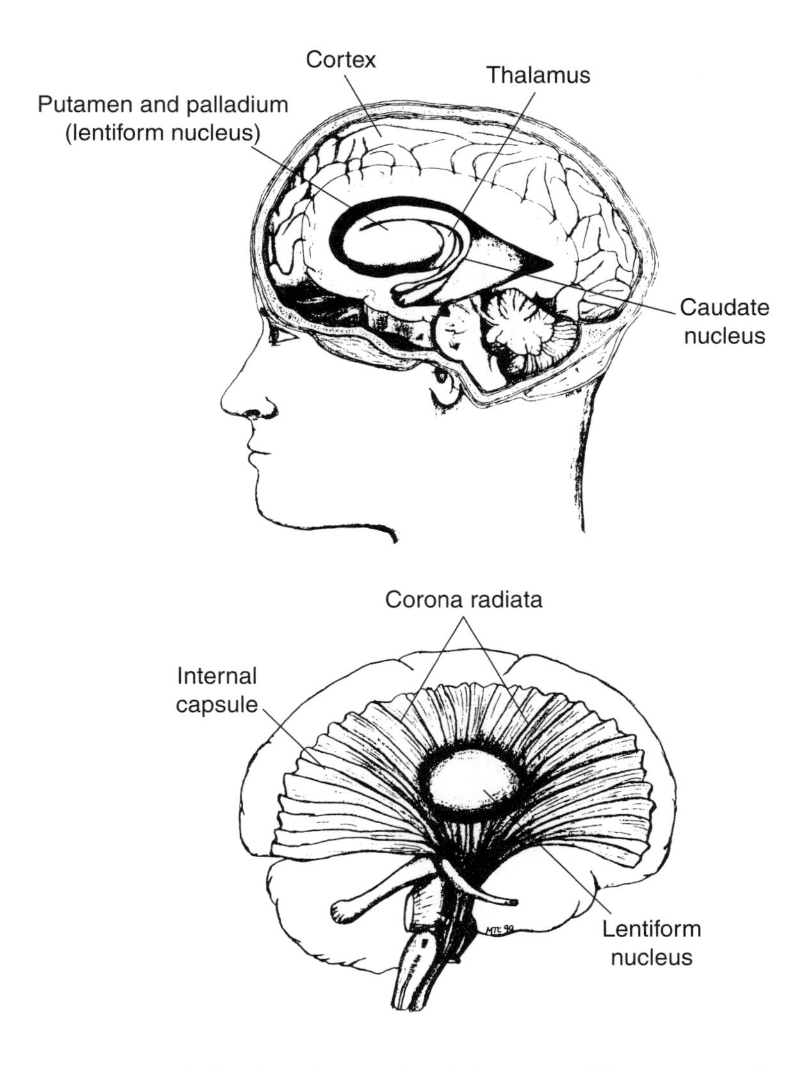

FIGURE 4.9 *The basal ganglia are subcortical structures. The putamen and globus pallidus (palladium) constitute the lentiform nucleus, which is cradled in the nerves descending from the neocortex that converge to form the internal capsule. The caudate nucleus is another primary basal ganglia structure.*

a motor circuit originating in the supplementary motor area, an oculomotor circuit with origins in the frontal eye fields, and three circuits originating in prefrontal cortex (dorsolateral prefrontal cortex, lateral orbital cortex and anterior cingulate cortex). The prototypical structure of all circuits is an origin in the

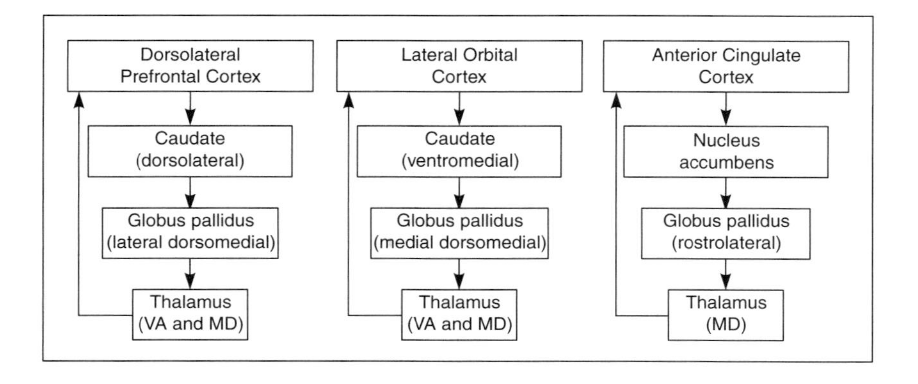

FIGURE 4.10 *Organization of three basal ganglia circuits that regulate various aspects of motor control, cognition, and emotion in human beings. The dorsolateral prefrontal circuit, for example, is implicated in speech motor programming, sentence comprehension, and some aspects of cognition. VA = the ventral anterior region of the thalamus, MD = the medial dorsal region of the thalamus. The diagrams are simplified and do not show the indirect connections of the substantia nigra (another subcortical structure) and other details. Damage to any of the neuroanatomic structures that support the neuronal populations of a circuit can result in similar deficits. (After Cummings 1993)*

> frontal lobes, projection to striatal structures (caudate, putamen, and ventral striatum), connections from striatum to globus pallidus and substantia nigra, projections from these two structures to specific thalamic nuclei, and a final link back to the frontal lobe. (Cummings 1993, p. 873)

As noted above, tracers attach themselves to the neural circuits formed by the outputs of neurons connecting to other neurons. Postmortem staining techniques then allow the neural pathways to be discerned under microscopic analysis. Therefore, traditional invasive tracer studies are limited to species other than humans. Tracer studies of monkey brains show that the striatum supports circuits that project to cortical areas associated with motor control and cognition (Alexander, DeLong, and Strick 1986; Middleton and Strick 1994; Graybiel 1995, 1997). However, diffusion tensor imaging (DTI) techniques, which are based on MRI technology, now reveal distinct cortical to striatal circuits in humans that are similar to those mapped out in monkeys (Lehericy et al. 2004). This is not surprising because the general topography of the striatum and associated structures such as the substantia nigra is quite similar in monkeys and humans.

Neurodegenerative Diseases

In short, damage to striatal and associated subcortical structures that support a cortical-striatal-cortical circuit can result in a behavioral deficit that was thought to derive from damage to the region of the cortex served by that circuit. Neurodegenerative diseases such as Parkinson's disease (PD) and progressive supranuclear palsy (PSP) result in major damage to subcortical basal ganglia, mostly sparing the cortex (Jellinger 1990). Independent studies of these neurodegenerative diseases have established the role of the basal ganglia in these circuits. The primary deficits of these neurodegenerative diseases are motoric; tremors, rigidity, and repeated movement patterns occur. However, these subcortical diseases also cause linguistic and cognitive deficits. Speech production, syntax, and cognitive deficits similar in nature to those typical of Broca's aphasia can occur in even mild and moderately impaired PD patients (Gotham, Brown, and Marsden 1988; Morris et al. 1988; Taylor et al. 1990; Harrington and Haaland 1991; Lange et al. 1992; Lieberman et al. 1992; Cools et al. 2001). In particular, deficits in the comprehension of and production of syntax have been noted in independent studies of PD (for example, Lieberman, Friedman, and Feldman 1990; Lieberman et al. 1992; Natsopoulos et al. 1993; Illes et al. 1988; Pickett 1998; Hochstadt and Lieberman 2000; Lieberman 2000; Grossman et al. 1991, 1992, 1993, 2001; Howard et al. 2001; Hochstadt 2004). As is the case for Broca's aphasia (Blumstein 1995), PD patients have difficulty comprehending sentences that have moderately complex syntax as well as long sentences that tax the brain's computational resources (Baum 1989).

In the final stages of PD, a dementia often occurs, different in kind from Alzheimer's dementia (Albert, Feldman, and Willis 1974; Cummings and Benson 1984; Xuerob et al. 1990). The afflicted patients retain semantic and real-world knowledge but exhibit "perseveration" and are unable to readily form or change cognitive sets (Flowers and Robertson 1985; Cools et al. 2001). In their daily life this manifests itself in cognitive inflexibility—the patients cannot alter the direction of a thought process or change plans when circumstances dictate a change. These seemingly unrelated deficits appear to derive from the "local" operations performed by the basal ganglia in the cortical-striatal-cortical circuits regulating these aspects of behavior.

Probable Basal Ganglia Operations

In the era before medication with Levadopa was prescribed to treat Parkinson's disease, thousands of operations were performed. The effects of these surgical interventions on motor control in humans and similar experimental lesions in monkeys were reviewed in a seminal paper by Marsden and Obeso (1994). They note that the basal ganglia appear to have two different "local" motor control functions:

> First, their normal routine activity may promote automatic execution of routine movement by facilitating the desired cortically driven movements and suppressing unwanted muscular activity. Secondly, they may be called into play to interrupt or alter such ongoing action in novel circumstances . . . Most of the time they allow and help cortically determined movements to run smoothly. But on occasions, in special contexts, they respond to unusual circumstances to reorder the cortical control of movement. (Marsden and Obeso 1994, p. 889)

Given the fact that the basal ganglia circuitry regulating motor control does not radically differ from that implicated in cognition,

Marsden and Obeso conclude that the basal ganglia have a second, similar "local" operation:

> the role of the basal ganglia in controlling movement must give insight into their other functions, particularly if thought is mental movement without motion. Perhaps the basal ganglia are an elaborate machine, within the overall frontal lobe distributed system, that allow routine thought and action, but which respond to new circumstances to allow a change in direction of ideas and movement. Loss of basal ganglia contribution, such as in Parkinson's disease, thus would lead to inflexibility of mental and motor response." (1994, p. 893)

The functional basal ganglia complex includes the putamen, caudate nucleus, globus pallidus, and close connections to the substantia nigra, thalamus, and other structures. The system essentially acts as a sequencing engine. Within the basal ganglia, information is transmitted by inhibitory and excitatory channels (Alexander, DeLong, and Strick 1986; Alexander and Crutcher 1990). The globus pallidus, to which the putamen and caudate nucleus project, is the principal output structure. The excitatory channel projects to the substantia nigra through the "internal" segment of the globus pallidus and essentially connects the information specified in a pattern generator with a cortical target. The inhibitory channel through the "external" output channel of the globus pallidus and subthalamic nucleus disconnects a pattern generator. Both motor and cognitive pattern generators, which respectively specify the submovements of motor acts and subelements of thought processes, including linguistic "rules," are activated and inhibited through these two routes.

Neuroimaging Studies—Set Shifting and Sequencing

Advances in brain imaging and behavioral studies of human subjects support this inference—that the basal ganglia perform cognitive sequencing functions. One of the first imaging studies that demonstrated basal ganglia switching cognitive sets in a linguistic task was a PET study performed at the Montreal Neurological Institute. Neurologically intact bilingual speakers of French and English were the subjects. The PET data showed that the putamen becomes hyperactive when bilingual French-English speakers switch to their second language (Klein et al. 1994).

Functional magnetic imaging (fMRI) data from neurologically intact subjects confirm the cognitive role of the dorsolateral-prefrontal-striatal circuit noted by Cummings (1993). The event-related fMRI study of Monchi et al. (2001) confirms the particular role of basal ganglia in the cortical-striatal-cortical circuits involved in performing the Wisconsin Card Sorting Test (WCST), which evaluates a person's ability to form and shift cognitive criteria. Brain activity was monitored in neurologically intact subjects in a version of the WCST that allowed the investigators to match stages of the task with the event-related fMRI signals. The subjects had to match test cards to reference cards based on the color, shape, or number of stimuli pictured on each card. Subjects were informed when they made either correct or incorrect matches and had to shift the matching criterion as the test progressed. Neural circuits involving prefrontal cortex and basal ganglia were activated during the test. Bilateral activation was observed in mesial and dorsolateral prefrontal cortex, including Broca's area (Brodmann areas 44 and 45), Brodmann area 46 (implicated in executive control and working memory), basal ganglia, and thalamus. (Figure 4.5 shows the cortical areas defined by Brodmann, 1912.)

Higher activation occurred in the right hemisphere, perhaps reflecting the visual material used in the test. Dorsolateral prefrontal cortical areas (Brodmann 9 and 46) were active at the points where the subjects had to relate the current match with earlier events stored in working memory. In contrast, a cortical to basal ganglia circuit involving the midventrolateral prefrontal cortex (areas 47/12), caudate nucleus, putamen, and thalamus was active when subjects had to shift to a different matching criterion. Increased activity occurred in the putamen, which showed activity throughout the task, during these cognitive shifts.

The behavioral study of Scott and his colleagues (2002) complements these findings. A comprehensive set of cognitive tests that assess frontal lobe functions such as planning, as well as tests of memory, were administered to PD patients who had undergone neurosurgery that produced precise bilateral lesions of the internal excitatory output pathway of the globus pallidus. The sole problem occurred on the Wisconsin Card Sorting Test, where the subjects were unable to shift the matching criterion as the test progressed. The bilateral lesions impeded the activation of a different cognitive process at the shift points, confirming the set-shifting role of the basal ganglia complex. Stowe et al. (2004) found similar linguistic set-shifting activity using PET to monitor neurologically intact subjects in a sentence comprehension task. When the subjects were presented with ambiguous sentences, activation of the basal ganglia to dorsolateral-prefrontal-cortex circuit occurred when they realized the ambiguity and changed their interpretation.

Both the role and time course of basal ganglia activity in linguistic tasks can be explored using fMRI techniques. Kotz and her colleagues (2003b) monitored neurologically intact subjects who were asked to judge whether spoken sentences conveyed positive, neutral, or negative emotion. A set of 108 spoken sentences were produced by a speaker who attempted to convey positive, negative, or neutral emotion by modulating the sentences' intonation. (The fundamental frequency of phonation, the amplitude of the speech signal, its spectral content, and temporal factors all enter into this process). The listeners' task was to identify the "emotion." The event-related fMRI study showed bilateral, left-accentuated activation of putamen and thalamus as well as temporal sites. When the spoken sentences were digitally processed to remove all acoustic energy above the third harmonic of the fundamental frequency of phonation, thereby removing most cues to a sentence's words, added frontal cortical and caudate nucleus activation occurred.

Rissman, Eliassen, and Blumstein (2003) explored the neural bases of semantic priming in an event-related fMRI experiment. Semantic priming, as I have already noted, is a straightforward procedure. When subjects are asked to decide whether a phonetic sequence that they hear or read is a word, they respond faster if they first hear or read a semantically related "prime." For example, you would respond to the word "mouse" faster if you first heard the word "cheese." Your response time would be longer if you instead heard the semantically unrelated word "table." You would need much more time to decide that [blig], which could be an English word, is in fact not. Subjects presumably have to search through their internal lexicon to come to that conclusion. The fMRI data obtained as subjects responded to English words shows activation of putamen, globus pallidus, anterior caudate, and cerebellum as well as the approximate locations of Broca's and Wernicke's areas and other cortical sites. Nonword identifications take longer and result in increased neural cortical and subcortical activity.

Experiments that make use of event-related-potentials (ERPs) can reveal the time course of basal ganglia activity as subjects listen to and interpret spoken utterances. Kotz et al. (2003) monitored subjects listening to short sentences and deciding whether they were grammatical on the basis of verb "argument structure"; some verbs can have objects, others not. For example, the sentence "the little boy grins" is grammatical, but "the little boy grins the old man" is not. Half of the subjects had basal ganglia (BG) brain damage. Half of the subjects had only cortical damage. In previous ERP studies of neurologically intact subjects, a negative electrical signal (the N400 response) occurred 400 msec after lexical-semantic processing of the words of a sentence and was followed by a positive signal occurring at 600 msec (the P600 response) that appears to reflect syntactic processing, such as taking into account a verb's argument structure. The BG subjects did not show the P600 response linked to syntactic processing and instead showed a delayed signal resembling the N400, consistent with the slow cognitive processing noted in many studies of BG damage. The cortically damaged subjects showed P600 but no N400 activity, a result that suggested lexical-semantic impairment.

These findings in experiments with human subjects match data from electrophysiologic studies of monkey brains (reviewed in Graybiel 1995, 1997). Basal ganglia perform similar functions in phylogenetically "simple" mammals. Studies of rodent brains in which basal ganglia structures were selectively destroyed or in which invasive, direct electrophysiologic recording was used again show motor sequencing activity which resembles that occurring in humans. When the rodents' basal ganglia were destroyed, they were able to execute the individual submovements that, when linked together, constitute a grooming sequence (Berridge and Whitshaw 1992), but they could not perform the grooming sequence itself. Electrophysiologic studies of the rodents' basal ganglia neurons showed firing patterns that sequentially inhibited and released submovements, thereby stringing them into a grooming sequence (Aldridge et al. 1993; Aldridge and Berridge 2003).

Implicit Associative Learning

The basal ganglia have other functions, including learning very specific patterns of action. Most of the instances in which humans and other species form cognitive categories and expectations are "unconscious." Huettel, Mack, and McCarthy (2002), for example, correctly point out that this is a central aspect of cognitive behavior:

> By identifying patterns, people form predictions about upcoming events. If, when waiting at a train crossing, one first feels the ground shake, then hears a low rumbling sound, and finally sees red flashing lights in the distance, the consistency of that pattern with past experience allows one to predict that a train will soon arrive. (Huettel, Mack, and McCarthy 2002, p. 485)

In their fMRI study, Huettel and his colleagues observed a pattern of prefrontal and basal ganglia activity similar to that seen in the

WCST when subjects formed and shifted cognitive sets in a visual task that involved no overt instructions regarding category formation. Subjects viewed a random binary sequence of squares and circles. Within the overall random sequence in which patterns occurred by chance, segments of up to eight repeating or alternating stimuli were presented. Prefrontal cortex was activated as the subjects unconsciously inferred the presence of a pattern of images, for example, a series of five circles followed by a square. Basal ganglia (putamen and caudate nucleus) also were activated when a series of repeated patterns was presented and followed by a shift to a different pattern. When patterns were alternated, no basal ganglia activity occurred—the subjects apparently treated the alternation itself as a pattern. These findings are consistent with electrophysiologic studies showing neuronal populations forming in the putamen as primates learn a task (Kimura, Aosaki, and Graybiel 1993; Graybiel et al. 1994).

The specificity of these learned behaviors also argues against the "procedural-declarative" model advanced by Ullman (2004) following Pinker's (1994) suggestions. They hardly constitute general "rules," even in the sense noted by phonologists such as Chomsky and Halle (1968), who codified the sound changes that convey the plural forms of the regular nouns of English, such as "book" and its plural "books." According to Ullman, the basal ganglia and frontal regions of the human brain control rule-governed processes such as the relationship that holds between the singular and plural forms of regular English nouns, while posterior regions are the seat of the dictionary, coding individual irregular nouns. While the basal ganglia clearly form part of the cortical-striatal-cortical circuits involved in comprehending syntax, sequencing motor pattern generators and so on, these circuits, as we have seen, also regulate emotion, attention, and other aspects of behavior. Their linguistic activity is not limited to the restricted role postulated by Pinker and Ullman. I will return to note the negative results of a specific test (Longworth et al. 2005) of the Ullman-Pinker theory after we review the findings of studies that reveal the role of cortical-striatal-cortical and other neural circuits in regulating motor, linguistic, and cognitive behavior.

Blobology

Before we review these studies, we must consider one caveat as we interpret data from fMRI and PET studies that make use of the "subtractive procedure." This procedure attempts to identify neural structures that are key factors in regulating a particular aspect of behavior by subtracting the pattern of activity associated with a presumed baseline condition from the total activity pattern associated with the experimental stimuli. For example, activity that is portrayed as a colored blob in an fMRI image occurs when the output from a person listening to acoustic white noise is subtracted from the output that occurs when that person listens to and comprehends words. This procedure presumably reveals the additional activity associated with listening to words, but it also wipes out the activity of neural structures that are active while a person listens to both noise and words. Nonetheless, the blobs produced in this way are often interpreted as proof of a word-comprehension organ.

Blobology overlooks the fact that circuits linking activity in many neuroanatomic structures generally carry out complex, observable behaviors. The neuroanatomic structures implicated in listening to acoustic signals must necessarily be activated when we interpret words. And there is no guarantee that the part of the brain that supposedly is the seat of word comprehension is not also activated in some other behavior that was not explored in the particular experiment. For example, the anterior cingulate gyrus, a structure of the paleocortex, is activated in virtually every fMRI or PET study on record. It appears to play a part in directing attention to a wide range of mental processes as well as in regulating speech, keeping infant mammals and their mothers in touch, and in maternal care. Clinical studies show that lesions in cortical-striatal-cortical neural circuits involving the anterior cingulate can result in mutism (Cummings 1993). Animal studies that make use of microelectrode techniques and surgical lesions show that the anterior cingulate also plays a part in regulating the laryngeal activity that yields the mammalian isolation call (a typical infant cry) and in maternal care (Slotnick 1967; Newman and MacLean 1982; MacLean and Newman 1988).

Over-reliance on the subtractive technique sometimes results in breathless press releases announcing the part of the brain that determines some newsworthy aspect of human behavior such as sex, fear, aggression, religion, and so on. Jerome Kagan's parody of the seat of tennis being located by subtracting brain activity when thinking about running, from brain activity while thinking about tennis, is not too far off the mark when we evaluate the claims of some PET and fMRI studies. Neuroanatomic structures that are activated in one activity in a particular experiment are generally activated in other behaviors—sometimes closely related, sometimes not.

Evidence for Basal Ganglia Activity in Cortical-Striatal-Cortical Circuits

As Charles Darwin realized in 1859, some measure of evidence must be provided to support any theory. The details noted in this section may suffice to flesh out the claims I have already made in this chapter and may also prove useful for readers who are not familiar with the findings of relevant clinical studies.

Broca's Syndrome

Broca's aphasia does not occur, absent subcortical brain damage. Indeed, Broca's syndrome may result from impairment of the basal ganglia. Crosson et al. (2005), for example, note the ineffectiveness of therapy directed at restoring language abilities after unilateral cortical brain damage when the basal ganglia are also damaged. A syndrome similar to that of Broca's aphasia occurs in neurodegenerative diseases and when brain damage has affected the basal ganglia and cerebellum. Broca's syndrome is real, though the site of brain damage apparently is not Broca's area. A brief discussion of the nature of the deficits of Broca's syndrome may be useful; the details are in the references cited.

Speech Sequencing

One of the primary speech deficits of Broca's syndrome and compromised basal ganglia function is a breakdown in the sequencing of the motor commands necessary to produce stop consonants. These speech sounds are produced by closing the lips, obstructing the flow of air from a speaker's mouth, and then abruptly opening the lips, thereby producing a burst of turbulent air that has distinct acoustic properties. At the same time, the speaker must adjust the muscles of the larynx to produce phonation subsequent to the burst. In order to produce a [b], phonation must occur within 20 msec of lip opening; longer delays yield the sound [p]. Similar temporal contrasts involving the muscles of the tongue and the larynx differentiate the sounds [d] from [t] (for example, "do" versus "to"), and [g] from [k] (for example, "god" versus "cod"). Lisker and Abramson (1964) coined the term "voice-onset time" (VOT) to describe this distinction, which appears to hold for all human languages studied to date. In brief, VOT is defined as the time that occurs between the burst that results from lip or tongue gestures and the onset of periodic phonation generated by the larynx. Figure 4.11 shows the waveforms of a [ba] and a [pa] with cursors superimposed that mark the onsets of the bursts and the phonation that define VOT.

Speakers must precisely control a sequence of independent motor acts to produce these sounds. Complex adjustments of the tension of at least six different laryngeal muscles must be initiated at least 60 msec before the onset of phonation (Atkinson 1973). Broca's aphasics are unable to maintain control of some of these sequential motor commands: their intended [b]'s may have long VOTs appropriate for [p]'s, their [t]'s may have the short VOTs of [d]'s, and so on (Blumstein et al. 1980; Baum et al. 1990). The problem is not limited to generally slower speech, though that also occurs; stop consonants such as [p] and [t], which should have long VOTs, may

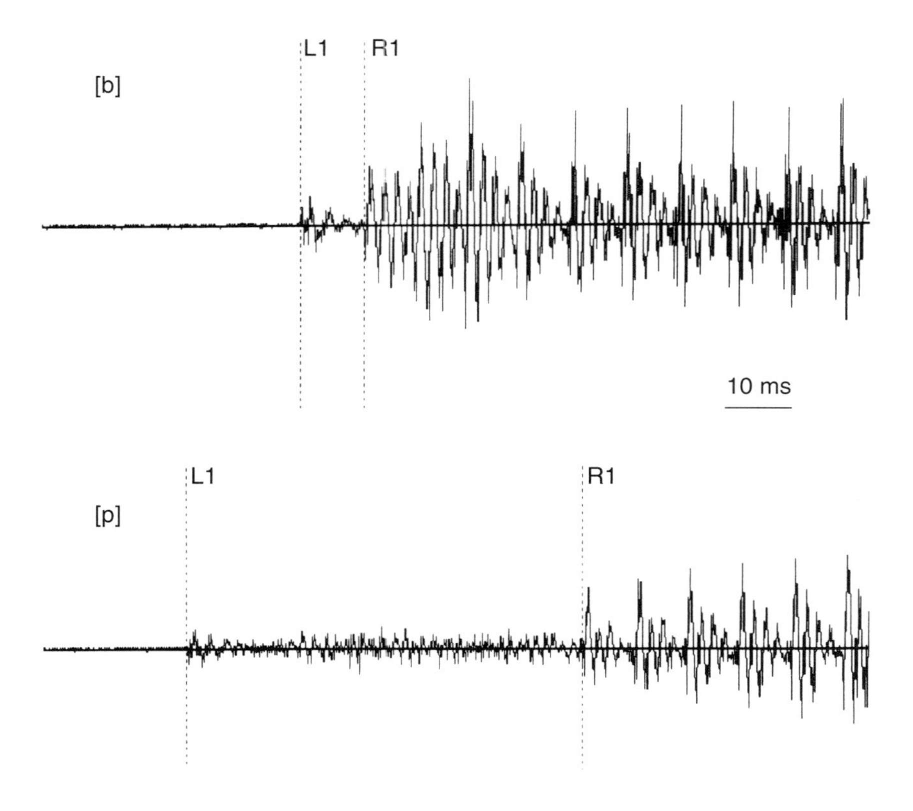

FIGURE 4.11 *Speech waveforms of the syllables [ba] and [pa]. Amplitude is plotted on the ordinate and elapse of time on the abscissa. "Cursors" L1 mark the beginnings of the "bursts" of the stop consonants [b] and [p] that occur when the lips open. Cursors R1 mark the onset of phonation that occurs when the vocal cords (folds) of the speaker's larynx start to produce periodic phonation.*

have short VOTs. The problem clearly does not involve a loss of control of duration. Broca's aphasics maintain the intrinsic durations that differentiate vowels (Baum et al. 1990); for example, the vowel of the word "bat" is three times longer than that of the word "bit." The relative distinctions in vowel duration that serve as cues to the identity of these vowels are maintained.

Surprisingly, Broca's aphasics maintain almost normal control of the magnitude and placement of tongue, lip, and laryngeal gestures;

no apparent loss of peripheral motor control occurs. The acoustic consequence is that the production of the formant frequency patterns that specify vowels is unimpaired in Broca's syndrome, though there is increased variability (Ryalls 1981, 1986; Kent and Rosenbek 1983; Katz 1988; Baum et al. 1990; Katz et al. 1990). Formant frequency patterns are determined by the configuration of the supralaryngeal vocal tract (tongue, lips, larynx height). Therefore, we can conclude that the control of these structures is unimpaired in Broca's aphasic syndrome. The speech deficit appears to involve the sequencing of independent motor commands. Studies of the neural circuits that control the larynx and tongue and lips in monkeys suggest that they are independent (Newman and MacLean 1982; Sutton and Jurgens 1988). Broca's aphasics also have difficulty executing oral nonspeech and manual sequential motor sequences (Kimura 1993).

The phonologic level, or the knowledge of the sound pattern that specifies the name of a word, is preserved in Broca's aphasics and in Parkinson's disease (PD). The problem appears to be limited to articulatory sequencing of certain motor gestures. For example, the acoustic cues and articulatory gestures that specify the stop consonants differ when they occur in syllable initial position or after the syllable's vowel. The speech-sound [t], for example, is signaled by a long VOT when it occurs in syllable-initial position. In contrast, after a vowel, the acoustic cues for [t] are reduced duration of the vowel that precedes it and increased burst amplitude. Broca's aphasics and PD subjects maintain normal control of these postvocalic cues although VOT sequencing is disrupted for syllable-initial [t]'s.

These distinctions are general. The duration of a vowel always is longer, taking into account other factors, before a [b], [d], or [g] than for a [p], [t], or [k]. The fact that Broca's aphasics preserve these durational cues indicates that the phonologic instruction set for producing stop consonants is intact. The preservation of these durational cues shows that the Broca's VOT deficit derives from the disruption of sequencing rather than impaired ability to control duration. Instrumental analyses of the speech of Broca's aphasics often reveal waveforms showing irregular phonation (Blumstein 1994). Speech quality is dysarthric—noisy and irregular phonation occurs, reflecting impaired regulation of the muscles of the larynx and alveolar air pressure. Similar problems can also occur in advanced stages of Parkinson's disease.

Linguistic Deficits of Broca's Syndrome

Contrary to the traditional Broca-Wernicke model that assigns speech production to Broca's area, higher-level linguistic deficits also occur in Broca's aphasic syndrome. They consistently show reduced auditory semantic priming effects, especially when the acoustic signal is degraded (Utman, Blumstein, and Sullivan 2001). The semantic priming phenomenon is quite evident in "normal" subjects, who respond faster to "dog" when they hear "cat" first.

Other linguistic deficits characterize Broca's syndrome. The utterances produced by Broca's aphasics often have been described as "telegraphic." In the days of telegrams, the sender paid by the word, so words were omitted whenever possible. English-speaking aphasics, who omit prepositions, articles, and tense markers, produce telegram-like messages such as "man sit tree" in place of "the man sat by the tree." These aphasic telegraphic utterances have generally been thought to be the result of the patients compensating for speech production difficulties by reducing an utterance's length, thereby minimizing the difficulties associated with speech production. The fact that Broca's aphasics have language comprehension deficits that involve syntax was established by studies starting in the 1970s. Broca's aphasics have difficulty comprehending distinctions in meaning conveyed by syntax (Zurif, Caramazza, and Meyerson 1972). Although agrammatic aphasics are able to judge whether sentences are grammatical, albeit with high error rates (Linebarger, Schwartz, and Saffran 1983), the comprehension deficits of Broca's aphasics have been replicated in many independent studies (for example, Baum 1989; Blumstein 1995). For example, higher error rates occur when subjects are asked to comprehend distinctions in meaning conveyed by passive sentences such as "the boy was kissed by the girl" than for the canonical sentence "the girl kissed the boy." High error rates often occur when subjects are asked to comprehend sentences containing embedded relative clauses such as "the boy who was wearing a red hat fell down." Long sentences generally present additional difficulty. Error rates exceeding 50 percent can occur for sentences that yield virtually error-free performance by neurologically intact control subjects (c.f. Blumstein [1995] for a comprehensive review).

Cognitive Deficits of Broca's Syndrome

Goldstein (1948) referred to the loss of the "abstract capacity"—or deficits in planning, deriving abstract criteria, and "executive capacity" generally associated with frontal lobe activity (Mesulam 1985; Stuss and Benson 1986; Fuster 1989; Grafman 1989). Goldstein noted the inflexibility of Broca's patients as they responded to the changing circumstances of everyday life. The cognitive deficits of Broca's syndrome appear to reflect the set-shifting problems that occur in PD and other disruptions of the basal ganglia "sequencing engine" as it attempts to shift cognitive sets. Baldo and her colleagues (2005) found deficits in set shifting on the Wisconsin Card Sorting Test (WCST) that were similar to those I noted for basal ganglia lesions (Scott et al. 2002).

Parkinson's Disease (PD)

VOT Sequencing Deficits

Voice-onset-time sequencing deficits, similar in nature to those of Broca's syndrome, can occur in the later stages of PD. Computer-implemented analysis reveals overlaps between the VOTs of stop consonants such as [t] versus [d] that exceed 19 percent for some PD subjects; the degree of VOT sequencing deficits depends on the severity of the disease state (Lieberman et al. 1992; Hochstadt 2004). The PD subjects, as is the case for Broca's aphasics, spoke at a slower rate, but they maintained control over the relative durations that differentiate vowels, other durational speech phenomena, and tongue and lip movements. Similar results occurred for subjects suffering degeneration of the cerebellum (Pickett 1998).

Sentence Comprehension Deficits

As is the case for Broca's syndrome, PD can result in sentence comprehension deficits (Grossman et al. 1991, 1993, 2001; Howard et al. 2001; Lieberman, Friedman, and Feldman 1990; Lieberman et al. 1992; Natsopoulos et al. 1993; Pickett 1998; Hochstadt 2004). The first study that associated grammatical deficits with PD was reported by Illes et al. (1988); their data showed deficits similar to those noted in Huntington's disease, a neurodegenerative disease deriving from degeneration of the caudate nucleus. The sentences produced by PD subjects often were short and had simplified syntax. However, Illes and her colleagues attributed these effects to the speakers compensating for their speech motor production difficulties by producing short sentences. A subsequent study of comprehension deficits of PD (Lieberman, Friedman, and Feldman 1990; Lieberman et al. 1992) showed that syntax comprehension deficits occurred that could not be attributed to compensatory motor strategies. The PD subjects in this study listened to a set of 100 sentences that had similar vocabularies but differed in both syntactic complexity and length. The subjects simply had to utter the number (one, two, or three) that identified a line drawing that best represented the meaning of the sentence that they heard. Deficits in the comprehension of distinctions of meaning conveyed by syntax occurred for long conjoined simple sentences as well as for sentences that had moderately complex syntax. The PD subjects' comprehension deficits noted clearly were not the result of any compensating strategy, because the motor component of their responses both to sentences that had complex syntax and high rates and sentences with simple syntax and low error rates was identical.

Nine of a sample of forty nondemented PD subjects had these comprehension deficits. Neurologically intact adult subjects made virtually no errors when they took this test, which was originally designed for use with hearing-impaired children (Engen and Engen 1983). The vocabulary was comprehensible to six-year-old children; the error rate for normal eight-year-old children rarely exceeds 5 percent. In contrast, 30 percent error rates occurred for some of the PD subjects. The PD subjects' comprehension errors typically involved repeated errors on particular syntactic constructions. Therefore, the observed syntax comprehension errors could not be attributed to general cognitive decline or attention deficits. The test battery used in this study included sentences having syntactic constructions that are known to place different processing demands in normal adult subjects, for example, center-embedded sentences, right-branching sentences, conjunctions, simple one-clause declarative sentences, semantically constrained and semantically unconstrained passives, and so on. The highest number of errors (40 percent) was made on left-branching sentences that departed from the canonical pattern of English having the form subject-verb-object (SVO). An example of a left-branching sentence is "Because it was raining, the girl played in the house." Thirty percent of errors occurred for right-branching sentences with final relative clauses such as, "Mother picked up the baby who was crying." Twenty percent of error rates also occurred on long conjoined simple sentences such as, "Mother cooked the food and the girl set the table." Similar sentence comprehension errors reflecting information conveyed by syntax were subsequently found in independent studies of nondemented PD subjects (Natsopoulos et al. 1993; Grossman et al. 1991, 1992, 1993, 2001; Howard et al. 2001) using procedures that monitored either sentence comprehension or judgments of sentence grammaticality.

The PD subjects studied by Grossman et al. (1991) were asked to interpret information presented in sentences in active or passive voices when the questions were posed in passive or active voices. Deficits in comprehension were noted when PD subjects had to shift cognitive sets, responding to a question posed in a passive voice concerning information presented in an active voice or the reverse. Higher errors, for example, occurred when the subjects heard the sentence "The hawk ate the sparrow" and were asked, "Who was the sparrow eaten by?" than when they were asked, "Who ate the sparrow?" Grossman et al. (1991) also tested PD subjects' ability to copy unfamiliar sequential manual motor movements (a procedure analogous to that used by Kimura (1993), who found deficits in this behavior for Broca's aphasics). Manual motor sequencing deficits have been noted in many studies of PD (for example, Cunnington et al. 1995). Deficits in sequencing manual motor movements and linguistic sequencing in the sentence comprehension task were correlated. The correlation between sequencing complex manual motor movements and the cognitive operations implicated in the comprehension of syntax is consistent with Broca's area playing a role in both verbal working memory and manual motor control (Rizzolatti and Arbib 1998) in circuits supported by basal ganglia.

Speech production sequencing deficits also have been found to correlate with syntactic comprehension deficits in PD (Lieberman et al. 1992) and bilateral damage to the head of the caudate nucleus and the putamen (Pickett et al. 1998), as well as for neurologically intact human subjects subject to hypoxia and subjects having cerebellar damage (Pickett 1998). In a series of studies of subject populations that included PD patients, subjects with cerebellar lesions, and "normal" old age-matched controls, VOT sequencing deficits and sentence comprehension deficits co-occurred 90 percent of the time. In contrast, no association was noted between the ability to comprehend distinctions in meaning conveyed by syntax and other motor functions, such as the rate of finger tapping, slow rising times from a sitting position, tremor, and other metrics often associated with PD (Parkinson's Study Group, 1989) and these subject populations (c.f. Lieberman 2000).

Grossman et al. (2001) interpreted the sentence comprehension deficits observed in Parkinson's disease as an attentional rather than a linguistic deficit. However, this is unlikely because, if it were simply a matter of not paying attention to the words of a sentence, it would be difficult to explain why both PD subjects and Broca's aphasics have far fewer errors when they are asked to interpret the meaning of a sentence such as, "The banana was eaten by the boy," than the sentence, "The clown was poked by the cowboy." They are clearly paying attention to the animacy constraints conveyed by words—in other words, the fact that an inanimate banana cannot eat a boy. Although anomia, or word-finding difficulties, are common in aphasics who have damage to the frontal areas of the cortex described by Paul Broca, as well as in other forms of aphasia, patients with Broca's syndrome are unable to name an object or a picture of an object though they appear to be fully aware of its attributes.[2]

Cognitive Set-Shifting Deficits

The Odd-Man-Out (OMO) test was crafted by Flowers and Robertson (1985) to test a frontal lobe function cognitive set shifting in Parkinson's disease. Cognitive behaviors involving frontal and prefrontal cortical regions deteriorate in PD because of dysfunctional basal ganglia activity in the neural circuits to these regions; positron emission techniques show that this is the case rather than intrinsic prefrontal dysfunction (Dirnberger, Frith, and Jahanshahi 2005). The OMO is a sorting test that measures two factors: the ability to derive an abstract criterion necessary to solve a problem, and sequencing, the ability to shift to a new criterion. Figure 4.12 shows a typical OMO test card. The subject is asked to pick out the odd figure. The decision can be based on either shape or size, and either decision is correct. The subject is then presented with a second card and again asked to pick out the odd image using the same criterion. If the alternative rule is instead used (for example, shifting to shape after starting with size), the subject is corrected. After selecting the odd figure on ten cards, the subject is asked to perform the task using the other criterion to select the odd figure on a set of ten additional cards. For example, if the subject starts by sorting pictures by their shapes, s/he must then switch to sorting them by size. The subject is then asked to sort the first ten cards using the criterion that s/he originally used. The sorting criterion is shifted up to six times using the same twenty cards.

FIGURE 4.12 *Two Odd-Man-Out test cards. The figure that "doesn't belong" can be chosen either by shape or size. For letter stimuli, shape corresponds to the letter's name, while size corresponds to case.*

Neurologically intact control subjects have almost no errors on the OMO test. Difficulties on the first and second sorts can reflect deficits in the ability to select a criterion, but subsequent errors are indicative of cognitive-shifting deficits. Moderate PD subjects consistently make more set-shifting errors on the OMO test, including the later trials (for example, Flowers and Robertson 1985; Parkinson's Study Group 1989; Lieberman et al. 1992). Impaired subjects have difficulty shifting to a different sorting criterion. The subject studied by Pickett et al. (1998), who had profound bilateral damage to the head of the caudate and putamen, was unable to shift criteria at all. Hypoxic impairment of the brain also results in OMO errors.

Parkinson's subjects who exhibit more profound speech motor deficits also make significantly more errors than less-impaired PD subjects on other cognitive tests involving working memory—in other words, tests of short-term and long-term recall; delayed recognition; the Digit Span Backwards test; and the Verbal Fluency test, in which a subject must name as many words as possible, starting with a particular letter of the alphabet, within one minute. Deficits on these tests often occur in PD (Parkinson's Study Group 1989). Performance on these tests was significantly correlated with performance on the RITLS sentence-comprehension test in the Lieberman et al. (1992) Parkinson's disease study.

Some of the syntax comprehension deficits of PD and Broca's syndrome subjects may reflect a general set-shifting deficit. These subjects may have difficulty shifting syntactic rules when they encounter embedded sentences in relative clauses. In processing a simple sentence such as "The boy ran," one general set of syntactic decoding rules, sequentially applied, will suffice to interpret the sentence. These rules traditionally are presented in linguistic studies in the form of tree structures, or equivalent sets of sequential rules, such as:

Sentence = Noun-phrase + Verb-phrase
Verb-phrase = Verb + Noun-phrase
Noun-phrase = Determiner + Noun

On encountering the word "who" in the sentence "I saw the boy who was fat fall down," the sequence of syntactic operations must be interrupted, the preliminary results stored in short-term verbal working memory, and the syntactic decoding process must be reset. I will return to this issue; it may reflect one of the primary local basal ganglia operations—set shifting, expressed in what might be aptly termed basal ganglia syntax.

Cognitive Planning

A series of experiments by Trevor Robbins and his colleagues (for example, Lange et al. 1992; Robbins 2000; Lewis et al. 2005) demonstrates that deficiencies in dopamine, the primary neurotransmitter modulating basal ganglia function, results in cognitive impairment in frontal lobe tasks mediated by cortical-striatal-cortical circuits. Lange et al. (1992) compared the error rates and thinking times of PD subjects when they were both unmedicated and medicated on a computer-implemented version of the Tower of London test, which assesses planning usually associated with cortical frontal lobe function. The subjects had all been on Levadopa treatment, which restores the level of the neurotransmitter dopamine, thereby ameliorating basal ganglia dysfunction. They were tested when they were

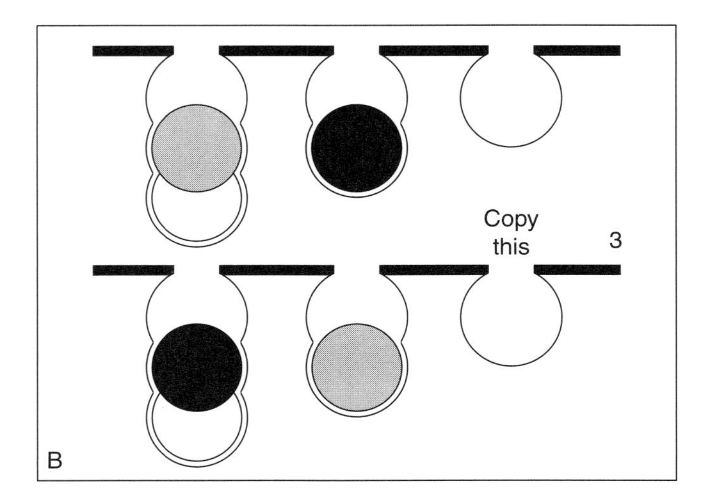

FIGURE 4.13 *The Tower of London test array. Subjects were shown images on a computer equipped with a touch screen and were asked to copy the pattern of balls in hoops shown on the upper part of the display. They could move the balls on the lower display by means of finger movements on the touch screen. (After Lange et al. 1992)*

medicated and when they had been off medication for several days. Their PD subjects saw a computer-generated target picture of three colored balls in two hoops. The third hoop is empty. The subject's task was to move the balls in the two hoops of a second computer-generated image presented below the target image and match the target configuration. Figure 4.13 shows a target configuration and the initial condition of the problem. The subject's task was to copy the ball positions shown in the upper picture.

The subjects moved the balls from hoop to hoop using a touch screen. Twelve tests that differed in complexity were presented. The main measure of accuracy of performance was the proportion of problems solved in the minimum number of moves. The time between the presentation of the problem and the first move was recorded as well as the total time needed to solve the problem. Yoked trials permitted computation of thinking time for each trial by taking account of the slow motor control that can occur in PD. After each problem a yoked trial was used to find the time involved in each move. The subject in the yoked trial moved the balls following a trace that duplicated his/her movements. Movement time was then subtracted from the execution time of the first move to determine initial thinking time. The movement time for all the moves was also subtracted from the total problem-solving time.

Figure 4.14 shows their initial thinking times and the proportion of "perfect" solutions as functions of problem complexity, in other words, the total number of moves that were needed for a perfect solution on and off Levadopa. Because the subjects served as their own controls, the only factor that differed in the two situations was the activity of basal ganglia and other dopamine-sensitive neural circuits. Using a similar paradigm, testing patients on and off Levadopa medication, Lewis et al. (2005) found that low dopamine

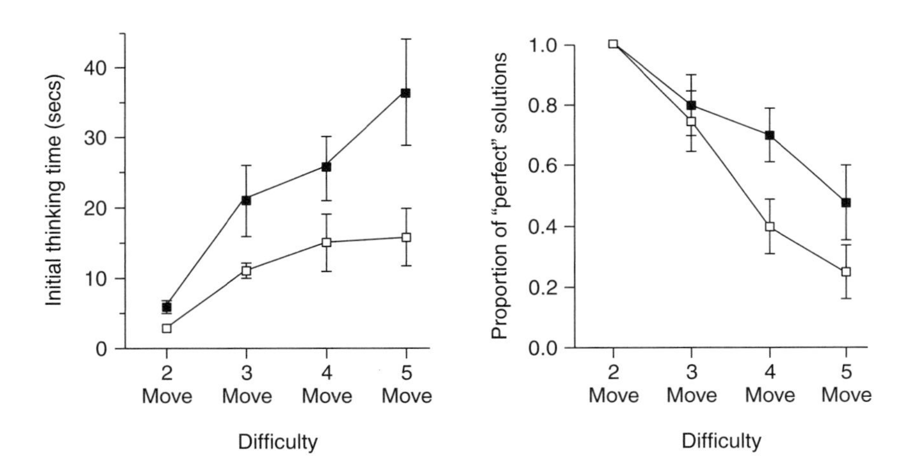

FIGURE 4.14 *Thinking time on the Tower of London test was calculated by subtracting the time that a Parkinson's disease subject took to move his or her finger on command on the touch screen from the time between the presentation of each ball pattern and the same subject's first move. The filled-in squares in A represent thinking time for problems that could be solved in 2, 3, 4, or 5 moves for subjects after they had received L-dopa medication. The open squares plot thinking time in the absence of L-dopa. The number of perfect solutions is plotted in B for the same subjects on and off L-dopa. (After Lange et al. 1992)*

levels resulted in deficits in tasks involving maintaining and reordering a series of words in verbal working memory. Given the involvement of verbal working memory in sentence comprehension, this may in part explain some of the findings noted above for the linguistic deficits associated with Parkinson's disease.

Mount Everest as an Experiment in Nature—Hypoxia

It is generally not possible to compare the behavior of a subject before and after injury to the brain occurs, nor is it ethically justifiable to place neurologically intact subjects in a situation that might degrade neural activity to test a theory. However, mountain climbers who are determined to reach the summit of Mount Everest provide a unique, ethically sound experimental situation for determining the effects of basal ganglia dysfunction on motor control, language, cognition and other aspects of behavior. The experimental design is unique insofar as the cognitive abilities of individual subjects can be assessed before and after exposure to an insult to the brain, allowing the assessment of subtle as well as profound impairment. Moreover, the techniques being developed, which link deficits on speech motor control to cognitive dysfunction, provide a means for monitoring and warning individual subjects of cognitive dysfunction that has demonstrably resulted in fatal accidents. These procedures have already been applied to the diagnosis and treatment of Parkinson's disease (Friedman et al. 1996) and are currently being applied to the study of verbal apraxia (difficulties in speech and oral movements) in children, which can result after births in which oxygen flow to the brain was insufficient.

Hypoxia (an oxygen deficit) commonly occurs in mountain climbers at extreme altitude. Histologic studies of the hypoxic brain identified regions of selective vulnerability in the hippocampus, cerebellum, basal ganglia, and layers III, V, and VI of the neocortex (Brierley 1976). Damage to all of these neural structures may have motor and cognitive consequences. However, motor cortex and basal ganglia appear particularly sensitive to hypoxia (Inoue et al.

1992; Burke et al. 1994; Azzarelli et al. 1996; Jansen et al. 1997). Imaging techniques (SPECT), which noninvasively monitor blood flow, show the effects of hypoxia in the cortical-striatal-cortical circuit linking basal ganglia and motor cortex. Significant increases in blood flow occur in basal ganglia and motor cortex because these structures are linked in a neural circuit (Pagani et al. 2000). Increased blood flow also occurred in the hippocampus, indicating that it too is metabolically active and thus sensitive to hypoxia (Pagani et al. 2000).

Other independent studies show that the globus pallidus (the principal basal ganglia output structure linking the striatum to cortex through the thalamus and other subcortical structures) is extremely sensitive to hypoxic damage (Laplane et al. 1984, 1989; Strub 1989; Kuoppamaki, Bhatia, and Quinn 2002). MRI imaging shows bilateral lesions localized to globus pallidus brought on by exposure to altitude that result in profound subcortical dementia and aphasia (Jeong et al. 2002; Chie et al. 2004). The deficits appear to be as severe as those noted for strokes that result in Broca's syndrome. Marked behavioral and cognitive disruption, similar to that noted for frontal lobe lesions, was noted by Laplane et al. (1984) for eight patients with bilateral globus pallidus lesions resulting from extreme hypoxia. Similar losses were noted by Strub (1989) in a study of a subject having bilateral globus pallidus hemorrhages resulting from exposure to extreme altitude. Repeated ascents to altitudes exceeding 8000 meters results in permanent cognitive deficits (Regard et al. 1989), similar in nature to the cognitive set-shifting inflexibility seen in Parkinson's disease (Flowers and Robertson 1985) and surgically induced bilateral lesions to the internal segment of the globus pallidus (Scott et al. 2002).

The ascent of Mount Everest by the normal route starting in Nepal involves establishing and frequenting a series of high camps over a three-month period. After reaching Everest base camp, which is sited on a glacier below the start of the ascent, climbers rest in order to acclimatize to the altitude. The 5400-meter altitude of base camp is the limit for permanent human habitation (Bouhuys 1974).

In a series of studies that started in 1993 (Lieberman et al. 1994, 2005; Lieberman, Kanki, and Protopappas 1995), speech samples were recorded and cognitive tests were administered (including the OMO test) to acclimatized volunteers after informed consent was obtained. From base camp, climbers ascend through a series of high camps—camp 2 at 6300 meters, camp 3 at 7150 meters, and camp 4 at 8000 meters. Supplementary oxygen usually is not used at camps 2 and 3. From camp 4, climbers attempt to reach and return from the 8600-meter summit. Communications from base camp with VHF radios and Palm Pilot PDAs, programmed with a cognitive test battery (Shepard and Kosslyn 2005) enabled speech and cognitive data to be obtained from camps 2 and 3.

In the 1993 experiment, a battery of speech, syntax, and cognitive tests similar to those first used to assess deficits in PD (Lieberman et al. 1992) was administered to five climbers as they reached each of these progressively higher camps. Testing at camp 4 was impossible, owing to equipment failure. As was the case for PD subjects, the control of VOT sequencing degraded, though to a lesser degree than that manifested in PD. The VOT ranges of their stop consonants converged. As noted before, the voiced consonants of English have short VOTs, and the unvoiced consonants have long VOTs. The minimal VOT separation width was determined by subtracting the longest voiced consonant's VOT from the shortest unvoiced consonant's VOT for each place of articulation ([b] versus [p], [d] versus [t], [g] versus [k]), and calculating the mean for all of the subjects at each high altitude camp. The minimal VOT separation width of all five climbers decreased from 26.0 to 6.4 ms. The time needed to comprehend spoken English sentences also increased. Response times on the RITLS were 54 percent longer at camp 2 for simple sentences that are readily comprehended by six-year-old children. Sentence response time and VOT decrements were highly correlated ($r = 0.74$). Obviously, in contrast to PD, other motor functions such as upright walking were not affected to any significant degree because the climbers were able to ascend Mount Everest.

Subsequent Everest studies in 2001, 2002, 2003, and 2004 involving more than sixty subjects have replicated these findings and have refined the experimental procedures. Computer-implemented analysis of hypoxic climbers' speech reveals motor control sequencing deficits manifested in VOT and other parameters as well as slower speech, similar to that noted in basal ganglia impairment in PD and focal lesions. These speech measures track cognitive impairment manifested in slower response times in tests of sentence comprehension, the OMO test, the Wisconsin Card Sorting Test (WCST), and other measures associated with frontal lobe dysfunction. The speech measures provide a means for predicting mild cognitive dysfunction—hit rates of 85 percent with errors of 10 percent have been achieved. The overall correlation between WCST set completion and vowel duration for nine subjects tested in 2004, for example, was $r = -.71$, $p < .05$. The correlation between vowel duration and response time in the sentence comprehension test was also highly significant ($r = .85$, $p < .05$).

Cognitive perseveration, evidenced in the OMO and WCST tests, is perhaps the best indicator of extreme dysfunction that can have fatal consequences. An extreme example involved a physically fit young climber. Oxygen deficits apparently resulted in his not being able to change plans in an obviously life-threatening situation (Lieberman et al. 2005). When tested at base camp after acclimatization to altitude, the climber in question made no errors whatsoever on the OMO test. His speech at base camp also was within normal bounds. A dramatically different pattern was seen five days later, when he was tested the morning after he reached camp 2. Whereas the climber's base camp performance on the OMO test had been error-free, at camp 2 he made errors similar to those of subjects with extreme frontal lobe or basal ganglia dysfunction. He was able to form a cognitive set: his first set of OMO sorts (done on the basis of size) was error-free. However, when he was asked to shift the sorting criterion, he had extreme difficulties. After starting to sort the second set of images by shape, he reverted to size. His error rate on each of the second, third, and fourth sorts was 40 percent. Difficulties on the first and second sorts can reflect deficits in the ability to select a criterion, but subsequent errors are indicative of cognitive-shifting deficits. The set-shifting impairment was similar to that occurring with bilateral surgical lesions of the globus pallidus (Scott et al. 2002). Acoustic analysis of the climber's speech at camp 2 showed profound sequencing deficits similar to those occurring in extreme cases of Parkinson's disease and in a subject with bilateral lesions in the caudate nucleus and putamen (Pickett et al. 1998).

The subject was informed of his cognitive impairment and advised to descend by the members of his climbing team who were present at camp 2. The weather was deteriorating. However, short of physical restraint, it was impossible to prevent him from following his original climbing plan. He ascended the next day through an intense storm to camp 3. After two nights there, he fell to his death while attempting to return to camp 2 on the fixed ropes. These fixed ropes are anchored at intervals to the mountain. Climbers are secured by attaching a safety harness to the fixed ropes with two carabiners (snap links). To descend safely, a climber must execute a sequence of motor acts at each anchor point, first unclipping one carabiner and then moving and securing it to the next rope, then following a similar sequence with the second carabiner. As noted above, basal ganglia dysfunction disrupts the execution of such internally guided, sequential, manual motor acts. This incident was an isolated occurrence; climbers will generally descend when informed of cognitive impairment.

Cognitive perseveration resulting from basal ganglia dysfunction is the probable factor leading to Everest climbers failing to change plans in this and other situations. The fatal errors in judgment that led to disaster on Everest in 1997 reflected a failure to change plans and descend when the weather abruptly deteriorated (Krakauer 1997). The technique being developed through these Everest experiments could lead to systems that would monitor the speech of persons exposed to hazards that could result in degraded cortical-striatal-cortical circuits involved in decision-making, comprehending the meaning of sentences, and other aspects of behavior.

The Everest studies have found that speech also slows down in hypoxic climbers who have cognitive deficits; the correlation coefficient between increased vowel duration (signifying slower speech) and response time to a MiniCog test of verbal working memory was ($r = 0.61$, $p < .05$). The tests of verbal working memory were part of a battery of cognitive and motor tests (MiniCog), which were implemented on handheld Palm Pilot PDAs. The correlation between working memory and slow speech is not unexpected in the light of the findings of Lewis et al. (2005), which show that low dopamine levels in Parkinson's disease result in deficits in verbal working memory and its role in sentence comprehension. The Palm Pilots were carried up Mount Everest so tests could be run at camps 2 and 3; subjects were not slower on a test of simple manual motor coordination or tests that did not involve short-term memory (Larkin 2003; Shephard and Kosslyn 2005; Lieberman et al. 2005). VOT sequencing errors also were significantly correlated with slower response times on a version of the Stroop test (responding to the number of digits displayed rather than their value, for example, four presentations of the number 5) ($r = -0.84$, $p < .01$). Elderly, neurologically intact subjects who had longer vowel durations also had higher error rates in a study of the comprehension of distinctions in meaning conveyed by syntax (Lieberman et al. 1989). Similar techniques—acoustic analysis of the speech and tests of sentence comprehension and cognitive set-shifting ability—have proven useful in evaluating different surgical procedures that are useful in alleviating tremor in PD (Friedman et al. 1996). Hypoxia also affects memory consolidation and retrieval, most likely reflecting hippocampal dysfunction (Lieberman et al. 2005).

Cerebellum

Less is known about the cognitive role of the cerebellum, a subcortical structure that is linked to the prefrontal and motor cortex as well as to the basal ganglia. fMRI and tracer studies show that it is active in motor learning, apparently acting in concert with prefrontal cortex (Thach et al. 1993; Thach 1996; Deacon 1997). More general linguistic and cognitive roles for the cerebellum, particularly the neocerebellum, which is disproportionately large in humans, have been proposed (Leiner, Leiner, and Dow 1993). However, it is unclear whether the cerebellum plays a role in sentence comprehension and linguistic tasks that do not involve modeling of motor activity (Thach 1996). However, Pickett (1998), who tested the ability of persons suffering cerebellar degeneration to comprehend distinctions in meaning conveyed by syntax, did not find deficits attributable to the neocerebellum (Pickett 1998). The linguistic and cognitive deficits noted in this study may have reflected damage to the neural pathways linking it to the basal ganglia and cortex (Pickett 1998). The cerebellum has been linked to the control of timing motor activity (Ivry and Keele 1989; Ivry and Gopal 1992); however, although VOT sequencing was degraded in some of the subjects studied in Pickett (1998), the intrinsic durations of English vowels was preserved. This might reflect the highly overlearned nature of the neural pattern generators that specify the motor gestures underlying human speech.

Verbal Working Memory

Imaging studies that monitor brain activity during different linguistic tasks consistently show activation of Broca's and Wernicke's areas of the cortex, as well as many other cortical areas. Broca's area clearly is involved in sentence comprehension in neurologically intact subjects. As I have already noted, it plays a role in regulating other aspects of behavior and is not a sentence-comprehension module. It is, for example, implicated in manual motor control (Kimura 1979). Recent data show that Broca's area and its homologue in monkeys support a neural system that generates and monitors grasping and manual gestures (Rizzolatti and Arbib 1998). Stromswold et al. (1996), following modular theory, concluded that

Broca's area is the brain's syntax organ. Using PET, they studied neurologically intact subjects whose task was to decide whether sentences were grammatical. The sentences varied in grammatical complexity; the greatest activation of Broca's area occurred in the sentences that were most complex, leading to the correct conclusion that Broca's area was implicated in analyzing the syntax of a sentence.

But Broca's area is not a syntax module, even if we were to limit our world view to sentence comprehension; a body of evidence that extends back thirty years shows that the meaning of a sentence involves recourse to the brain's neural dictionary as well as to short-term storage and operations in verbal working memory, a short-term neural memory buffer (Baddeley 1986; Gathercole and Baddeley 1993). Scores of independent studies show that the words of a sentence are held in verbal working memory by means of a process of phonetic rehearsal, silent speech that makes use of the neural mechanisms that also control overt speech. The data of Awh et al. (1996), for example, show that neurologically intact subjects use neural structures implicated in speech production to subvocally rehearse letters of the alphabet, maintaining them in working memory. Subtractions of PET activity showed increased metabolic activity (rCBF values) in Broca's area (Brodmann area 44) as well as premotor cortex (area 6), supplementary motor area, cerebellum, and anterior cingulate gyrus when PET data from a task involving verbal working memory were compared with a task that had a substantially lower working memory load. These brain regions are all implicated in speech motor control. Electrophysiologic data from nonhuman primates, for example, show that the anterior cingulate gyrus is implicated in regulating phonation (Newman and MacLean 1982) as well as in attention (Peterson et al. 1988). Left hemisphere posterior (Wernicke's area) and superior parietal regions also showed greater activity as working memory load increased. These PET data are consistent with the results of studies of patients with lesions in these cortical areas: they show deficits in verbal working memory that appear to reflect impairment to phonological knowledge, that is, the sound pattern of words (Warrington, Logue, and Pratt 1971; Vallar et al. 1997).

Imaging studies confirm that Broca's area and these cortical areas are involved in overt speech as well as in silent reading. The PET study of Peterson et al. (1988), in which neurologically intact subjects were asked either to read or repeat spoken isolated words, showed activation of primary motor cortex, premotor cortex, supplementary motor cortex in the subjects' left hemispheres, and bilateral activation of areas near Broca's area and its right hemisphere homologue. Bilateral activation of areas near Broca's region also occurred when subjects were asked simply to move their mouths and tongues. This finding is consistent, to a degree, with the data of many studies of patients with cortical lesions, because lesions confined to Broca's area often result in oral apraxia—deficits in motor control instead of the deficits in motor planning associated with aphasia (Stuss and Benson 1986; Kimura 1993).

Frontal and posterior regions of the cortex also activate when people listen to speech and talk. A series of PET studies performed at the Montreal Neurological Institute consistently showed increased activity in Brodmann's areas 47, 46, 45 and 8 in the left frontal region as well as activity in the subcortical left putamen and posterior secondary auditory cortex. (Klein et al. 1995; Paus et al. 1996). As noted earlier, similar activation patterns were noted in an event-related fMRI study in which subjects listened to and interpreted spoken sentences (Kotz et al. 2003b). These studies demonstrate the presence of pathways from motor to auditory cortex. Signals transmitted from neural structures regulating speech motor control result in increased activity in regions of posterior temporal cortex associated with speech perception when a person talks. Broca's area is also activated when musicians listen to a performance and make decisions concerning its musical structure (Maess et al. 2001). Broca's area thus does not constitute a localized speech production module or a syntax comprehension module. Posterior parietal regions, anterior cingulate gyrus, premotor cortex, and supplementary motor area are all implicated in these processes.

Dynamic Systems: Throwing More Resources at the Problem

It is often necessary to throw in more resources when confronted with a difficult problem. This also seems to be the case for the brain. The neural system that carries out sentence comprehension is dynamic, recruiting additional resources as task demand increases. The fMRI study of Just et al. (1996) made use of the same "subtraction" technique as Stromswold et al. (1996). Neural metabolic activity was monitored as subjects read sentences that expressed the same concepts and had the same number of words but differed with respect to syntactic complexity. The sentences all had two clauses. The sentences with the simplest syntactic structure were active conjoined sentences (type 1) such as, "The reporter attacked the senator and admitted the error." The same information was conveyed by the subject relative clause sentence (type 2), "The reporter that attacked the senator admitted the error" and the object relative clause sentence (type 3), "The reporter that the senator attacked admitted the error." These three sentence types differ with respect to syntactic complexity by several generally accepted measures. Progressively longer reading times and higher comprehension error rates occur in these sentence types. Neurologically intact subjects read sets of exemplars of each sentence type while activity in their brains was monitored by means of fMRI. Measures of comprehension were also obtained, as well as mean processing time and error rates. Activity in the left temporal cortex, the superior temporal gyrus and superior temporal sulcus, and sometimes the middle temporal gyrus, Wernicke's area (Brodmann's areas 22, 42 and sometimes 21), increased as the subjects read the sentences with increasing syntactic complexity. Similar increases in activity occurred in the left inferior frontal gyrus—in other words, Broca's area (Brodmann's areas 44 and 45).

The novel finding was that the three sentence types resulted in increased activity in areas that were spatially contiguous or proximal to the areas activated while the subjects read simpler sentences. Furthermore, the right hemisphere homologies of Broca's and Wernicke's areas became activated, though to a lesser degree, as syntactic complexity increased. Moreover, dorsolateral prefrontal cortex (generally not associated with language) showed bilateral activation for three of the five subjects who were scanned in an appropriate plane (coronal scans). Activation levels in the dorsolateral prefrontal cortex also increased with sentence complexity for these subjects. Dorsolateral prefrontal cortex is implicated in executive control, visual working memory, tasks requiring planning, deriving abstract criteria, and changing criteria in cognitive tasks (Grafman 1989; Paulesu et al. 1993; D'Esposito et al. 1995).

It is clear that the neural bases of language are complex and appear to involve many different neural circuits (Mesulam 1990). Moreover, our knowledge is imperfect. For example, cortical-striatal-cortical circuits linking prefrontal cortex to other neural structures do not appear to be implicated in the linguistic deficits associated with Wernicke's syndrome, in which fluent, often meaningless speech containing neologisms is produced (c.f. Blumstein 1995). While PET studies show prefrontal hypometabolism in patients with Broca's syndrome, this is not the case for Wernicke's (Metter et al. 1989). The neural bases of Wernicke's syndrome are still unclear. What is also unclear is the degree to which cortical-striatal-cortical circuits are anatomically and functionally independent. Is there some "special" relationship between circuits that regulate speech production and syntax, or is the correlation noted in many studies between sequencing deficits in speech motor control and the comprehension of distinctions in meaning conveyed by syntax simply the result of speech production being very complex? It takes more than seven years for even simple aspects of speech motor control to be established in children (Smith 1978; Smith and Goffman 1998; Nittrouer et al. 2005). Fully adult speech production capabilities are not attained until after puberty—sometime after age fourteen years (Lee, Potamianos, and Narayanan 1999). Is the fact that chimpanzees can neither talk nor comprehend complex syntax the result of some commonality in the circuits regulating speech and syntax, or does it simply reflect similar levels of complexity? These are open questions.

The Brain's Dictionary

At present the neural bases of the brain's dictionary are even less clear than the bases of speech and syntax. However, the available evidence suggests again that complex neural circuits are involved. It clearly is impossible to comprehend the meaning of a sentence without identifying its words and determining their meanings and syntactic constraints. For example, the argument structures of verbs determine, among other things, whether they can refer to animate subjects (Croft 1991). And, in fact, a growing body of psycholinguistic research based on interactive-activation models of linguistic representation and processing indicates that sentence processing is lexically driven and takes into account probabilistic, semantic, and syntactic knowledge coded in the lexicon (Bates and Goodman 1997; MacDonald 1994). As noted earlier in Chapter 2, the neural structures that "define" the meaning of a word appear to be the ones that are relevant in real life. Neuroimaging studies show that when we think of a word, the concepts that are coded by the word result in the activation of the brain mechanisms that concern the real-world attributes of the word in question. For example, the PET data of Martin et al. (1995b) show that the primary motor cortex implicated in manual motor control is activated when we think of the name of a hand tool. Primary visual cortical areas associated with the perception of shape or color are activated when we think of the name of an animal. Neurologically intact subjects who were asked to name pictures of tools and animals activated the ventral temporal lobes (areas associated with visual perception) and Broca's area. Martin and Chao (2001) review the results of independent studies that show that regions of "sensory" temporal cortex that are active during perception store the information that characterizes words and concepts in the brain's dictionary. These effects are not limited to native speakers of English. A review of independent neuroimaging studies in subjects speaking a variety of different languages shows similar results. As the studies reviewed in Chapter 2 noted, progress is being made toward understanding how we access semantic "knowledge" through the sound pattern of a word (Damasio et al. 1996).

The neural bases of memory, including the lexicon, appear to be complex and involve the nature of the information that is stored (Kirchoff et al. 2000), whether the memory is old or new (Wheeler and Buckner 2003), as well as the degree of attention directed at retrieving a memory (Velanova et al. 2003). It is becoming clear that the brain bases of memory are complex, dynamic neural networks linking areas of prefrontal cortex—including Broca's language area, Brodmann's areas 44, 45—with areas of temporal cortex and subcortical hippocampus. The pattern of cortical activity appears to reflect the nature of the memory. Findings consistent with those of Martin and his colleagues show that prefrontal areas associated with verbal working memory are predominantly left-hemisphere activated when words are involved, that pictures activate both hemispheres, and temporal areas of the cortex and hippocampus were consistently activated (Kirchoff et al. 2000). The hippocampus plays a critical role in forming and retrieving memories in humans and other species. It was thought that the hippocampus's role was limited to spatial memory in species other than humans—coding the locations of objects and paths. However, electrophysiologic studies of rats show that the neurons of the hippocampus code behavioral cues in these primitive mammals (Wood, Dudchenko, and Eichenbaum 1999). Whether fundamental differences exist between the neural circuits coding memory in humans and other mammals is unclear at this juncture.

These findings add to the fund of knowledge that rules out neophrenological models such as that proposed by Ullman (2004), who claims that the frontal regions of the brain and basal ganglia in themselves constitute a procedural system that regulates syntax, with the brain's posterior regions constituting the neural lexicon. Ullman claims that frontal regions play no essential role in retrieving memories. In his model, posterior temporal lobe regions of the brain, in themselves, constitute the brain's dictionary. As we have seen, this is not the case—frontal "motor" regions of cortex are activated for words relating to motor activity. Moreover, frontal regions and circuits involving the hippocampus are critical elements in the neural processes implicated in retrieving both episodic memories and the semantic—real-world neural referents of words from the brain's dictionary. Ullman also does not take into account the neurophysiologic and behavioral studies I have previously described that establish the particular role of basal ganglia in set shifting, the complexity of the basal ganglia sequencing engine (Ullman localizes striatal cognitive activity in the caudate nucleus, but diffusion tensor fiber tracking (Lehericy et al. 2004) and clinical findings (Pickett et al. 1998) also show that the putamen is involved), or the fact that cortical-striatal-cortical circuits also regulate affect and emotion (Cummings 1993).

The Dog Who Did Not Bark: When Procedural Knowledge Becomes Declarative Knowledge

In the case of *Silver Blaze,* published in 1892 by Sir Arthur Conan Doyle, the key to the mystery is captured in Sherlock Holmes's remark about "the curious incident of the dog in the nighttime." The dog in question did not bark because of his "declarative knowledge" gained by observing procedures carried out repeatedly. The dog did not bark because he recognized the perpetrator of the nefarious deed committed in the dead of night; the absence of overt motor activity was a sign of knowledge.

The basal ganglia are involved in learning procedures contingent on cognitive inferences. The net result is declarative knowledge, coded in the brain's memory (Kimura, Aosaki, and Graybiel 1993; Graybiel et al. 1994; Huettel, Mack, and McCarthy 2002), such as whether you should come to a halt before a railroad track because you can hear a locomotive approaching. The sign of knowledge concerning trains and cars is what you do not do—you do not continue to drive onward. The act of stopping could be viewed as procedural knowledge, but it is contingent on more general concepts concerning the probable consequences of locomotives and cars colliding. However, it is also declarative knowledge—specific information concerning what the sound of a train approaching on a railway line that intersects a road signifies. In short, the distinction between procedural knowledge and declarative knowledge inherent in the Pinker-Ullman model is, at best, tenuous. Neurophysiologic data reinforce this view. According to Ullman (2004), procedural knowledge is gained through basal ganglia activity; a direct test of Ullman's theory by Longworth et al. (2005) showed that PD (which as we have seen involves basal ganglia dysfunction) does not affect the subjects' ability to command rule-based past-tense morphology (for example, "jump" versus "jumped"). This study again showed that PD affects higher-level syntactic processes that involve sequencing and the inhibition of competing alternatives—basal ganglia operations noted in the independent studies discussed above.

Cortical Plasticity

One of the surprising findings of current research on the brain is cortical plasticity. The supposition of Damasio et al. (1996) regarding the phenotypic acquisition of the circuits that neurally instantiate words is well founded. Neurophysiologic studies indicate beyond reasonable doubt that many of the neural circuits that regulate human and animal behavior are shaped by exposure to an individual's environment. For example, the visual cortex develops in early life in accordance with visual input; different visual inputs yield different neural connections (Edelman 1987; Hata and Stryker 1994). Moreover, cortical regions that normally respond to visual stimuli in cats respond to auditory and tactile stimuli in visually deprived cats. Rauschecker and Korte (1993) monitored single-neuron activity in the anterior ectosylvian visual cortical area of normal cats and cats that had been vision deprived. Neurons in this area in normal cats had purely visual responses. In young cats that had been deprived of vision from birth, only a minority of cells in this area responded to visual stimuli: most responded vigorously to auditory and to some extent somatosensory stimuli. Imaging and behavioral data indicate that similar processes account for the formation of such basic aspects of vision as depth perception in children.

PET data show that both primary and secondary visual cortex in persons blinded early in life are activated by tactile sensations when they read Braille text (Sadato et al. 1996). Children who have suffered large lesions to the classic language areas of the cortex usually recover language abilities that cannot be differentiated from other normal children (Bates, Thal, and Janowsky 1992). Sign language is "heard" in the auditory cortex of deaf people (Nishimura et al. 1999), which most likely explains the fact that hearing-impaired persons can communicate in a linguistic mode using manual signs. The cortical representation of the tactile finger receptors involved in playing stringed instruments is a function of the age at which musicians started musical lessons (Pantev et al. 1998).

The details of the motor programs instantiated in motor cortex likewise are phenotypically acquired (Edelman 1987; Nudo et al. 1996; Sanes and Donoghue 1994, 1997). The traditional view of the primary motor cortex was as a somatopic map of an animal's body that charted a one-to-one correspondence between specific cortical areas and the parts of the body that they control. Individual muscles were supposedly activated by particular cortical neurons. For example, the thumb supposedly was controlled by neurons in one area, the fingers in another, and the wrist in a pattern that mirrored the shape of the hand. Many texts still represent the organization of human primary motor cortex by means of an upside-down cartoon of the body (toes, feet, hands, fingers, lips, tongue, and so on) in which different areas of motor cortex each control a given part of the body. The somotopic organization of motor cortex implicitly was similar for all human beings and was presumed to be innate.

However, the functional architecture of motor cortex is neither somotopic or innate. Although regions of primary motor cortex control head, arm, and leg movements, discrete innately specified locations controlling individual fingers, the wrists, tongue, and lips do not exist (Sanes et al. 1995). A series of studies (reviewed in Barinaga 1995) that started in the late 1970s show that individual muscles are influenced by neurons in several separate locations in motor cortex. Moreover, individual cortical neurons have branches linking them to other cortical neurons that control multiple muscles. The complex assemblage of neurons appears to be a functionally organized system in which circuits consisting of neuronal populations work together to coordinate groups of muscles that carry out particular actions. Sanes et al. (1995) used functional magnetic resonance imaging (fMRI) to study human subjects performing motor tasks. The fMRI findings were similar to data obtained using microelectrode arrays in monkey brains (Sanes and Donoghue 1996).

Multiple representations, or circuits, controlling finger and wrist movements in human primary motor cortex were charted during different voluntary movements. As Sanes and his colleagues note, the hypothesis that best explains their data is one in which "neurons within the M1 [primary motor cortex] arm area form a distributed and cooperative network that can simultaneously and optimally control collections of arm muscles . . . Furthermore, it is likely that neural elements or ensembles that make up the functional processing units continually recombine while forming new output properties to produce normal adaptive behavior." In brief, neuronal populations are shaped that regulate learned motor control patterns (Donoghue 1995; Pascual-Leone et al. 1995; Sanes et al. 1995; Nudo et al. 1996; Karni et al. 1995, 1998; Classen et al. 1998). The potential for the neural control of complex motor activity is part of the genotype. However, the neural mechanisms that regulate particular movements in animals and humans are not necessarily genetically specified. Some activities, such as breathing, must be in place at birth and necessarily must be genetically specified; a timetable that schedules the development of breathing in the first days of life is evident in normal human infants (Bouhuys 1974). However, most aspects of human motor control, including walking (Thelen 1984), appear to be the result of many attempts involving trial and error. Skilled performance is the end result of the process of automatization by which means animals, including humans, learn to rapidly execute skilled motor control programs (MCPs) without conscious thought. In fact, primary motor cortex appears to be adapted for learning and storing these MCPs (Evarts 1973). Repeated trials shape neuronal circuits.

Automatization is apparent when a novice driver learns to shift the gears of a car that has a manual transmission. At first, shifting requires total concentration and is slow and inaccurate. After repeated trials, gear-shifting becomes automatic and rapid. A similar process can account for our learning to walk, catch balls, and talk. Although data on the automatization of speech motor activity are presently not available, it is unlikely that the role of primary motor cortex in speech differs from that of any other motor act. In the absence of compelling contrary evidence, we can also state that the rules of syntax are not innately specified. The architectural similarity of the cortical-striatal-cortical circuits implicated in motor control and cognition is evident (Cummings 1993; Kimura et al. 1993; Graybiel et al. 1994; Marsden and Obeso 1994; Middleton and Strick 1994; Mirenowicz and Schultz 1996). Therefore, it is improbable that the cognitive pattern generators (Graybiel 1997, 1998) that specify syntactic operations (Lieberman 2000) are innately specified.

Basal Ganglia Grammar

Human languages generally make use of regular operations that signify meaning; for example, in English the suffix "ed" signifies the past tense. Theoretical linguists formulate rules to describe these grammatical operations, but the nature of these rules varies from one theory to the next. Chomsky's theories have taken a very twisting path from the 1957 version of generative linguistics. The grammatical rules presented in his 1995 minimalist theory bear no resemblance to those proposed in 1957. Nor has any attempt been made to relate grammatical rules to any known neurophysiologic mechanism.

Basal ganglia activity in the neural circuits that regulate motor control and cognition may provide an answer. Basal ganglia sequencing can be described by two rules. The basal ganglia essentially operate as a switch; stored pattern generators are successively connected (activated) and disconnected (inhibited) to cortical targets to enable a rat to execute a grooming sequence (Aldridge et al. 1993; Aldridge and Berridge 2003) or to allow me to write this sentence on the keyboard of my computer. The functional architecture of the basal ganglia includes inhibitory and excitatory channels (Marsden and Obeso 1994), which either (A) inhibit or (B) activate pattern generators, disconnecting or linking them to cortical targets. A string of motor pattern generators thus can be successively transferred to cortical circuit targets by successively (A) inhibiting the ongoing motor subcomponent and then (B) activating a new motor component. Similar inhibition and activation of cognitive pattern generators would generate sentences and comprehend distinctions in meaning conveyed by syntax.

The rules of syntax of a particular language or dialect would, in my proposed theory, be acquired by general cognitive processes. As is the case for automatized motor control "algorithms," they would be stored in the brain along with the phonological shapes and semantic extensions of words. As I have noted, evidence from many sources shows that neural motor pattern generators are acquired as an animal or human learns to perform a task, and they are stored as automatized routines. For example, neurons in the prefrontal cortex of monkeys signal the start of the elements of motion sequences (Fujii and Graybiel 2003). Given the fact that neural structures that support the circuits regulating motor control and language overlap, there is no reason to suppose that the basic neural operations of the brain differ for syntax and motor control. Syntactic rules would be transferred into the computational space of working memory (most likely Broca's area in an intact brain) by the basal ganglia acting as a switching mechanism–activating and inhibiting cognitive pattern generators that constitute the rules of syntax. In some ways, Chomsky's (1995) minimalist theory is compatible with this view, insofar as most of the details of syntax are in the lexicon.

Motor Control Programs and the Rules of Syntax

Some further inferences concerning syntax follow from this line of thought. The neural mechanisms implicated in motor control are massively parallel (Alexander et al. 1986; Alexander and Crutcher 1990; Marsden and Obeso 1994; Sanes and Donoghue 1996, 1997) and cannot usefully be described by means of sequential algorithms similar to those commonly used by linguists to describe the rules of syntax (Alexander, Delong, and Striker 1992). As Croft (1991) notes, the inability of formal linguistics to describe the sentences of English, arguably the most intensively studied language on earth, may derive from an overreliance on algorithmic procedures that do not take account of graded semantic and real-world knowledge (MacDonald 1994). Even formal linguists committed to generative grammars such as Jackendoff (1994) note that it has not been possible to describe the syntax of English by these procedures. As Marsden and Obeso (1994) pointed out, it is possible to describe basal ganglia functions, but how they achieve these ends is still a mystery. Distributed parallel processes seem to be the means whereby local operations are instantiated; but having said that, how brains work is still a mystery. The devil is in the details. The problem for linguistic research rests in the fact that language is the product of a biologic brain that does not resemble a digital computer programmed by means of sequential algorithms (for example, Pinker 1994).

Brain Lateralization and Language

The human brain is lateralized, and the left hemisphere in about 90 percent of the present human population has a greater role in regulating both motor control and language. Lenneberg (1967), in one of the first modern studies of the biology of language, believed that brain lateralization was the key to human language; many subsequent studies have sought to establish brain lateralization in extinct hominids. However, lateralization is a primitive feature (Bradshaw and Nettleton 1981). The brains of some species of frogs are lateralized; the left hemisphere of their brains regulates their vocalizations (Bauer 1993). Moreover, studies of many mammalian species show that paw movements are under lateralized neural control (Denneberg 1981; MacNeilage 1991). And current neural imaging studies show that, although one hemisphere generally is more active during linguistic tasks, both hemispheres of the brain are activated (Just et al. 1996; Kotz et al. 2003b; Rissman, Eliassen, and Blumstein 2003). Theories that identified asymmetric development of the traditional language areas of the neocortex with linguistic ability have not stood the test of time. The planum temporale of Wernicke's area was thought to be symmetric in apes, in contrast to asymmetrically larger planum temporale in the human dominant hemisphere. However, further study shows that apes and humans both have similar asymmetric planum temporale (Gannon et al. 1998). Broca's area likewise has been found to be asymmetric in apes (Cantalupo and Hopkins 2001). Because apes, which have asymmetric neocortical language areas, lack human language, it is clear that asymmetry in itself does not confer linguistic ability.

Take-Home Messages

- The studies I reviewed demonstrate that the neural bases of human language are not localized in Broca's and Wernicke's areas of the cortex. Our knowledge of the brain is imperfect and much is unclear. However, it is clear that Broca's and Wernicke's areas are not the brain's language organs, devoted to language and language alone.
- Cortical-striatal-cortical neural circuits are implicated in sentence comprehension and cognitive sequencing as well as in speech and other aspects of motor control. Local operations

appear to take place in particular subcortical structures or regions of the cortex, but these processes do not necessarily constitute an observable behavior. Local processes in different parts of the brain linked in a circuit carry out motor and cognitive acts. Basal ganglia play a critical role in these circuits and carry out at least three motor and cognitive control functions:

- They are involved in learning activities that yield a reward.
- They play a part in sequencing the individual elements that constitute a motor or cognitive "pattern generator." The cognitive acts include both forming syntactically complex utterances and comprehending distinctions in meaning conveyed by syntax. They can reiterate a finite set of elements to form a potentially infinite number of outputs.
- They confer adaptive motor and cognitive ability by interrupting an ongoing motor or cognitive sequence, contingent on external events and prior knowledge. They work in concert with other subcortical structures and areas of the cortex in circuits that allow us to change plans and rapidly adjust our behavior to changing circumstances. The subcortical neuroanatomic structures that support the neuronal populations that constitute these circuits also play a part in regulating emotion. We also do not know the degree to which circuits are linked. Is there a closer link between circuits regulating speech and syntax than, for example, manual motor control and syntax?

- Human beings possess a verbal working memory system that allows us to comprehend the meaning of a sentence, taking into account the syntactic, semantic information coded in words as well as pragmatic factors. The neural substrate that instantiates verbal working memory appears to be a dynamic distributed network that recruits neural "computational" resources as task demands increase—for example syntactic complexity and sentence length. The neural network links activity in posterior, temporal regions of the neocortex including Wernicke's area with frontal regions such as Broca's area (Brodmann areas 44 and 45), frontal regions adjacent to Broca's area, premotor cortex (area 6), motor cortex, supplementary motor area, the right hemisphere homologies of Wernicke's and Broca's area, and prefrontal cortex. Frontal regions of the cortex generally associated with nonlinguistic cognition are activated as task difficulty increases. Anterior cingulate cortex, basal ganglia, and other subcortical structures such as thalamus and cerebellum also are implicated.

- The brain's dictionary appears to be instantiated by means of a distributed network in which neuroanatomic structures that play a part in the immediate perception of objects and animals as we view them or the gestures associated with tools as we use them are activated. The lexicon appears to connect real-world knowledge with the sound-patterns by which we communicate the concepts coded by words. Like other neural structures implicated in language, it is plastic and is shaped by life's experiences. Frontal cortical regions as well as the hippocampus play critical roles in both consolidating and retrieving the knowledge conveyed by words. Simplistic models such as Ullman's (2004), which claim that posterior regions of the brain constitute the dictionary while a frontal-striatal system regulates procedural processes such as syntax, are not valid.
- The motor patterns that generate the complex articulatory gestures that produce human speech (c.f. Perkell 1969; Lieberman 1984; Stevens 1992) appear to be learned, as is the case for other acquired motor patterns. They are not fully in place until age ten years (Smith 1978; Hulme et al. 1984). Cognitive, procedural knowledge, including the syntactic operations specific to a particular language, also appear to be learned by children (Bates, Thal, and Janowsky 1992; Bates and Goodman 1997; Deacon 1997). The cerebellum is implicated in motor learning and may

play a part in cognitive and linguistic tasks involving motor imagery. The brain's dictionary likewise involves activity in different parts of the brain linked in neural circuits. Frontal regions, including Broca's area, are activated together with motor cortex in the case of words referring to objects and activities involving motor activity and posterior cortical regions concerned with visual perception, color, and shape, as well as the hippocampus. The phylogenetically primitive hippocampus appears to be a crucial element in consolidating and retrieving memories—including those coded as words.

- Lateralization clearly is not the "key" to human linguistic ability. The brains of most, perhaps all, living animals appear to be lateralized.
- Linguistic theorists could profit by taking account of the fact that the circuits regulating motor control and syntax share neuroanatomic structures performing similar local operations. Moreover, it is improbable that basic neurophysiologic processes differ for motor control and syntax. Motor control involves distributed, parallel processes that fundamentally differ from the serial, algorithmic computations typically employed by linguists to describe syntactic operations. Moreover, the local operations by which the basal ganglia sequence pattern generators are similar in nature to that proposed by linguistic models (Chomsky 1995).

CHAPTER 5

Motor Control and the Evolution of Language

HUMAN BEINGS ARE the single living species that possesses complex linguistic ability. If it were possible to study the brains, bodies, and behavior of extinct hominids who may have possessed intermediate stages of language, we would be able to provide definitive answers to such questions as when and how the several components of fully human language evolved. As Chapter 2 noted, linguists since the time of the Sanskrit grammarians have realized that the ability to communicate and think by means of language involves different elements. Speech, syntax and the lexicon, the store of concepts conveyed by the words of the brain's dictionary, all make human language possible.

Here We Must Proceed with Trepidation

We are not in a position to form "real" theories—that is, theories that can be refuted—for the evolution of these capabilities. Although current research has yielded insights on the neural circuits and mechanisms that code, store, and retrieve words, the brains of extinct hominids cannot be examined. All that remains are skulls, bones, and a few things that they made, used, or threw away. However, I believe that the present state of knowledge allows us to move beyond sheer speculation toward refutable theories on some aspects of the evolution of speech, syntax, and cognition. The studies reviewed in Chapter 4 reveal the mark of evolution on the human brain, and they show that the basal ganglia constitute a "sequencing engine."

In species as far removed from humans as frogs, the basal ganglia sequence motor pattern-generators to produce acts that contribute to the survival and propagation of a species, though that act may be as deceivingly simple as snapping up a fly. The evolutionary root of human cognitive ability may be observed when the frog's feast of flies is abruptly terminated as an inquisitive person approaches, sending the frog into a nearby pond. In short, the proposal that I will explore for the evolution of human capabilities, such as talking and commanding complex syntactic and cognitive ability, is that adaptive motor behavior is one of the keys to understanding how we came to be. In the hominid line leading to us, the basal ganglia became a superb sequencing engine, regulating acts that make us human. As I stated at the beginning of this book, those acts include such activities as dancing as well as the capacities generally rewarded in academic life.

The starting point may have been upright walking. Recent evidence points to the evolution of neural circuits that confer completely voluntary speech, complex syntax, and cognition appearing only 100,000 or so years ago, coincident with the appearance of anatomically modern human beings in Africa. Speech production may have been the driving force for this last phase of neural development, together with many of the "infinitely complex" factors that to Charles Darwin's mind guided the course of evolution.

As I hope to show, the story that the FOXP2 gene (the so-called language gene) tells is that a string of mutations resulted in a cortical-subcortical system capable of learning and freely sequencing both motor and cognitive pattern-generators, yielding human motor and cognitive ability, including complex syntax. Fully human linguistic capability in these domains, as well as related "creative" abilities such as dancing, composing music and works of art, appears to be species-specific. Walking (upright bipedal locomotion), dancing, manual dexterity, talking, singing, complex social interaction, mate selection—all may have been factors in the evolutionary trajectory that produced a human language-capable brain.

Reiteration and the FOXP2 Gene

At the beginning of this book I took issue with the song that many linguists and philosophers have sung for hundreds of years: human language is so unique that it bears no relation to the manner in which any other animal communicates or thinks. Chomsky has consistently argued that human linguistic ability involves some unique feature whose scope is restricted to language and language alone. The most recent candidate is a narrow "faculty of language" (NFL) that is species- and domain-specific (Hauser, Chomsky, and Fitch 2002). The domain is, as usual for Chomsky, syntax, which to him is the putative essence of language. The NFL hypothetically provides humans with recursion—the ability to recombine and reorder a finite set of words and phrases using a finite set of linguistic rules, so humans can form and convey a potentially infinite number of different sentences. Chomsky and his colleagues are correct, insofar as recursion, a restricted form of reiteration, is a key element of human linguistic ability.[1] However, as we have seen, the ability to reorder a finite set of elements to form an infinite set of actions is a key feature of motor control.

Forming different grammatical sentences entails more than simply changing the order of words. The semantic-syntactic constraints on the words in any dictionary, including that in your brain, must be considered. Different verbs, for example, have particular argument constraints (the linguistic term often used is "argument structure"). For example, the sentence "I wished Ann" violates an argument structure constraint because the verb "wish" cannot refer to an object. The sentence thus is not grammatical, whereas "I kissed Ann" is acceptable. Motor control entails similar constraints. As the basal ganglia release and inhibit successive pattern generators, these constraints come into play. Consider walking, which involves a sequence of ordered, constrained submovements. One submovement is heel strike, which contributes to the efficiency of upright bipedal locomotion: the back of the heel makes first contact with the ground as you walk. You cannot sequence the motor act that yields heel strike before your lower leg swings forward. The motor pattern generator that swings the lower leg forward can be followed with heel strike. The pattern generator that locks your legs in place while you stand still cannot be followed with heel strike. In short, the motor pattern generators involved in walking have argument structures.

I present the case for the evolution of reiterative ability in the domain of cognition, including language, as an example of Darwinian preadaptation. Neural mechanisms that initially evolved to facilitate adaptive motor control took on a more general cognitive role. In essence, this proposal for the evolution of the neural bases of syntax is not original. Karl Lashley in 1951 suggested that neural mechanisms that were initially adapted for motor control are the basis for syntax and some aspects of human cognition. Lashley was following in the footsteps of Charles Darwin, who noted that "an organ might be modified for some other and quite distinct purpose" (Darwin, 1859, p. 190).

Lashley clearly believed that syntactic capability transcended the domain of language. In previous chapters I have presented the evidence for neural cortical-striatal-cortical circuits playing a part in regulating the syntax of speech motor-control and cognition. These same circuits play a part in operations that linguists would clearly consider syntactic. Areas of the brain linked to the striatum such as Brodmann's cortical area 47 play a part in these operations (Monchi et al. 2001), as does the supplementary motor area (Cunnington et al. 1995; Lehericy et al. 2004). Given the redundancy that is evident in the brain, it also would not be surprising to find other neural structures that are involved in sequencing motor or cognitive acts. However, it is clear that the basal ganglia constitute a general-purpose sequencing engine. The basal ganglia can link the sequence of motor pattern-generators that permits me to tap away at my computer's keyboard. They also link the sequence of cognitive pattern generators (specifying syntactic rules) that allows you to comprehend these sentences. And they can interrupt and switch to a different sequence of motor or cognitive pattern generators when that is appropriate. In short, the human basal ganglia sequencing engine is one of the critical neural bases of language and thought. It also allows us to play the piano, dance, paint, and continually change our minds.

FOXP2 and the Subcortical Brain

Molecular genetics has dramatically changed evolutionary biology and the study of human evolution. The FOXP2 gene, which was identified by Simon Fisher (Fisher et al. 1998), provides an insight into the evolution of the human sequencing engine. FOXP2 undoubtedly is not the only regulatory gene involved in the evolution of human language. Moreover, it is not a language gene because it governs the expression of genes that specify neural structures that also regulate motor control and other aspects of cognition. Moreover, apart from the brain, it governs the development of the lungs and other structures. However, FOXP2 provides solid evidence that allows us to move beyond speculation on what aspects of the brain confer the qualities that distinguish human beings from other species.

The discovery of FOXP2 is the result of a sustained study of the KE family. Orofacial motor, speech production, syntactic comprehension, and cognitive deficits occur in some members of the KE family because of a genetic anomaly. The finding was described in some quarters as proof for Noam Chomsky's views on the innate nature of human syntactic ability; the claim that the human brain instantiates an innate, genetically transmitted Universal Grammar (UG) (for example, Gopnik 1990; Gopnik and Crago 1991; Pinker 1994). No person would dispute that human beings have an innate capacity to acquire language. It is clear that neurologically intact infants and children raised under "normal" circumstances have the biological capacity to learn any language. However, as I noted earlier, Chomsky (1966, 1976, 1986) goes further: he claims that the detailed syntax of all human languages is an innate attribute of the human brain.

For example, English syntax has a regular plural rule that predicts the plural form of most nouns; dog-dogs, car-cars, and so on. This example usually is cited in introductory linguistic tests because it is one of the few undisputed linguistic rules of English. Children raised in an English-speaking environment acquire this knowledge without explicit tutoring. Specialists in child development, such as Elizabeth Bates and Patricia Greenfield, have argued that the processes that allow children to master other aspects of cognitive and social behavior can account for this and other aspects of the acquisition of language (for example, Greenfield 1991; Bates, Thal, and Janowsky 1992; Elman et al. 1997; Lieberman 1984, 1991, 2000). However, the Chomskian claim is that the UG, an "organ" of the human brain, instantiates innate knowledge of this rule, in effect triggering the genetic program if a child is exposed to regular English plural nouns early in life. In short, the UG constitutes an innate store of the detailed knowledge of the syntax of all possible human languages. According to this argument, other genetically transmitted components of the UG specify the rules governing the formation of yes-no questions, others confer the ability to form passive sentences, and so on.

It is clear that the vocal and gestural signals that many species use to communicate are genetically specified and need only triggering stimuli. For example, ducklings require very limited exposure to duck calls as they hatch to develop normal duck calls months later (Gottlieb, 1975). In effect, the implicit claim of UG is that human beings are superducks who possess a vastly more elaborate set of genetically transmitted linguistic information. The hypothetical innate blueprint allows children who receive limited exposure to the utterances of a language to master syntax. The appropriate rules are triggered by appropriate limited stimulation (some variants of this theory claim that innate constraints exist that filter out inadmissible rules, but the net effect is similar). The hypothetical UG must, of course, encode the different syntactic schemes that occur in the world's languages. Therefore, UG must contain many detailed syntactic rules.

Because diseases such as diabetes, which have a strong genetic component, result in specific deficits, one source of evidence for UG would be a genetic anomaly that prevented afflicted individuals from mastering a specific aspect of English syntax while retaining other aspects of normal linguistic ability. This was reported to be the case for the afflicted members of a large extended family (KE) who suffer from a genetically transmitted anomaly. Gopnik (1990) and Gopnik and Crago (1991) claimed that these individuals' linguistic deficits were specifically limited to their being unable to master the regular past tenses of English verbs and regular plural nouns. Pinker (1994) repeated and publicized these claims. The afflicted individuals' mastery of other aspects of English syntax as well as their cognitive and motor behavior supposedly were similar to those of normal members of family KE.

However, this is not the case. Intensive study of family KE reveals a syndrome, a suite of severe speech and orofacial movement disorders, cognitive deficits, and linguistic deficits that are not limited to specific aspects of the syntax of English (Lai et al. 2001; Vargha-Khadem et al. 1998, 2005; Watkins et al. 2002). Major orofacial movement errors occur in the afflicted persons. The afflicted members of family KE are not able to stick out their tongues while closing their lips; they have difficulty repeating two-word sequences. In a filmed interview, subtitles were necessary because the speech of afflicted family KE children was scarcely intelligible (BBC broadcast, 1994). On standardized intelligence tests, afflicted members of family KE have significantly lower scores than their nonafflicted siblings, a result that rules out the presence of environmental factors affecting intelligence. Some of the afflicted individuals had higher nonverbal IQ scores than unaffected members, which has led some investigators erroneously to conclude that FOXP2 does not affect intelligence. There is a significant difference in means: the mean for the affected members was 86 (a range of 71–101) versus a mean of

104 (a range of 84 to 119) for unaffected family members. Moreover, intelligence derives from the interaction of many neural systems and life experiences. The range of nonverbal IQs for the nonafflicted members of the KE family varies, and it is impossible to know what the nonverbal IQs of an afflicted individual would have been, absent the FOXP2 genetic anomaly.

MRI imaging of afflicted family members showed that the caudate nucleus is abnormally small bilaterally while the putamen, globus pallidus, angular gyrus, cingulate cortex, and Broca's area are abnormal unilaterally; functional abnormalities were found in these and other motor-related areas of the frontal lobe (Vargha-Khadem et al. 1995, 2005). PET imaging showed overactivation of the caudate nucleus in two afflicted individuals during a simple word repetition task (Vargha-Khadem et al. 1998). The affected neural structures include those supporting cortical-striatal-cortical circuits implicated in both talking and thinking (see Chapter 4). The reduction in caudate nucleus volume was "significantly correlated with family members' performance on a test of oral praxis, non-word repetition, and the coding subtest of the Wechsler Intelligence Scale" (Watkins et al. 2002). Watkins and her colleagues conclude that these "verbal and non-verbal deficits arise from a common impairment in the ability to sequence movement or in procedural learning." However, it's not clear that the volume of particular neural structures is, in itself, the neural basis for the observed motor and cognitive deficits. fMRI studies that compare afflicted members of the KE family with both their "normal" siblings and age-matched controls show that underactivation occurs in the putamen, Broca's area, and its right homologue (Liegeois et al. 2003), a finding that is consistent with circuits connecting the striatum and Broca's area (Lehericy et al. 2004). The behavioral deficits appear to derive from degraded cortical-striatal-cortical circuit activity.

Embryonic Development

The constellation of behavioral deficits noted in afflicted family KE subjects results from a dominant point mutation mapped to chromosome 7q31 in the FOXP2 gene (Fisher et al. 1998; Lai et al. 2001). Lai and her colleagues determined the neural expression of FOXP2 during early brain development in both humans and mice (Lai et al. 2003), mammalian endpoints separated by 75 million years of evolution (Mouse Genome Sequencing Consortium 2002). The FOXP2 gene encodes a forkhead transcription factor, a protein that regulates the expression of other genes during processes that are active during embryogenesis such as signal transduction, cellular differentiation, and pattern formation. Mutations to other forkhead transcription factor genes have been implicated in a number of developmental disorders. In the case of family KE, the mutation changes a single amino acid in the DNA-binding region of the protein, and that single change apparently leads to protein dysfunction. The areas of expression in both the human and mouse brain are similar and include the structures directly involved in sequencing in the human cortical-striatal-cortical circuits that regulate both motor control and cognition—the thalamus, caudate nucleus, and putamen as well as the inferior olives and cerebellum.

Independent evidence shows that FOXP2 is expressed in the putamen as well as in the caudate (Takahashi et al. 2003)—the behavioral deficits of afflicted family KE members clearly cannot be localized to the caudate nucleus. These structures are all intricately interconnected. The cerebellum, which receives inputs from the inferior olives (another subcortical structure), is involved in motor coordination. The cortical plate (layer 6) is also affected by the FOXP2 mutation. As Lai et al. (2003) point out, their "data are consistent with the emerging view that subcortical structures play a significant role in linguistic reasoning."

As the studies reviewed in Chapter 4 show, cortical-striatal-cortical circuits also regulate cognition. The striatal components of these circuits—the basal ganglia that connect to the cortex through the thalamus—constitute the sequencing engine that successively activates and inhibits cognitive pattern generators. Cognitive flexibility as well as adaptive motor programming derive from basal ganglia activity; basal ganglia are the neural switch that can shift to a different set of pattern generators. The close relationship between extreme verbal apraxia and the syntactic deficits of afflicted family KE members suggests a close link between the neural bases of speech production and syntax. The KE family members who cannot stick out their tongues and round their lips or repeat two words have cognitive deficits that appear to derive from being unable to shift sets, as well as syntax comprehension deficits that may reflect a similar problem. Other motor control problems may exist in afflicted family KE members, but they have not been documented. fMRI studies that compare afflicted members of the KE family with both their "normal" siblings and age-matched controls extend earlier limited PET imaging (Liegeois et al. 2003). Underactivation occurs in the putamen, Broca's area, and its right homologue—results that are consistent with circuits connecting the striatum and Broca's area (Lehericy et al. 2004).

In Chapter 4, I noted a similar correlation between speech motor deficits and the comprehension of distinctions in meaning conveyed by syntax for diverse groups including hypoxic climbers on Mount Everest, aged people, and young children. In addition to the studies I reviewed in the previous chapter, Kimura and Watson (1989), in a study of aphasic patients with focal brain damage, found that their patients, as is the case for the afflicted members of family KE, had coordinate oral sequencing and speech production deficits. Other studies of verbal apraxia, or problems in sequencing speech motor acts, have shown impairment of subcortical neural structures (le Normand et al. 2000). Children with verbal apraxia have been shown to also have cognitive and language comprehension problems (Dewey et al. 1988).

A pilot study of verbal apraxia in young children shows correlations similar to those seen in PD and hypoxia between speech motor-sequencing deficits and cognitive sequencing (Young and Lieberman, forthcoming). One child suffered oxygen deprivation during birth, which usually results in damage to the metabolically active basal ganglia and hippocampus (Burke, Franklin, and Inturrisi 1994; Robertson et al. 1999). Consistent with the findings of studies of PD and hypoxia on Mount Everest (see Chapter 4), deficits in forming cognitive sets and shifting on the WCST occurred, as well as VOT sequencing deficits. VOT sequencing deficits were correlated with perseverative errors on the WCST ($r = 1.0$, $p < 0.008$). Vowel durations also were significantly slower for this group than for age-matched normal controls (for example, 437 msec for a group of four children for the vowel of the word "bad" versus 279 msec for a group of thirty-two five-year olds. The apraxic children's cognitive deficits are consistent with the data of Monchi et al. (2001) and Scott et al. (2002), which confirm the role of the basal ganglia in cognitive set shifting. Three siblings, for whom no evidence of oxygen deprivation during birth exists, have similar, even more profound speech and cognitive deficits than the child known to have suffered hypoxic damage. A genetic basis for these deficits, similar in nature to that demonstrated for the afflicted members of family KE, is quite likely. Genotyping should establish whether an anomalous version of FOXP2 or some other regulatory gene is involved.

Speech, Cognition, and Neanderthals

The correlation between speech motor sequencing and cognitive set-shifting impairments (including the comprehension of distinctions in meaning conveyed by syntax) may reflect the fact that speech motor control is exceedingly difficult, providing a more sensitive measure of basal ganglia dysfunction. The correlation instead may reflect a link between neural circuits regulating speech motor activity and syntax, reflecting a stage in the evolution of the human brain in which selective pressure for enhancement of speech may have played a part in selection for the human version of FOXP2. The next chapter discusses the evolution of human specialized speech-producing anatomy. The last stage in the evolution of the human brain may reflect the adaptations that enhanced the reiterative nature of human speech motor control, with the concomitant enhancement of human cognition. Neanderthals constitute a hominid species that diverged from humans some 500,000 years ago (Krings et al. 1997; Ovchinnikov et al. 2000; Adcock et al. 2001). We may wonder whether disabilities in articulate speech, complex syntax, and cognitive flexibility played a part in their extinction. I return to this contentious issue in the next chapter.

The dating of the human form of FOXP2 thus is consistent with superior speech, syntactic and cognitive capabilities playing a major role in humans replacing Neanderthals. A key finding of the Lai et al. (2003) study is the similarity between the mouse and human FOXP2 expression pattern in the developing brain. FOXP2 clearly is implicated in the development of neural circuits regulating motor control in mammals. However, despite the high degree of similarity, there are important distinctions between the mouse and human versions: foxp2 and FOXP2. The dating of the human form of this gene may not reflect that of the proteins that are directly involved in embryogenesis. However, the techniques of molecular genetics indicate that the human-specific changes in the FOXP2 protein sequence underwent positive selection sometime about 100,000 years ago, which is consistent with independent estimates (Stringer 1992, 1998, 2002; Templeton 2002; Clark et al. 2003; White et al. 2003) for the appearance of anatomically modern *Homo sapiens*—us.

These estimates are based on techniques used to track the divergence of species and the dispersal of populations. These techniques provide estimates of the date at which hominids first diverged from the common ancestor of present-day apes and humans (6 million years ago), the date when modern human beings evolved (100,000–200,000 years ago), and the date when Neanderthals diverged from the evolutionary trajectory that resulted in modern humans (500,000 years ago). The "clock" is the rate at which random mutations could account for the differences in DNA that can be observed in the mitochondrial DNA transmitted from females (Ingman et al. 2000) and recovered from chimpanzees, humans, Neanderthal fossils, and different contemporary human populations. Evolution always involves the individual; a particular man and woman were the forebears of all present-day human beings. In time, successive random mutations yielded the genetic variation that can be observed in human populations separated by time and space. The genetic clock-rate, the pace at which mutations occur, can be established by comparing the genetic variation that exists between humans that are the descendants of individuals who emigrated from one geographic location to another at a particular date, leaving relatives behind. For example, the genetic differences between the indigenous inhabitants of Borneo who emigrated from the Asian mainland and their relatives who remained behind have provided reasonable estimates of the genetic clock. Using these techniques, Enard et al. (2002) estimate that the human form of FOXP2 appeared in the last 100,000 years during the period that marks the emergence of anatomically modern *Homo sapiens.* Studies of the DNA of creatures as far removed from humans as mice and fruit flies generally support these inferences (Carroll, Grenier, and Weatherbee 2001). The degree of similarity of FOXP2 in the modern human populations studied by Enard and his colleagues suggests that it was subjected to intense selective pressure, as we might expect if it played a significant role in conferring human linguistic and cognitive ability.

The Antiquity of foxp2

Studies of songbirds point to a time depth transcending the evolution of mammals for foxp2's role in the evolution of syntactic processes. Avian foxp2, which is 98 percent identical to the human version, is also expressed in striatal nuclei. The avian striatum is a key element of the neural circuits that allow birds to learn songs (Brainard and Doupe 2000). As noted in the preceding chapter, neuronal populations form in the primate putamen that code and initiate learned motor acts (Kimura, Aosaki, and Graybiel 1993; Graybiel et al. 1994; Mirenowicz and Schultz 1996). As stressed earlier, the basal ganglia are a key element in learning both motor acts (Graybiel 1995) and patterns of action having implicit "cognitive" bases (Graybiel 1997; Huettel, Mack, and McCarthy 2002). Thus, foxp2/FOXP2 may be a key element governing the striatal bases of learning. Molecular genetic studies of birds show that foxp2 is localized in a region of basal ganglia present only in species that can learn different songs. Moreover, avian foxp2 expression peaks at the times associated with song acquisition, in zebra finches at post-hatch days 35 and 50. In adult canaries, which relearn songs seasonally, peak foxp2 expression occurred in those periods when songs became unstable (Haesler et al. 2004). The evolutionary antecedents of the human basal ganglia sequencing engine clearly extend back to the amphibian and reptilian species ancestral to birds and mammals.

It is unlikely that FOXP2 is the only gene involved in the evolution of the human brain. However, it clearly is involved in the development of the neural circuits that regulate aspects of behavior that are species-specific human attributes. The fact that foxp2 expression and basal ganglia activity are factors governing the acquisition and regulation of syntactic processes in other species opens up an avenue of research that may establish the functional distinctions that result from bird, mouse, and chimpanzee versions of the gene. Ultimately, we may be able to establish the genetic links to the neural substrates that regulate motor control and cognition, as well as their evolutionary history.

Walking and Thinking

One of the major behavioral deficits of Parkinson's disease (PD) is a breakdown in executing the complex sequential movements involved in upright bipedal locomotion and in maintaining upright balance. The Hoehn and Yahr (1967) diagnostic scale for the severity of PD essentially is a measure of upright balance and locomotion. The basal ganglia clearly play a major role in regulating walking and maintaining upright balance. It has become evident that the neural bases of locomotion in humans differ from those in most quadrupeds; we do not come equipped with an innate walking reflex. At one time, my wife and I used to jog in the pastoral setting of the University of Connecticut's School of Agriculture. As we paused to watch young foals, we could see that, after a few shaky moments, they started to trot minutes after birth; their neural systems clearly were "preloaded" with an innate quadrupedal locomotion "program." Studies of the acquisition of human walking suggest that we have a similar phylogenetically primitive quadrupedal locomotion reflex. Children have to learn to walk upright (Thelen 1984). Toddlers, for example, are toddlers because they lurch about for more than a year before they learn to control heel strike. The apparent bipedal walking reflex that we observe in human infants is an artifact resulting from an infant's shoulders and arms being immobilized as she or he is held upright.

The subcortical structures whose expression is regulated by FOXP2, the basal ganglia and cerebellum, play a critical role in motor learning. The process of learning to execute a highly automatized motor sequence involves activity in these subcortical structures as well as prefrontal cortex (for example, Kimura, Aosaki, and Graybiel 1993; Thach 1996). The evolutionary processes that enabled early hominids to perfect upright locomotion, starting from the base apparent in present-day chimpanzees, which can walk and run for limited periods with great effort, could have shaped a bipedal human walking reflex. Natural selection could, in time, have yielded innate human motor reflexes for walking. But that doesn't appear to have occurred. As Jesse Hochstadt (personal communication) has suggested, the hominid solution to walking may have involved enhancing the power of the subcortical-cortical circuits involved in motor learning—the prefrontal cortical and subcortical structures of the basal ganglia and cerebellum that are implicated in motor learning. In short, selection for walking may have been the starting point for the evolution of our human cognitive capacities.

The subsequent evolution of the genus *Homo* was marked by adaptations for endurance running (Bramble and Lieberman 2004), which places still further demands on the basal ganglia sequencing engine. A transfer to speech motor control in early *Homo erectus* could easily have taken place. Because no videotapes or movies of any extinct hominid are likely to be found outside the pages of fantasy, indirect tests of this hypothesis can only hint at whether it is correct. Developmental studies that track the onset of walking and running in children with speech production might provide some insight.

Theories for the Evolution of Language

It would be disingenuous to leave you with the impression that my argument in this chapter represents the only theory for the evolution of human language or that other scholars universally accept the motor control theory. Although discussion of the evolution of language was dormant until the 1970s, biannual international conferences on the topic now occur. The approach to addressing the evolution of some of the derived features of human language, speech production, and complex syntax that I have presented differs from many, draws on others, and is essentially disjoint from still others. But before presenting differing views on the evolution of human language, it is essential to differentiate between identifying behavioral factors that may have played a part in driving natural selection and the biological mechanisms that confer human language and cognition.

Biological Determinants of Human Language and Cognition

The Einstein-Photo—Big Brains

The human brain is about three to four times larger than that of the chimpanzee, depending on how you factor in body size (Jerison 1970, 1973; Deacon 1997; Semendeferi and Damasio 2000; Semendeferi et al. 2002), a fact that has led to many attempts to track the evolution of language by estimating the brain size of extinct hominids. The underlying premise is that bigger brains are markers of "bigger" mental capacities. Again, current technology may motivate this approach; computers with larger memory capacities generally have greater power insofar as they can store and execute longer and more complex programs. Comparative evidence supports this theory. Because chimpanzees possess both lexical and syntactic ability in reduced degree, natural selection that gradually increased the size of the brain's mental store of words and syntactic rules most surely has played a part in the evolution of these aspects of human language. Moreover, although folk physiology is supposedly removed from scientific inquiry, the belief that big brains and heads show genius is ubiquitous. As a result, we have the "Einstein look"—a photographer conveys the genius of his subject by a subtle camera angle, tilting the forehead toward the camera. Bushy, even wild, hair and a decrepit, baggy sweater help but are not essential.

However, even extremely large brains do not necessarily indicate that a creature possesses human language. Whales are an example. Bigger-brained hominids probably had "better" language ability than ones with small ape-like brains; but just what linguistic capabilities were more advanced in big-brained than in earlier, small-brained hominids is unclear. The extent of lexical ability might well be connected, to a degree, with overall cortical volume because memory does appear to be instantiated in the areas of the cortex involved in perception and motor control. The expansion of hominid brain volume that may be linked to the ASPM gene about 2 million years ago (Evans et al. 2004; Zhang 2003) could account for many of the cognitive differences that are apparent when we compare chimpanzees and humans. However, there may be more to human lexical capacity than simple cortical volume. It is unclear what aspects of cognition or language could derive from an increase in brain size. Can we say that a brain volume of 1000 cc is an absolute measure of syntactic abilities capable of uttering and comprehending sentences in the passive voice?

Nonetheless, many papers and books make specific claims about when fully human language appeared, based on estimates of brain volume reconstructed from fossilized fragments of the skulls of extinct hominids. Schoenemann (in press) is a recent example. However, a big-brained person is not necessarily smarter than a small-brained one. Schoenemann's own work supports that statement. Schoenemann and his colleagues (2000) used MRI imaging and a battery of cognitive tests to study brain size and cognition within the same families. When socioeconomic status (SES), education, and the countless differences that occur in the course of life are controlled in this manner, the correlation between brain size and the battery of cognitive tests was zero.

Increases in brain size that most likely bear no relation to language are evident in the fossil record. These findings surely must temper claims for specific linguistic capacities based solely on brain size. Consider the evolution of wolves and sheep. Although wolves and sheep communicate, it's a stretch to say that they possess language. But it is clear that wolves are cleverer than sheep. And, not surprisingly, wolves have bigger brains than sheep. One of the first studies that showed that selection for bigger brains occurred in the struggle for existence, and thereby enhanced biological fitness, was published only ten years after Darwin's *On the Origin of Species.* Lartet (1868) studied the increase in brain size in closely related mammalian species over the course of time. Lartet's measure of time was provided by the depth at which fossils were found in the strata of the river Seine. The depth at which a fossil skull was found marked relative time, with deeper depths signifying an earlier period. Lartet attempted to take into account the effects of body size—for example, the huge brains of whales compared to those of apes. Lartet observed that "the further back that mammals went into geological time, the more was the volume of their brain reduced in relation to the volume of their head and to the overall dimensions of their body." None of these mammals possessed language, but selection for increased brain size clearly had adaptive value.

Jerison (1970, 1973) replicated Lartet's findings; a "race" towards bigger brains had occurred over a 60 million-year period. Jerison attempted to account for the effects of body size on the brain to discern increases that might enhance cognitive behavior. As the studies of neural circuits reviewed in the previous chapters indicate, that cut is not an easy one, because the circuits that regulate mood and motor control include structures that also support circuits conferring cognitive ability. Jerison's findings, nonetheless, show a gradual increase in brain size over time in most species. Most animals got bigger brained as time went on. The beasts that were on the menu got bigger brains, but the carnivores doing the dining maintained an edge in brain size. In any given epoch, carnivores have had larger relative brain sizes than their prey. Some small-brained species evolved who appeared to elude capture by running faster and some little-brained carnivores with big teeth also were successful. However, bigger brains generally paid off. Because wolves, sheep, and tigers do not talk, it is difficult to see how brain size, in itself, can date the appearance of any particular aspect of human language.

However, the fossil hominid record clearly shows that brain size abruptly increased in early members of the genus *Homo* some 2 million years ago. Genetic studies indicate that regulatory genes such as ASPM may have been factors in this process (Evans et al. 2004; Zhang 2003), and the survival of these bigger-brained members of the *Homo* species (bigger brains have high metabolic costs) suggests that they had cognitive abilities superior to earlier, small-brained hominids. However, the open question is the nature of these cognitive capacities.

More Is Different—A Neural Legend

There is more to the hunt for brain size than simply brain size itself. The trophy apparently is the "special" part of the brain that would yield human language if it were plugged into a chimpanzee brain. If you were trapped overnight in a university library, you could pass the hours reading urban legends documented in sociological journals—alligators cruising the storm drains of New York City, Elvis alive in Topeka or Newark, and so on. A parallel neural legend is "more is different," meaning that bigger brains must *necessarily* have structures specialized to carry out different aspects of behavior. The basis for this claim rests in the fact that in small-brained animals such as hedgehogs, it is difficult to distinguish sensory from motor cortex. In humans, posterior occipital and temporal regions traditionally have been identified with processing sounds, visual displays, colors, and so on. The motor areas of the cortex that control movement are frontal regions. However, this distinction is becoming very, very fuzzy. For example, Brodmann's area 17 in occipital cortex is involved in processing incoming visual information and hence is sensory cortex. However, area 17 also is activated when people think of colors (Kosslyn et al. 1999). The distinction between sensory, cognitive, and motor functions becomes even fuzzier when we take into account neuroimaging studies that explore the semantic properties of words. Regions of both sensory and motor cortex appear to be the neural instantiation of the referents of words; Martin and Chao (2001) review much of the evidence.

The distinction between motor and sensory cortex becomes still murkier when we consider the fact that many of the mirror neurons in the monkey homologue of Broca's area discharge when a monkey hears the sound of a piece of paper being torn apart. Other loud sounds do not trigger these mirror neurons, which bind actions with the sounds that result from these actions. Other mirror neurons discharge both when a monkey observes someone eating and when the monkey eats.

Is Any Part of the Human Brain Disproportionately Larger Than a Chimpanzee's?

If "more is different" were true, some special part of the human brain could be the seat of language; but the relevant comparison must be with chimpanzee, not hedgehog, brains. Granting the fact that different tasks might be carried out in a particular part of the brain, any disproportionately larger part of the human brain might, hypothetically, carry out chimpanzee tasks plus language—with the added task becoming an explanation for the increase in size. Therefore, some studies have searched for a part of the human brain that is disproportionately larger than that of a chimpanzee, looking for a seat of language or some particular aspect of language. In making the case for the bigger part of the brain that yields language, the prefrontal cortex is an obvious candidate because, as we have seen, prefrontal regions of the cortex are involved in language and thinking. Deacon (1997) proposed that human cognitive and linguistic ability derives from our prefrontal cortex working in concert with our cerebellum. According to Deacon, the human prefrontal cortex is disproportionately larger than in apes and archaic hominids. Other theorists have proposed that the cerebellum plays a critical part in higher cognition and language (Schmahmann 1991; Leiner, Leiner, and Dow 1993). However, Deacon's theory is not consistent with MRI imaging studies that have determined the volumes of different parts of the brain in living monkeys, apes, and humans or with studies of the effects of damage to the human cerebellum. MRI volumetric studies show that the cerebellum is disproportionately smaller in human beings (Semendeferi and Damasio 2000; Rilling and Seligman 2002). Moreover, the frontal regions of the brain are not disproportionately larger in humans (Semendeferi et al. 1997; Semendeferi and Damasio 2000) than in apes. If human prefrontal regions were disproportionately larger, other frontal regions would have to be disproportionately smaller, and this is not evident. The cerebellum is involved in learning motor acts (Thach 1997) and may have other cognitive functions. However, a controlled study of human subjects with brain damage localized to the cerebellum did not reveal deficits in cognition and sentence comprehension (Pickett 1998).

In this vein, Rilling and Seligman (2002) propose that enhanced human lexical and semantic ability derives from our having a disproportionately larger temporal cortex. Because many, but not all, aspects of phonologic and syntactic-semantic information appear to be coded in temporal cortex (Damasio et al. 1996), this supposition is reasonable. Rilling and Seligman computed the expected size of various parts of the human cortex using complex "allometric" procedures that attempt to account for brain size increases which follow from increased body size rather than from cognition. In this approach they followed the principles proposed by Lartet (1868) and Jerison (1970, 1973). They concluded that the temporal regions of the human brain are about 24 percent larger than one might expect if we simply had a very large chimpanzee brain. However, their claim rests on the complex scaling procedures used in their study. When the actual measured volumes of temporal cortex and total brain volumes are compared, the human temporal cortex is no larger than one would expect for a big ape brain. Semendeferi and Damasio (2000), who measured the same sample of ape brains as Rilling and Seligman, found that no particular area of the human cortex, including the temporal regions, is disproportionately larger than an ape's. This conclusion is consistent with Stephan, Frahm, and Baron's (1981) anatomic studies, which directly measured the volume of many structures of the brains of dead primates. Stephan concluded that no part of the human cortex was disproportionately larger than an ape's.

However, all of these studies are based on exceedingly small sample sizes of monkeys, apes, humans, and other species. Given the variation in human brain size, it would be necessary to show that the volume of some hypothetical, disproportionately larger structure of the human brain was significantly larger than that of apes. For example, the "disproportionately larger" temporal lobe volume of humans, compared to a scaled-up ape brain inferred by Rilling and Seligman (2002), was 40 mL. This difference does not appear to be significant because the difference in volume between the smallest and largest brains of the 72 subjects studied by Schoenemann and his colleagues (2000) was 400 mL (1015 versus 1419 mL); the average within-family difference between siblings was 62 mL. No part of the human brain appears to be disproportionately larger than that of a chimpanzee brain scaled up to human size. It is therefore most unlikely that a disproportionate increase in any part of the brain occurred in the course of hominid evolution.

Brain Lateralization

If we return to global theories that attempt to account for the evolution of human language, one recurring candidate is lateralization. The human brain is lateralized, and the left hemisphere in about 90 percent of the present human population has a dominant role in regulating both motor control and language. Right-handedness usually corresponds with left-brain dominance. Lenneberg (1967), in one of the first modern studies of the biology of language, claimed that lateralization was the key to human language. Subsequent comparative studies of brain lateralization for other species have disproved this claim: lateralization is a primitive feature found in species as far removed from human beings as frogs. The left hemisphere of the brains of some species of frogs regulates their vocalizations (Bauer 1993). Moreover, studies of many mammalian species show that paw movements are under lateralized neural control (Bradshaw and Nettleton 1981; Denneberg 1981; MacNeilage 1991). Some mice are right-pawed, others left-pawed, but they all survive and have viable progeny—left-paw dominance does not increase biological fitness. Corballis (2002) has claimed that lateralization (particularly left-brain) is a key factor for the evolution of human language, but he does not explain how this can be the case in light of evidence for lateralization in many other species.

Why Do Humans Tend to Be Right-Handed?

Although lineages of laboratory rats and mice are predominantly right- or left-pawed, some in the wild can be either. The mouse and rat strains that are consistently pawed are the descendants of a few ancestors that all had the same paw preferences. We humans most likely are the descendants of a small group of hominids who happened to be predominantly right-handed. Stone tools flaked more than 2 million years ago appear to have been produced by right-handed hominids (Toth and Schick 1993). We may not be directly related to these particular small-brained toolmakers, but hominids who were mostly right-handed surely are in our line of ancestry.

Language Is Not Lateralized

Moreover, language capacities are *not* isolated in the dominant hemisphere of the human brain. As noted earlier, current neural imaging studies show that, although one hemisphere generally is more active during linguistic tasks, both hemispheres of the brain are activated (Just et al. 1996; Kotz et al. 2003b; Rissman, Eliassen, and Blumstein

2003). Theories that equate asymmetric cortical language areas with linguistic ability have not stood the test of time. The planum temporale—a sector of Wernicke's area—was thought to be symmetric in apes, in contrast to the asymmetrically larger planum temporale in the human dominant hemisphere. However, further study shows that apes and humans both have similar asymmetric planum temporale (Gannon et al. 1998). Broca's area likewise has been found to be asymmetric in apes (Cantalupo and Hopkins 2001). Despite popular accounts of human language being in the left brain, no apparent linguistic or cognitive deficits or decrements in biological fitness have been traced to right-brain dominance. The claim that art and music reside in the right hemisphere likewise may fall in the category of neural mythology—were Rembrandt and Mozart southpaws?

Behavioral Factors Driving the Evolution of Human Language and Cognition

Syntax Is Useful, But?

In Molière's seventeenth-century play, *Le Bourgeois Gentilhomme,* the central character learns that he speaks prose. Although the focus of Chomsky's work is syntax, he provided the same lesson in a documentary film about his linguistic theories. Chomsky did not present any evidence for syntactic universals that supported his position; instead, he stated that languages employ nouns and verbs. Molière might as well have been interviewed. Linguists now have accepted the premise that human language must have evolved (for example, Pinker and Bloom 1990; Pinker 1994; Hauser, Chomsky, and Fitch 2002). Accordingly, the proceedings of recent conferences on the evolution of human language are heavy with studies that attempt to prove the utility of syntax. In the absence of access to extinct hominids, the method of choice generally involves computer modeling rather than comparative studies of animal communication.

The computer-modeling studies reported by Nowak and Krakauer (1999) and Nowak, Plotkin, and Janson (2000) have been widely cited. Nowak and his colleagues model the organization of a system that relies on words to communicate information. As the number of words and concepts that must be communicated increase, syntactic devices that make use of a finite set of words to communicate a potentially infinite number of propositions become useful. On the basis of the computer simulation, they make two claims: (1) that syntax becomes necessary as a language acquires many words, and (2) that the rules of syntax must be innately determined. The first claim is consistent with studies of the development of lexical ability and syntax in young children. Bates and Goodman (1997) show that, as children mature, syntax develops as vocabulary size increases. The second claim, which reiterates Chomsky's claim that the rules of syntax are innately specified, has little to do with the computer simulation. It relies on Gold's theorem—an exercise in mathematical logic in which Gold (1967) demonstrated that Chomsky's system of linguistic rules (ca 1966) could not be acquired by a child with 100 percent accuracy, unless "negative" evidence—corrections to the child's ungrammatical sentences—were continually provided. Linguists who were following the general line of Chomsky's work noted that parents rarely corrected their young children. Gold and these linguists ignored virtually everything that influenced a child's acquisition of language (c.f. Karmiloff and Karmiloff-Smith 2001; Tomasello 2004a for comprehensive reviews of this issue). The conclusion of Nowak's studies was that knowledge of syntax must be innate. However, even if we ignore the evidence that shows that children are not little logicians who ignore everything that happens around them, the problem only exists if the child has to acquire each and every rule of a "perfect" grammar shared by every speaker of a given language. Horning (1969), using mathematical logic similar to that employed by Gold, showed that negative evidence was *not* necessary if a close approximation of the grammar would suffice. For example, if 95 percent of the child's rules were similar to his parents', negative evidence was not needed. And studies of the transmission of language from one generation to the next show that children do not preserve the language of their forebears. A short interlude with the *Oxford English Dictionary* will establish this fact.

Horning's (1969) position is generally accepted by specialists in computer analysis of natural languages and is noted in textbooks on computer processing of natural language (for example, Manning and Schutze 1999, pp. 386–387). However, Nowak and his colleagues apparently were unfamiliar with Horning's refutation of Gold's theorem. Nowak and his colleagues disregard or are unfamiliar with the body of evidence that suggests that children acquire words and syntax by means of associative learning, imitation, and the subtle social cues that parents and caretakers use to convey displeasure to children or to correct errors (for example, Lieberman 1984; Meltzoff 1988; Greenfield 1991; Bates and Goodman 1997; Elman et al. 1997; Karmiloff and Karmiloff-Smith 2001; Saffran, Aslin, and Newport 2001; Tomasello 2004a). Human beings are intensely social animals and normal children are sensitive to these cues at early ages. Language, moreover, continually changes. We don't speak early English or even preserve the linguistic forms of our parents.

Social Factors

Social factors have been cited as driving natural selection that increased neocortical size, thereby enhancing lexical and syntactic ability. It is hard to disagree with the general premise. Clearly, supportive social interaction enhances biological fitness; Charles Darwin discussed many examples in *On the Origin of Species.* Speech and language play a part in obtaining mates in contemporary and historical times and there is no compelling reason to think that humans acted in very different ways in prehistoric epochs.

In this vein, Dunbar (1993) claimed that grooming, gossip, and the cohesion of large groups of hominids resulted in increases in brain size that ultimately led to human language (though the brain size of solitary orangutans does not fit his model). I (Lieberman 1984) cited collective insight, the sharing of information among the members of a group so as to enhance problem solving. Falk (2003) proposes that language evolved as hominid mothers attempted to keep in touch with their helpless infants. Hunting, tool use, and tool fabrication as factors leading to language have been noted time and time again. They may have played a part in the evolution of human language, though it is apparent that chimpanzees hunt and make and use tools (Goodall 1986).

But it is difficult to single out any single behavior as *the cause* for the evolution of human language. It surely is the case that language enhances the conduct of every aspect of human behavior that could enhance biological fitness. Paraphrasing Charles Darwin, in the infinitely complex relations of hominids with each other and external nature, language would have been an asset. Virtually all aspects of human behavior that are enhanced by language, except for suicide, misanthropy, and total warfare, increase the survival of progeny, yielding biological fitness. So virtually all of the factors noted here could have played a part in the evolution of human language.

Date-Stamping the Stages for "Fully" Human Language

Again, because we lack any direct evidence of the vocalizations or gestures by which extinct hominids may have communicated, or for that matter the language of early human populations, it is difficult to ascertain the precise dates at which the derived features of human language came into being. But many attempts to date-stamp these attributes of human language have been made.

The presumed causal relationship between left-brain dominance and language is the basis for studies that attempt to establish the date at which fully developed human language appeared. The paper published by Wilkins and Wakefield (1995) is typical. Wilkins and Wakefield claimed that they could discern asymmetric, hence lateralized, Broca's and Wernicke's areas by examining endocasts (casts of the inside surfaces) of fossil skulls. They concluded that these extinct hominids possessed an innate 1995 version of Chomsky's Universal Grammar 2 million years ago. Knowledgeable biological anthropologists such as Ralph Holloway (1995) do not believe that it is possible to identify these areas by looking at fossil skulls. Holloway likened the exercise to reading tea leaves. But the presence of Broca's area or its homologue in an archaic hominid would not be a proof of fully developed human linguistic ability, much less an innate syntax organ: as we have seen, it is only part of the complex neural circuits that regulate human language and play a role in other aspects of human and nonhuman primate behavior. If it were possible to determine that Broca's area (there is, by the way, some dispute as to what constitutes Broca's area) was present in an extinct hominid's brain, about all that we could state with certainty is that Broca's area was present.

Gestural Language and the Stages of Language

Gordon Hewes over the course of many years was the principal advocate of a stage in the evolution of language in which manual gestures communicated linguistic distinctions. The very use of the word "stage" suggests there was a series of abrupt transitions, which may or may not have been the case. The evolutionary history of human linguistic ability most likely involved both gradual changes through natural selection on an existing range of genetic variations and mutations on regulatory genes that yielded branch-point events. Hewes' (1973) proposal did not specify either the degree or date at which manual gestures played a greater role than vocal signals in language. Hewes, moreover, did not claim priority for this theory; he traced similar theories back to the nineteenth century.

Somewhat different suggestions for mixed vocal-gestural systems playing a part in the evolution of human language have been presented by Greenfield (1991), Kimura (1993), Lashley (1951), Lieberman (1975, 1984), and others. Rizzolatti and Arbib (1998) proposed that the natural selection that initially enhanced manual motor control played a part in the evolution of Broca's area and human language; moreover, they fleshed out this theory with the discovery of mirror neurons. Theories proposing a mixed manual-vocal system leading into predominantly vocal communication are consistent with the fact that cortical-striatal-cortical circuits regulate all aspects of motor control, as well as the apparent greater difficulty of speech versus manual motor control. The actual presence of this (or any) stage in the evolution of human language is difficult to test, absent living archaic hominids.

Other proposals call for a stage in which manual gestures were the sole medium for language (for example, Burling 1993; Corballis 2002). However, recent comparative studies of animal communication, such as Slocombe and Zuberbuhler (2005), show that monkeys and chimpanzees use a limited number of vocal signals to transmit referential as well as emotional information; a "pure" manual language stage is unlikely. Moreover, we also must pose a question: If manual language was, as Corballis (2002) claims, so prevalent and efficient, why is the present default mode primarily vocal? Children also typically start to "talk" using a mix of gestures and speech (Greenfield 1991). The neural bases for all complex manual gestures, manual sign language, and vocal language are also conjoined; strokes that result in Broca's syndrome in deaf persons who communicate using American Sign Language result in a breakdown in their ability to produce any complex manual gestures (Kimura 1979).

Manual gestures continue to play a part in linguistic communication, even in hearing individuals (McNeill 1985), and adaptations that enhanced manual gestures may have played a part in the evolution of language. As Donald (1991) notes, mimetic imitation may have played a role in the evolution of language; gestures can mime action and generally complement spoken language (McNeill 1985). Gestures are often the medium of communication when speech fails. However, speech is the default medium for human language and the human brain is singularly adapted to regulate speech production. Speech sounds apparently yield a longer working memory span (Boutla et al. 2004). Any theory for the evolution of human language must account for these facts. Speech also allows us to use our hands for purposes other than communication. The occurrence of tools in the archaeological record going back to the epoch of Australopithecines (with new discoveries the dates are continually being pushed back in time) also argues against manual language ever being the sole phonetic medium of human language.

The Archaeological Record and the Evolution of Language

Corballis (2002) elaborates the speculation on the initial stages of human language that I presented in Lieberman (1975, p. 159): "the initial language of the Australopithecines may have had a phonetic level that relied on both gestural and vocal components." However, Corballis places the evolution of spoken language at 50,000 years before the present (BP), correlating it with the appearance of "modern" behavior at that time. The evidence for the sudden appearance of modern behavior consists of the artifacts unearthed in western Europe. However, virtually all of the artifacts found in Europe can be found in Africa at earlier epochs: red ochre, for example, which is thought to have been used for symbolic ends, can be traced back in Africa to 90,000 years BP (McBrearty and Brooks 2000). Moreover, humans had reached Asia and most likely Australia by that date, and any child from those regions can effortlessly acquire any language on earth. The biologic substrate that confers human language and cognition must have been in place before humans left Africa.

But a greater problem precludes equating implements, weapons, or most anything that has survived 50,000 years with human cognitive ability. The aggregate endeavors of humans over thousands of generations yield these artifacts. Does the fact that George Washington's official conveyance was a coach drawn by horses signify that his linguistic and cognitive ability was inferior to that of George Bush, who travels in a Boeing 747? Cultures also lose technology. European seafarers observed that the native population of Terra del Fuego was unclothed; they also had lost the art of producing sophisticated stone tools, but their cognitive and linguistic abilities were and are intact. There is no connection between the level of human technology and human brains.

Motor Control and the Timetable for the Evolution of Language

Having demonstrated the folly of presenting a timetable for the evolution of language and cognition, I now present one. Such is the folly of the human condition. In light of the evidence that I have reviewed throughout this book and this brief survey of approaches to the evolution of human language, which has omitted many stories based on how humans behaved 150,000 years ago that cannot be tested, and absent time travel, what can I conclude regarding the antiquity of language?

Although a slight increase in neural activity occurs in one hemisphere of the human brain in many linguistic tasks, lateralization is ubiquitous in mammals and birds and occurs in species as far removed from humans as frogs. Increased brain size surely contributes to human cognitive and linguistic ability. Genes such as ASPM may have been factors in the enlargement of the brain 2 million years ago (Zhang 2003; Evans et al. 2004). Other genes that result in neural restructuring, perhaps in a similar way to FOXP2, have yet to be identified. To me, the strongest evidence that neural mechanisms initially adapted for motor control are key elements in the evolution of human language is the mark of evolution in the human brain.

Walking may have been the initial starting point, and studies of the development of cognitive ability and walking may provide useful insights. But the verdict is still out. As the neural bases of behavior and the genetic expression of neural circuits and structures become evident, hopefully we will be able to move from speculation toward understanding of our evolution and nature. Chapter 6 presents evidence that humans have anatomy specialized for speech production. That finding paradoxically suggests that speech communication occurred at an earlier epoch in the extinct hominids ancestral to *Homo sapiens.* That probability is supported by studies of the vocal communications of other species and the anatomy involved in generating their calls.

Take-Home Messages

- No direct evidence on the evolution of human language exists because the intermediate species are extinct.
- Lexical ability, syntax, and speech are elements of human linguistic ability, and it is unlikely that any single factor suddenly conferred fully human language.

- Because chimpanzees possess limited lexical and syntactic abilities, it is most unlikely that any protolanguage devoid of syntax ever existed. The evolution of lexical ability may have been a gradual process and some syntactic ability was most likely present in the common ancestor or ancestors of humans and apes. Because nonhuman primates communicate referential information using a limited number of vocalizations, some form of speech most likely was present in the earliest stages of hominid language.
- Brain lateralization, in itself, cannot account for human linguistic ability. Nor does any part of the human brain appear to be disproportionately larger than a chimpanzee's, so relative size does not account for human linguistic ability either.
- Bigger brains go with being cleverer, but cleverness does not necessarily equal language.
- Adaptive motor control and the subcortical sequencing engine may provide one avenue for exploring the evolution of human cognitive and linguistic ability. The starting point for the gradual evolution of enhanced human motor and cognitive ability may have been upright bipedal locomotion at the dawn of hominid evolution.
- The FOXP2 gene points to neural circuits that confer fully human reiterative abilities (and human speech, syntax, and cognition) appearing in the timeframe associated with the appearance of anatomically modern *Homo sapiens.*

CHAPTER 6

The Gift of Tongue

As Julius Caesar commented, "All Gaul is divided into three parts." This chapter, likewise, is divided into three parts.

I first attempt to place human speech into the broader evolutionary framework of vocal communication. Most aspects of the anatomy involved in the production of human speech are "primitive" features that we share with other species. Comparative studies of living animals show that they make use of this anatomy to communicate vocally, so such studies help illuminate the possible early stages of hominid language. Next I focus on one apparent derived feature, our species-specific tongue. Other species, such as birds, have evolved alternative means for generating useful vocal signals. The particular shape of the human tongue enables us to produce the full range of human speech sounds, including the "supervowel" [i] (the vowel of the word "see"), while conserving articulatory effort. As Chapter 3 noted, being able to say [i] enhances the robustness of the encoding-decoding process that yields the high information transfer rate of human speech. I also discuss studies of the vocal abilities of other species and recent studies on the development of the human tongue and SVT in infants and children, which cast light on the probable evolution of human speech capabilities. The third part of this chapter discusses the speech production anatomy and phonetic repertoires of extinct hominids, particularly Neanderthals, in the light of these studies. I reassess the claims of previous studies, including my own. The findings of these studies also reflect on the possible evolution of the neural bases of other aspects of language and cognition.

As we have seen, speech is a key element of human linguistic ability. Most discussions of human language trivialize the role of speech. At normal speaking rates, twenty to thirty phonemes per second are transmitted from a speaker to listeners. This rate exceeds the temporal resolving power of the human auditory system; individual nonspeech sounds merge into a buzz at rates in excess of fifteen per second (Liberman et al. 1967). If we were to communicate vocally using other signals, the rate of transmission would be so slow that it would be difficult to keep track of the words that constitute a sentence. The species-specific human tongue—the "gift" of tongue recognized in creation myths, fairy tales, and children's stories—contributes to the robustness of this process. But the evolution of vocal communication extends far back into time, hundreds of millions of years before the appearance of human beings or any primates, to the ancestors of present-day frogs. These amphibians possessed anatomic specializations for vocal communication. Vocal signals convey both referential and emotional information in mammals. And vocal anatomy that plays a part in producing human speech is present in many other living species. Indeed, mechanisms adapted for vocal communication that differ from those of humans have evolved in other species, including other primates. The ubiquitous nature of vocal communication and anatomy suited to that purpose perhaps addresses the question of why the sounds of speech, not gestures, are the default medium for human language. It reflects the mark, the stamp, of evolution.

Primitive Features of Speech-Producing Anatomy

Larynx

Human speech involves much more than the larynx. Nonetheless, the larynx is a key element in the production of speech. Although laryngeal anatomy and physiology have been studied since the early years of the nineteenth century, it is to a British surgeon, Victor Negus, to whom we owe the first systematic study of its evolution. We build on Negus's masterful works, *The Mechanism of the Larynx* (1928) and *The Comparative Anatomy and Physiology of the Larynx* (1949).

Negus traced the evolution of the larynx from its appearance in extinct fish, who, like similar living fish such as the African lung fish and the mud fish of the Amazon, could breathe air (Negus 1949, pp. 2–8). Negus pointed out the selective advantage that the larynx confers to these air-breathing fish compared to "primitive" air-breathing fish like the climbing perch, which lack a larynx. The simple larynges of these air-breathing fish prevent water from entering their lungs. The larynx in these creatures is a sphincter, a slit that acts as a valve, positioned in the floor of the pharynx. When the fish is immersed, the laryngeal sphincter closes; when the fish is out of the water, the larynx opens and allows air to be swallowed and forced into the fish's lung. The shared evolutionary basis of the development of these features, now the subject of renewed attention by the field of evolutionary-development biology, or evo-devo (Alberch 1989; Arthur 2002), is apparent in Negus's observation "that the human embryo, when it is about 5 mm long, shows a slit in the pharyngeal floor 'much like that of the Lung Fish'" (Negus 1949, p. 6).

The larynx's initial adaptive function was to facilitate terrestrial life in fish that *already* could breathe—lung fish that could survive when they were stranded in dry river beds. The evolutionary sequence that ultimately resulted in human speech anatomy would not have occurred, absent this ecological singularity hundreds of millions of years ago. The evolutionary process involved preadaptation: "the highly important fact that an organ originally constructed for one purpose, namely flotation, may be converted into one for a wholly different purpose, namely respiration" (Darwin 1859, p. 190). The swim bladders of fish, which are homologous with the lungs of vertebrates, evolved to facilitate flotation (see Chapter 3). Fish can extract dissolved air from water through their gills. The air can then inflate their swim bladders, which are elastic sacs that expand to increase the volume of water displaced by the fish. Air can be removed from the swim bladders, thereby reducing the fish's size. In a manner similar to the blimps that float overhead trailing banners, a fish can displace a volume of water equal to its weight at a specific depth by changing its size. Whether the swim bladders of fish may have evolved from lungs is irrelevant in this context because fish neither need nor possess lungs.

Branch Points

Air-breathing fishes fill their swim bladder/lung by swallowing air, which then is transferred to the bloodstream. We can regard the change in function of the swim bladder and the evolution of the larynx as a branch point. The new ecological state, terrestrial life, provided a setting disjoint with that of fish. Thereafter, the challenges and opportunities present in life out of water drove the further evolution of the larynx. In short, the evolution of the larynx yielded the possibility for further changes.

Figure 6.1 illustrates hypothetical "branch points" in the evolution of the speech-producing anatomy. A branch point marks a point at which the course of evolution can change through selection that enhances fitness in a new setting or, as recent evo-devo studies show, through a mutation, generally one involving some regulatory gene that restructures a creature's biology substrate, suddenly changing or enhancing existing modes of behavior. In time, profound differences can occur through a chain of branch points, ultimately yielding a new species. The concept of branch points is implicit in the theory that Darwin proposed in 1859. Classic Darwinian theory posits abrupt changes as well as gradual natural selection, and can account for the uneven tempo of evolution noted by Gould and Eldridge (1977). The selective value of small steps is apparent in the gradual changes that occur within a functional continuum, such as the differences in the muscles and cartilages that open the larynx (Negus 1949, p. 7), which gradually became more

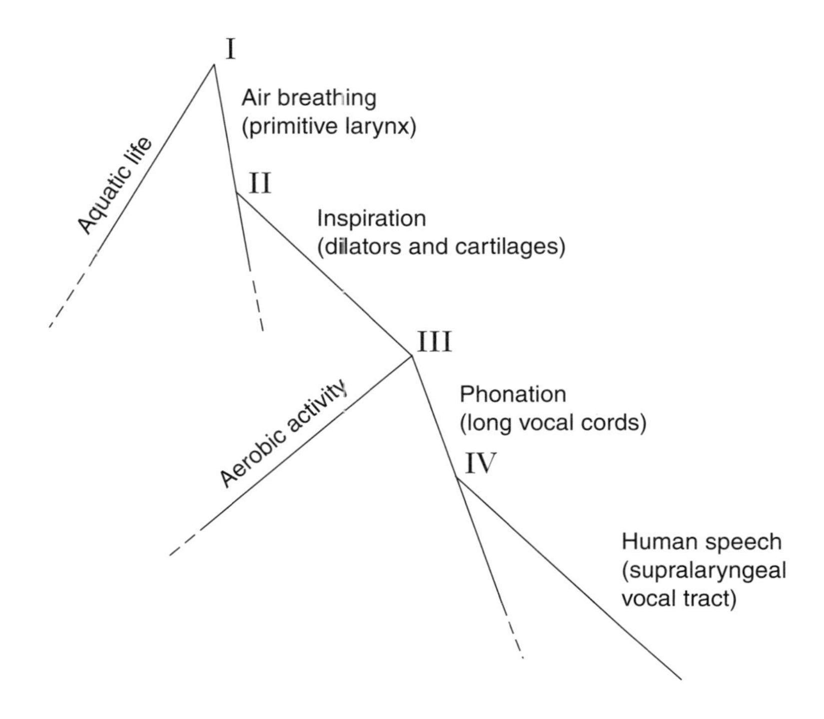

FIGURE 6.1 *A functional branch-point diagram charting some steps in the evolution of the upper respiratory system.*

effective over the course of time. Darwin's comment on the value of small changes still holds true:

> For as all the inhabitants of each country are struggling together with nicely balanced forces, extremely slight modifications in structure or habits of one inhabitant would often give it an advantage over others; and still further modifications of the same kind would often still further increase the advantage. No country can be named in which all the native inhabitants are now so perfectly adapted to each other and to the physical conditions in which they live, that none of them could anyhow be improved. (1859, p. 82)

I do not advocate a neo-Darwinian model for economic development, but it is obvious that branch point models hold true for the marketplace. For example, in the "struggle for existence" in car sales, small differences can yield "selective" advantages. Slight perceived differences in cost, performance, or reliability often have profound effects. The changes that have occurred in the American marketplace in the past decades derive from these small perceived advantages because car makers are "struggling together with nicely balanced forces." In contrast, in the early years of the twentieth century, a branch point change occurred when automobiles began to compete with horse-drawn carriages. The nature of the contest was inherently different. Finely designed carriages could not compete, no matter how well made or how durable they were. Automobiles were qualitatively different; a functional branch point had occurred, and new selective forces entered into the competition.

A twenty-fifth-century archaeologist studying the evolution of cars would see a long period extending over centuries in which gradual changes yielded better horse-drawn carts and carriages. The period of horse-drawn conveyances would appear to be a static period compared to the sudden introduction and improvement of the automobile. However, preadaptation clearly played a part in the sudden evolution of automobiles. Automobiles were made possible by gradual improvements in engines spanning several centuries. The invention of cars was triggered by a set of technological advances—internal combustion engines originally intended for boats, tires designed for bicycles—and an ecological change: smooth, surfaced "metaled" roads that had been built for horse-drawn conveyances. Here is a set of unrelated events that, in concert, profoundly changed the way we live.

In contrast, the sudden introduction of digital computers was a branch point event triggered through an innovation that enhanced an existing "behavior." Digital computers were at first expensive, cumbersome devices that used tens of thousands of hot, glowing vacuum tubes. Computers served as space heaters. Computer centers generally had no heating systems; air conditioning instead was necessary to prevent meltdowns. The invention of the solid state transistor, which in relation to computer technology was a chance event, and the subsequent development of integrated solid state circuits made the computer revolution possible. No conceptual changes occurred, but ever more refined and inexpensive small computers followed from a branch point triggered by the transistor. Scholars unfamiliar with the manner in which complex machinery is designed and built might argue that these analogies are irrelevant because neither cars nor computers are living organisms characterized by genetic heritability. However, constraints similar to those of genetic heritability are inherent in the machine tools and assembly line methods by which cars and computers are built and in the minds of their designers.

Laryngeal Modifications at Branch Points

A branch point theory accounts for the tempo of evolution. It incorporates both the slow, gradual process of change within a functional modality and the abrupt shifts triggered by either behavioral openings that yield the opportunity for rapid change or chance events that enhance an ongoing behavior. Within an ongoing functional modality, slight structural changes yield selective advantages. At a branch point, small *structural* changes, such as the alterations of the cartilages of the larynx that enhance phonation, can produce an abrupt, great *functional* advantage by virtue of the selective advantage. Negus (1949) explored the anatomic bases of the first three stages of laryngeal evolution diagrammed in Figure 6.1.

As Negus pointed out, the initial stage in the evolution of the larynx was the development of dilators that can pull the larynx open to allow more air into the lungs during breathing. A second functional branch point accounts for the larynx's becoming a sound-generating device. The process of phonation, in which the vocal cords move rapidly inward and outward to convert the steady flow of air from the lungs into a series of "puffs" of air, occurred in animals similar to present-day frogs. The anatomic modifications that yield more efficient, controlled phonation and/or "better" breathing occur at branch point III in the figure. Negus (1949, pp. 40–42) showed that the larynges of some animals are specialized for phonation at the expense of respiration while other species have taken the opposite course. For example, horses have a larynx that is "designed" to maximize the flow of air to and from the lungs. In contrast, the human larynx is designed to enhance phonation for the process of vocal communication. Canids such as dogs and wolves, which are social animals that communicate vocally but run down their prey, represent an intermediate solution, balancing the competing selective forces of respiratory efficiency and phonation.

Figure 6.2 shows the relative area of the trachea, or windpipe, which connects the larynx to the lungs, compared with the maximum opening of the larynx for a horse, a dog, and a human being

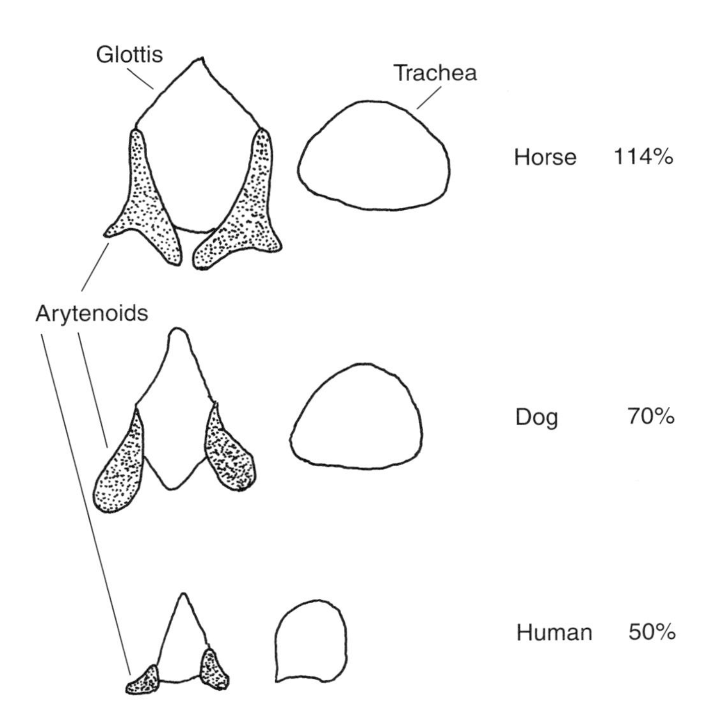

FIGURE 6.2 *Relative opening of the larynx and trachea in horse, dog, and human being. The arytenoid cartilages are stippled in these transverse views looking down on the larynx. Note that the maximum opening of the larynx is smaller than that of the trachea in human beings, where it is 50 percent of the tracheal area. The human and dog larynges thus obstruct the airflow to the lungs even when they are wide open.*

(Negus 1949, p. 31). The larynx acts as a valve constricting the tracheal flow of air to the lungs. A valve that can open wider offers less resistance to the flow of air through the airway from the lungs and allows greater airflow than one that cannot open as wide. You can model this effect by closing and opening the faucet of your kitchen sink. Negus observed that animals that run long distances under aerobic conditions have to transfer enough oxygen to their lungs to maintain the metabolic rate of sustained running. Therefore, horses that can escape from predators by running steadily over long distances have larynges that open wide to yield minimum airflow resistance.

Figure 6.2 shows the laryngeal morphology that determines maximum opening. Horses have a larynx that has long arytenoid cartilages. These cartilages swing apart to open the larynx for respiration. The sketch shows the two arytenoid cartilages abducted (opened) to their maximum opening. The cross-section of the trachea immediately below the larynx, which leads to the lungs, is sketched to the right of the diagram of the section of the open larynx. Note that the opening of the horse larynx is larger than the cross-sectional area of the trachea. The horse larynx thus does not impede the airflow to the horse's lungs. Horses are "designed" to maintain sustained long-distance running. Their hooves and legs as well as hidden details such as their laryngeal arytenoid cartilages have evolved with running driving natural selection.

Phonation

In contrast, social animals such as dogs possess different laryngeal specializations. Negus's sketch of the laryngeal opening of a dog (a staghound) in Figure 6.2 shows that its larynx presents an obstacle to airflow, even when it is at its maximum opening. The airway to the dog's lungs is restricted to 70 percent of the tracheal cross-section. Figure 6.2 also shows that the human airway is restricted to 50 percent of the tracheal cross-section. The deficiencies of the dog and human larynges with respect to respiration follow from the short length of their arytenoid cartilages.

The larynx opens by swinging the arytenoid cartilages outward from their posterior pivot point. The optimum length of the arytenoid cartilages for maximizing the opening of the larynx relative to the trachea, while maintaining the possibility of phonation, is about 0.7 times the diameter of the laryngeal opening (Negus 1949, pp. 40–42). The nearest approach to this optimum length occurs in the Persian gazelle. The Persian gazelle can sustain speeds of 60 miles per hour. Human beings, in contrast, have almost the shortest arytenoid cartilages of any mammal relative to their tracheal cross-section. The probable reason is phonation. Short arytenoid cartilages yield long, pliable vocal "cords" or "folds" (the two terms refer to the same structures) that enhance controlled phonation.

Phonation is possible with long arytenoid cartilages, but more force is necessary to rapidly move these massive, inflexible bone-like structures. High-speed motion pictures and acoustic analyses show that vocal cord motion involving the arytenoid cartilages yields low-frequency "fry" phonation that has a "creaky" sound quality (Timcke, von Leden, and Moore 1958; Lieberman 1967). The average glottal (laryngeal) opening also is large in fry phonation; airflow thus is higher, impeding sustained phonation. Moreover, given the high mass of the arytenoid cartilages, the fundamental frequency of phonation is not as readily controlled as it is in the normal modes (the phonetic term is a "register") for human phonation. Normal phonation in the "chest register" during human speech generally involves only the anterior, soft-tissue portion of the vocal cords, which run from the ends of the arytenoid cartilages to the thyroid cartilage (Van den Berg 1958). (The traditional term, "chest register," has nothing to do with the chest.) The short human arytenoid cartilages swing the anterior vocal cords, which consist of the vocal ligaments, and the thyroarytenoid and cricoarytenoid muscles to their adducted, constricted position, which sustains efficient, controlled phonation (Timcke, von Leden, and Moore 1958; Lieberman 1967). The ability of some humans to run long distances may reflect adaptations similar to those that mark Tibetan populations—perhaps efficient transfer of oxygen in the lungs (Bouhuys 1974; see also Chapter 7).

It is possible to speak without phonation by whispering. Whispering involves exciting the supralaryngeal vocal tract with turbulent noise generated by high airflow through an opened larynx. But the sound produced is feeble and the airflow is about ten times higher than in normal phonation. The larynx is a transducer that efficiently generates audible acoustic energy: this function explains its role and the evolutionary adaptations that enhanced phonation. The trade-off, therefore, is between respiration and phonation. Animals that took the right branch point at level III in Figure 6.1 retained changes that yielded smaller arytenoid cartilages and more efficient phonation. These animals, as Negus notes, are social animals that rely on vocal communication. Phonation plays a major role in human and animal communication, both in the domain of language and in the expression of emotion.

In most of the world's languages (for example, the Chinese languages), fundamental frequency (F0) tones signify different words in the same manner as the phonemes /d/ and /t/ distinguish the different English words "bad" and "bat." In both tone languages and languages such as English, which do not use F0 patterns to differentiate words, F0 patterns can signal syntactic boundaries (for more discussion on this topic, see Chapter 7). A speaker's identity and emotion can be conveyed by F0 variations and the spectral content of the glottal source (Jones 1932; Lieberman 1961, 1967; Lieberman and Michaels 1962; Protopappas and Lieberman 1997; Coster 1986; Kagan, Reznick, and Snidman 1988; Chung 2000). High F0 and F0 variations also characterize child-directed "motherese" and serve as directing signals to infants (Fernald and Kuhl 1987; Fernald et al. 1989). The link between human prosody and animal communication is apparent in studies of mammalian calls that convey information using F0 patterns (Heffner and Heffner 1980; Herman and Tavolga 1980; Goodall 1986; Owren and Bernacki 1988; Cheney and Seyfarth 1990; Owren 1990; Sayigh et al. 1990; Hauser and Fowler 1991; Hauser 1996; Fischer, Cheney, and Seyfarth 2000; Fischer et al. 2002). The leftward branch at level III takes the direction of selection for more efficient respiration as part of a total behavior complex that stresses sustained aerobic activity.

A Derived Human Feature—Our Supralaryngeal Vocal Tract

Many anatomic features differentiate human beings from earlier extinct hominids. Some of these differences play key roles in the evolution of the derived feature that makes fully human speech possible.

Tongues and the Knights Who Said "Ni"

In the film *Monty Python and the Holy Grail,* Roger the Shrubber encounters the Knights of Ni who utter the syllable [ni], which has magical properties. Using English orthography, [ni] would be spelled "nee." The Pythons, perhaps intuitively, had the knights produce the supervowel of human speech. As Chapter 3 noted, one of the most effective sounds for SVT normalization is the vowel of the word "see"–[i] in phonetic notation. It is less often confused with other sounds (Peterson and Barney 1952; Hillenbrand et al. 1995). Words formed with the vowel [i] are more often correctly imitated by young children as they acquire their native languages (Olmsted 1971). The vowel [i] with its cousins, the "quantal" vowels [u] and [a], are among the few attested universals of human language (Greenberg 1963; Maddieson 1984). Virtually all languages make use of these vowels, though they may add other vowel distinctions. A few languages do not make use of all of these vowels to signal different words, but that does not detract from their special status. A biologic propensity to use these vowels clearly exists. Human beings are creatures of culture and can act in ways that are not optimum. Suicide at an early age, for example, does not enhance biologic fitness.

The vowel [i] is less susceptible to misidentification because it yields an optimal reference signal from which a human listener can determine the length of the supralaryngeal vocal tract of a speaker's voice (Nearey 1979; Fitch 1997). The vowel [i], the other quantal vowels, and consonants have properties that make them superior acoustic signals for speech communication (Stevens 1972; Beckman et al. 1995). Speech communication clearly would be possible if human beings were unable to produce the vowel [i]; listeners can estimate the length of a person's SVT from other sounds (Gerstman 1968; Rand 1971; May 1976; Strange et al. 1976; Verbrugge, Strange, and Shankweiler 1976) as well as by listening to a long stretch of speech (Ladefoged and Broadbent 1957). If the phonetic content of the message is clear, as is the case for the stylized openers for conversation, such as "hello," "hey," "hi," "ni hao," a listener can reverse-engineer and estimate the speaker's SVT length. However, it is clear that natural selection can act on small distinctions (Mayr 1982); [i] and the sounds that the human SVT adds to the primate phonetic repertoire apparently provide selective advantages for vocal communication. The branch point at level IV thus denotes the modifications of the supralaryngeal vocal tract that typify anatomically modern *Homo sapiens.* The cost has been a greater probability of choking to death and perhaps less chewing efficiency.

Tongues and Airways of the Human Snoutless Monsters

All of the anatomic structures—lungs, larynx, tongue, lips, and mouth—involved in generating human speech have other, more basic vegetative functions. We must breathe, drink, and eat. However, these structures have been modified in such manner that they do not optimally carry out these basic life-supporting acts. Charles Darwin in *On the Origin of Species,* for example, noted "the strange fact that every particle of food and drink that we swallow has to pass over the oriface of the trachea, with some risk of falling into the lungs" (1859, p. 191). Darwin recognized the fact that the human supralaryngeal airway differs from that of other primates and most mammals; it is an inferior swallowing machine. He may also have known that human infants at birth have an SVT similar to an ape's. The features that distinguish human SVTs from those of other primates were noted in studies published decades ago (for example, Negus 1949; Lieberman, Klatt, and Wilson 1969; Lieberman and Crelin 1971; Bosma 1975). Recent studies show that the process that yields the human SVT begins before birth. The shape of the human fetal supralaryngeal airway can be seen in Jeffrey and

Spoor (2002), and Jeffrey (in submission). Major changes continue until age six to eight years (Lieberman and McCarthy 1999; Lieberman, Ross, and Ravosa 2000; Lieberman et al. 2001; Vorperian et al. 2005).

Different biologic processes appear to have been involved in the evolution of the human SVT. The genetic bases for these processes are presently unknown and progress in this direction is necessary if we are to fully understand the evolution of human speech. One factor is the length of the oral cavity. Apes and most other primates and mammals have long snouts. Their long mouths support tongues that cannot produce the SVT shapes necessary for generating the formant frequency patterns of the full range of human speech sounds. Some change in the regulatory genes that govern the development of the skull occurred in the transition to *Homo erectus.* The ASPM gene (Zhang 2003; Evans et al. 2004), which may foster cranial expansion, perhaps played a part in this process. But another, later process played a role in shaping the human skull. The unique retraction of the human face may have played a role in the initial migration and reshaping of our tongues (McCarthy and Lieberman 2001). Mutations on regulatory genes most likely are responsible for this phase of facial retraction as well as for the full course of the human tongue migration down into the pharynx, which yielded the tongue shape, laryngeal position, and shared food-airway passage that Darwin recognized was not particularly well designed for the very necessary act of eating.

The schematic midsagittal section of the head of an adult chimpanzee in Figure 6.3 shows the typical nonhuman "standard-plan" supralaryngeal airway recognized by Negus (1949). The chimpanzee's supralaryngeal vocal tract (SVT) and food passages constitute a breathing and eating machine that is better adapted to these tasks than a human's. The chimpanzee's tongue is long and thin compared with that of an adult human being, and it is positioned almost entirely within the animal's long protruding mouth. The ape's larynx is positioned behind the tongue and is close to the roof of the nasopharynx, which leads into the nasal cavity. The roof of the

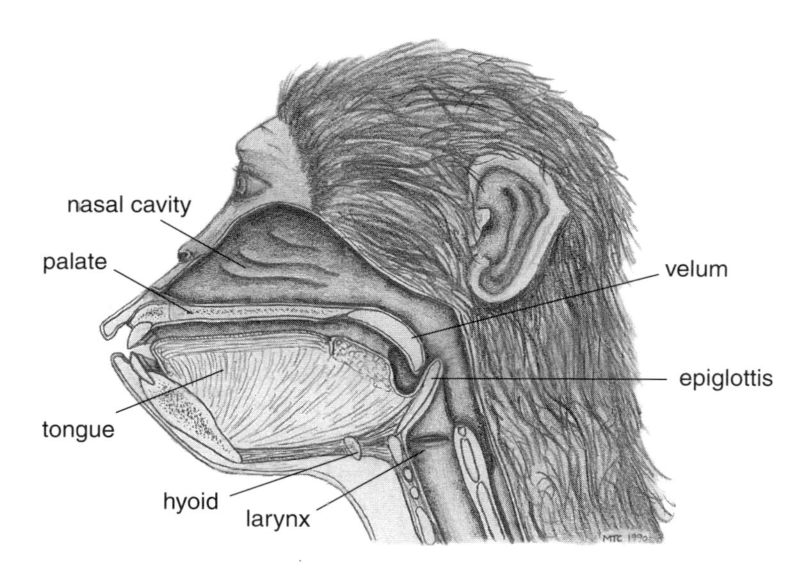

FIGURE 6.3 *Vocal tract of a chimpanzee. The tongue is positioned entirely within the oral cavity; and the larynx is positioned high, close to the opening to the nose. The epiglottis and velum overlap to form a watertight seal when the larynx is raised, locking into the nose during feeding. The hyoid bone is connected to the larynx, jawbone, and skull by means of muscles and ligaments; it is part of the anatomic system that can raise the larynx.*

nasopharynx consists of the bones that form part of the base of the skull. The hard palate, which is a bony structure that is often present in the remains of extinct hominid fossils, forms the anterior (front) part of the roof of the animal's mouth. The roof of the posterior (back) part of the mouth, often termed the oropharynx, is the soft palate, or velum. The velum can be pulled upward and backward by levator and tensor muscles, isolating the nasal airway from the rest of the SVT to produce nonnasal sounds (Bell-Berti 1971).

Human newborns retain most features of the nonhuman SVT. The functional "logic" for the morphology of the standard-plan supralaryngeal airway is apparent in Figure 6.4 when we consider the position of the larynx during respiration. In human newborns, the larynx moves upward through the oropharyngeal region of the mouth and locks into the nasopharynx leading into the nose during quiet respiration. The high position of the newborn larynx relative to the nasopharynx permits this maneuver. Note that the epiglottis and soft palate overlap and form a double seal. The newborn's larynx is small and can fit into the space between the foramen magnum, which is the base of the spinal column, and the end of the bony hard palate, which defines the oral cavity. The larynx, like a little periscope, pokes through the flow of mother's milk passing on either side.

The result is a sealed air path extending upward from the trachea into the nose, isolated from the flow of milk or other fluids. Newborn infants, therefore, can simultaneously swallow fluids while they breathe through their noses. Fluids enter their mouths, pass to either side of the elevated larynx, and enter the pharynx and esophagus positioned behind the larynx (Truby, Bosma, and Lind 1965). Liquids rarely fall into the larynx and trachea, so an infant does not choke as it suckles. The neural mechanisms that control respi-

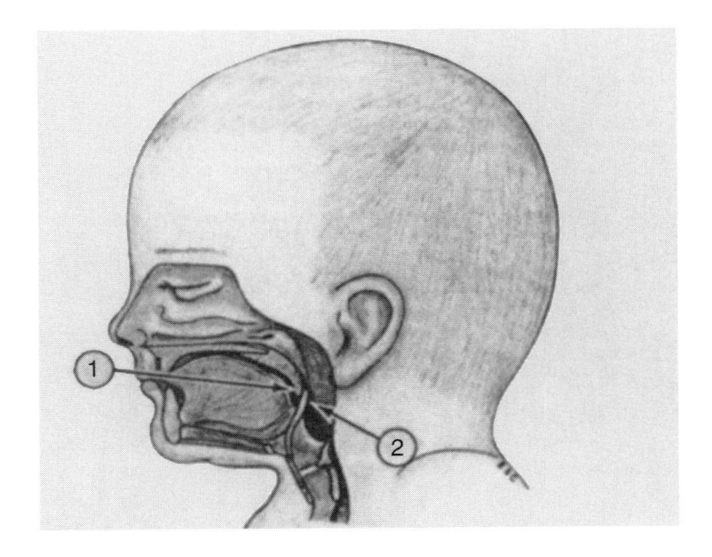

FIGURE 6.4 *Diagram of a human newborn showing the larynx locked into the nasopharynx during quiet respiration. The soft palate "1" and epiglottis "2" overlap, permitting suckling while the baby is breathing. (After Laitman, Crelin, and Conlogue 1977)*

ration appear to be matched to the standard-plan supralaryngeal morphology and human newborns are obligate nose breathers. As Crelin notes, "Obstructions of the nasal airway by any means produces an extremely stressful reaction and the infant will submit to breathing through the mouth only when the point of suffocation is reached" (Crelin 1973, p. 28).

Figure 6.5 shows the risky food and air pathway of an adult human being. We cannot simultaneously breathe and drink. As Darwin realized, compared to the upper airways of apes, the adult human supralaryngeal airway yields an increased risk of choking to death while swallowing. The pharynx in humans is a common food-and-air pathway (Negus 1949, pp. 176–177; Bosma 1975; Laitman

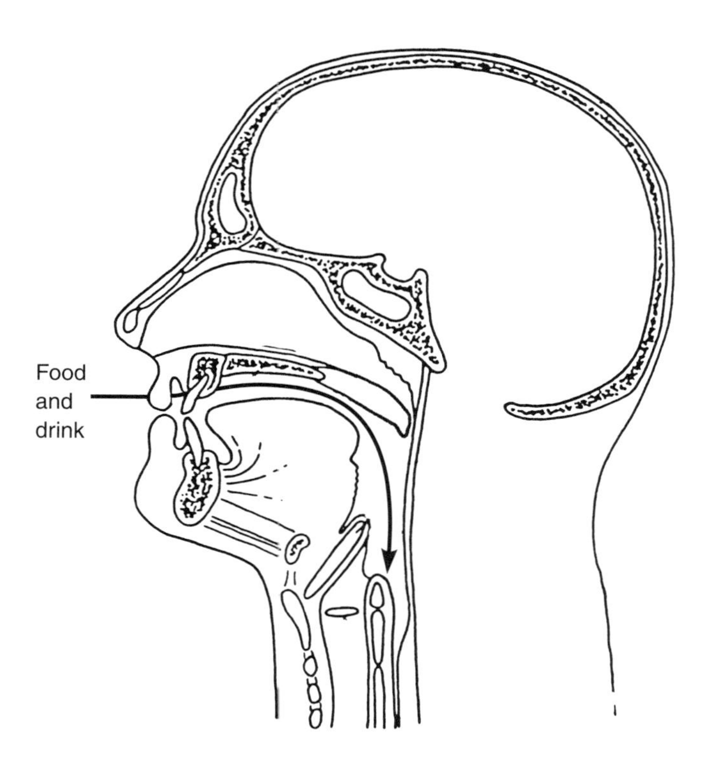

FIGURE 6.5 *Diagram of adult human being showing pathway for the ingestion of food. Solid food and liquids must pass over the opening of the larynx, increasing the risk of choking.*

and Crelin 1976; Palmer et al. 1992; Hiiemae et al. 2002). Successful swallowing requires precise coordination of the tongue, the hyoid-larynx complex, the epiglottis, and the esophageal sphincter (the pathway to the stomach), which is positioned behind the resting position of the larynx (Bosma 1992; Palmer et al. 1992; Hiiemae et al. 2002). As we swallow, we must execute a set of precisely timed and coordinated maneuvers. The tongue first thrusts food backward in the oral cavity, with the epiglottis folding backward. In order to avoid having food lodge and obstruct our larynx, we also must pull the larynx forward and upward, tucking it behind the pharyngeal section of the tongue, and opening the upper esophageal sphincter as we simultaneously push food toward the open esophagus. The powerful pharyngeal constrictor muscles then squeeze and propel food down the pharynx into the esophagus.

Unfortunately, part of the adult human pharynx is also part of the respiratory airway, owing to the ontogenetic descent of the tongue root carrying the larynx into the pharynx. An error in timing can propel a mass of food into the larynx, with results that sometimes are fatal. Jeffrey Palmer and his colleagues (1992) reaffirm Negus's observations. They note that, in contrast to nonhuman mammals and human newborns: "new findings suggest that normal humans are at risk for inadvertently inhaling food particles both before and after swallowing. Indeed, obstruction of the airway by inhaled food is a significant cause of morbidity and mortality in otherwise healthy individuals." Studies of swallowing show that tens of thousands of incidents of fatal choking have occurred (Feinberg and Ekberg 1990). In the absence of an autopsy, death resulting from a blocked larynx often is attributed to other causes. That was the case for a noted speech scientist whose death was at first attributed to a heart attack. An autopsy revealed a piece of steak blocking his larynx. These risks may account for the fact that newborn humans retain the standard-plan system in which they can isolate the pathway for swallowing from the breathing airway.

The nonhuman supralaryngeal airway is better adapted for swallowing in any position; chimpanzees habitually swallow food while they sit upright (Goodall 1986). Thus, claims that archaic hominids could not have had supralaryngeal airways with laryngeal positions close to those seen in chimpanzees because they would not have been able to swallow while they were upright (Falk 1975) can be refuted by a visit to a zoo. In fact, the pattern generator—in other words, the set of neural-activated motor commands that execute these complex maneuvers—appears to be similar in humans and other primates (Palmer et al. 1992) and is most likely an innate "reflex" regulated by the brainstem (Jean 1990). As is the case for breathing, there is little time available for a newborn infant to learn to swallow—the motor pattern generator is innate.

The adult human supralaryngeal airway has other life-supporting deficiencies compared to that of apes and, most likely, that of many of the extinct hominids. The nonhuman primate and human newborn larynx exits directly into the nose. Compared with the adult human supralaryngeal airway, there is less of a bend in the airway. As Negus (1949, p. 33) notes, the bend increases the resistance to airflow in humans. We most likely also are less efficient chewers compared to apes. Apes and these extinct hominids have long jawbones and mouths. Studies that were directed at perfecting dentures show that one important determinant of chewing efficiency is "swept tooth area," the moving surface area that comes into contact as the jaws move (Manley and Braley 1950; Manley and Shiere 1950). Longer jaws yield the possibility for a chewing machine with longer sweeps bringing more tooth surface area into contact. In Neanderthal hominids who also had long mouths, a long span exists behind the third molar (Howells 1976, 1989), whereas in humans our molars are positioned in a shorter upper jaw.

Chewing softer processed food has resulted in our using our jaw muscles to a lesser extent, which has reduced further the length of our jaws, crowding our teeth even more. From time to time, I survey the students in my classes concerning their molars; and I find that about 20 percent usually have had their molars removed before the age of 20. Crowded, impacted molars are susceptible to infection. Extracting an impacted, infected molar was virtually impossible before the introduction of anesthesia in the early nineteenth century and most likely would have resulted in death. Along with the lack of sanitation, this condition may be a key reason why early death occurred before the era of anesthesia. The demographic records of pre-Revolutionary France show that few adults survived past age 45 (Darnton 1985). As Darnton also notes, the French peasants ate soft mush—boiled cereal grains—for virtually all of their meals. That detail argues against the Industrial Revolution and processed food being branded as the cause of shorter jaws and impacted teeth.

Mouths, Tongues, and the Human Supralaryngeal Vocal Tract

To Negus, we owe the insight that shorter mouths are linked to human speech capabilities. Negus oversimplified the evolutionary process because he lacked data on the ontogenetic development of the supralaryngeal vocal tract in children. Negus suggested that the human supralaryngeal vocal tract evolved to enhance speech communication through the reduction of the oral cavity and the concomitant movement of the tongue down into the pharynx. Figure 6.6 shows an adult human supralaryngeal vocal tract in greater detail. Negus presciently focused on the human tongue and the proportions of the oral cavity and pharynx. After examining the fossil hominids that were available during the epoch in which he worked and the tongues and airways of humans and many other living species, Negus concluded that:

> As the oral cavity in Man has diminished in size, so the pharynx has become bigger. The large size of the tongue may be a relic from ancestors with a prognathous jaw, but there would seem to be no good reason why it should remain large if its main function were the propulsion of food. It would rather appear that its value for purposes of speech may have led to its retention in an exaggerated size. The alteration in relative size between the pharynx and the oral cavity is to the vocal advantage of Man, as it is possible for double resonance to be effected. This is of considerable use for variations in quality of sound. (1949, p. 198)

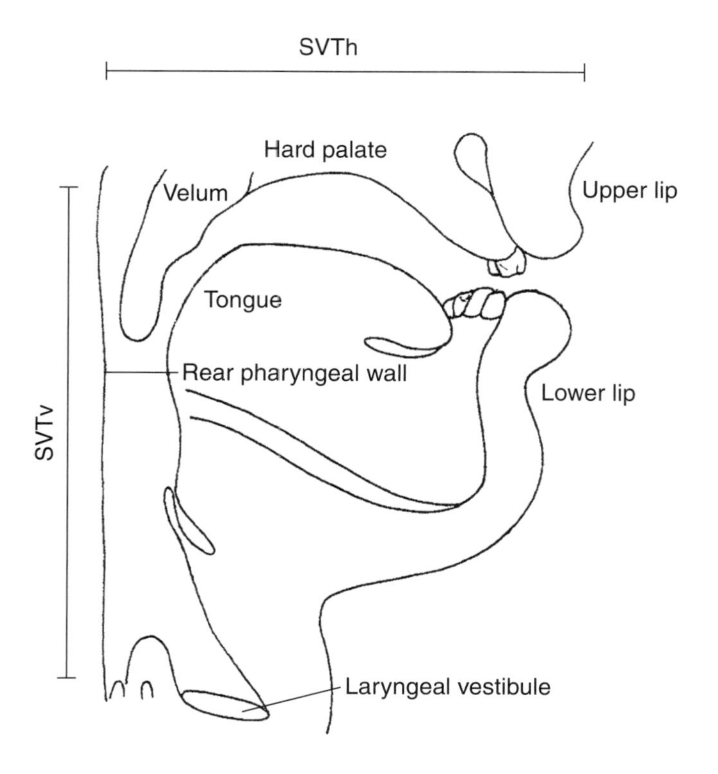

FIGURE 6.6 *Midsagittal view of the adult human supralaryngeal vocal tract. Note the 1:1 SVTh to SVTv proportions that follow from the shape and position of the human tongue.*

The key that to Negus appeared to unlock the evolutionary mystery was that human beings do not have long mouths. Negus failed to differentiate between prognathism—the forward projection of the lower face relative to the upper face that yields the snouts of apes—and the retraction of the entire face. However, he did recognize the linkage between shortness of the face and jaw and tongue restructuring. According to Negus:

> There is a gradual descent [of the larynx] through the embryo, foetus and child . . . The determining factor in man is recession of the jaws; there is no prognathous snout . . . The tongue however retains the size it had in Apes and more primitive types of

> Man, and in consequence it is curved, occupying a position partly in the mouth and partly in the pharynx. As the larynx is closely approximated to its hinder end, there is of necessity descent in the neck; briefly stated, the tongue has pushed the larynx to a low position, opposite the fourth, fifth and sixth cervical vertebrae. (Negus, 1949, pp. 25–26)

Recent studies of human and primate development confirm Negus's general insight: human beings and nonhuman primates follow two different paths after birth. In humans the tongue descends into the pharynx, taking the larynx down with it. However, it has become clear that the process entails more than the recession of the jaws. The rotation of the facial block is one factor implicated in the restructuring of the human face and the human SVT. Figure 6.7 illustrates this process. The anterior cranial base (ACB) angle and the line marked "PM" are always nearly at a 90° angle owing to the basic structure of the skull. The facial block, or the bones that support the face, rotate backward in humans during the first two years of life, thereby reducing the nasopharynx length; in contrast, such a rotation does not occur in apes and archaic hominids such as Neanderthals. This process does not appear to be linked to the expansion of the human brain (Lieberman and McCarthy 1999; Lieberman, Ross, and Ravosa 2000). As Daniel Lieberman and his colleagues note, "The human cranial base *flexes rapidly* after birth, almost entirely prior to 2 years of age, and well before the brain has ceased to expand appreciably . . . In contrast, the nonhuman primate cranial base *extends gradually* after birth throughout the neural and facial growth periods, culminating in an accelerated phase during the adolescent growth spurt" (Lieberman, Ross, and Ravosa 2000, p. 130).

Figure 6.7 illustrates the rotation of the facial block that restructures the human face. As a result, the majority of the human face lies beneath the braincase and we have less projecting brow ridges than archaic hominids. The rotation also reduces the space between the back of the hard palate and the foramen magnum (Mc-

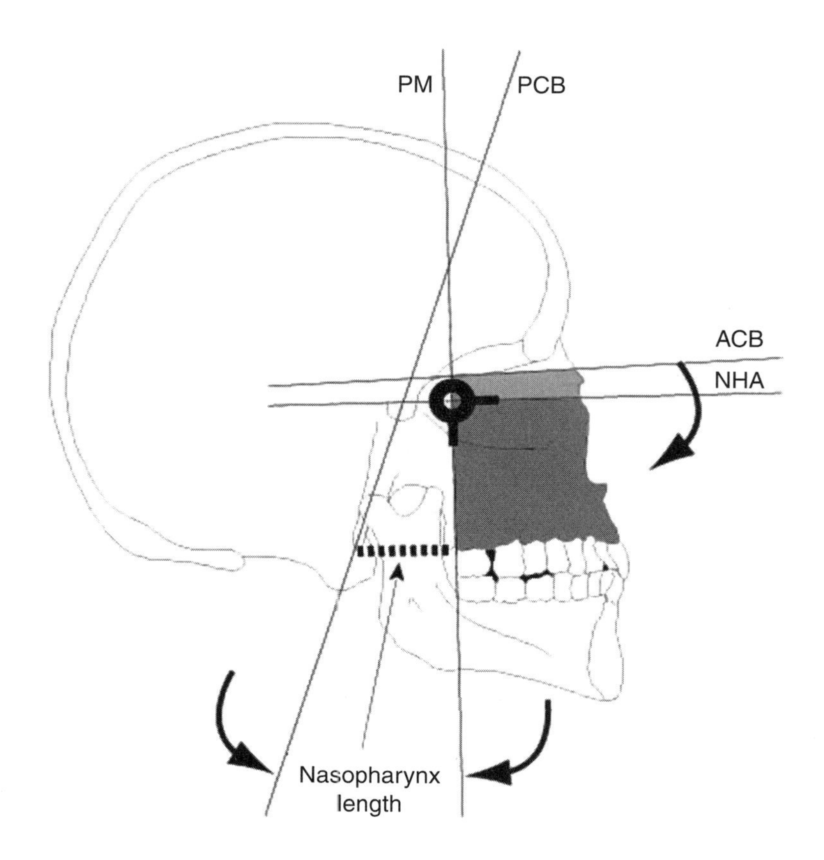

FIGURE 6.7 *In humans the "facial block" (the bones that support the face) rotates backward during the first two years of life around the axis noted in this sketch. This shortens the length of the nasopharynx, which forms part of the horizontal SVTh component of the vocal tract (adapted from Lieberman, Ross, and Ravosa 2000).*

Carthy and D. Lieberman 2001). The relevance of facial block rotation to human speech is that it shortens the space into which the larynx must fit in order to lock into the nose in the standard-plan SVT (Lieberman and Crelin 1971; Lieberman et al. 2001; Lieberman, Ross, and Ravosa 2000; McCarthy and Lieberman 2001). There is insufficient room for the adult human larynx to fit into this space, the oropharynx. The total length of the mouth is reduced and the tongue begins to descend down into the pharynx, carrying the larynx down to form the human SVT.

The Proportions of the Human SVT

However, other evidence precludes linking the form of the human upper airway solely to reduction of mouth length. The two-tube human SVT can be functionally divided into two sections: a horizontal section (SVTh) defined by the oral and oropharyngeal length, and a vertical SVTv section defined by the vertical pharyngeal length. The human tongue delimits both the horizontal oral (SVTh) and vertical pharyngeal (SVTv) sections of the human SVT. Daniel Lieberman and Robert McCarthy (1999) studied the ontogenetic development of the human SVT using a series of cephalometric radiographs (X-rays usually used by orthodontists) of individuals taken at intervals, almost from birth to adulthood. Analysis of these radiographs (the Denver series; Maresh and Washburn 1938; Maresh 1948; McCammon 1970) showed that between birth and age three years, the proportions of the SVT change from an SVTh/SVTv ratio of 1.5/1 to a ratio of 1.25/1. The restructuring of the basicranium and proportionate reduction of the mouth are essentially complete by this age. Yet as children mature, the proportion of the tongue body in the vertical pharyngeal section continues to increase until the fully quantal adult SVTh/SVTv 1:1 ratio is attained between ages six and eight years.

This ratio is subsequently retained as the skull and face attain their adult proportions (Lieberman et al. 2001). The independent studies of Fitch and Giedd (1999) and Vorperian et al. (2005), using MRI imaging of children, replicate these findings. In contrast, as apes mature, their faces gradually project forward, preserving the standard-plan SVT in which the tongue rests almost entirely within the oral cavity and oropharynx, which together constitute the mouth.

Factors Relevant to the Evolution of Human Speech Anatomy

For many years, a misconception about the sounds human infants produce led to the incorrect conclusion that the cranial angle could, in itself, serve to predict the shape of a fossil's SVT and its speech capabilities. The error derived from the difference between what we "hear" infants producing (our auditory perception) and the actual formant frequencies of the vowels that they produce. Early studies that relied on a linguist's ear noted that infants did not at first produce vowels such as [i], or the vowels [u] and [a], of the words "boo" and "ma." However, by the end of the second year of life, parents and linguists generally are convinced that they can hear infants producing these sounds (Irwin 1948). George (1976, 1978), who studied the same Denver series of radiographs I already mentioned, noted that by age two, the cranial base angle's flexure had reached its adult value. Because, as we shall see, independent studies had shown that an adult-like human SVT is necessary to produce these sounds (Stevens and House 1955; Lieberman, Klatt, and Wilson 1969; Lieberman and Crelin 1971; Lieberman, Crelin, and Klatt 1972; Stevens 1972; Carre, Lindblom, and MacNeilage 1995), the logical conclusion was that the descent of the tongue root and larynx in children resulted in adult-like SVT proportions. George thus concluded that the cranial base angle was a good predictor of SVT shape.[1]

The Perceptual "Magnet"

However, acoustic analyses show that, despite what our ears "tell" us, young children do not produce the formant frequency patterns that specify these vowels. Buhr (1980) derived the formant frequency patterns of the vowels produced by children in the first years of life; they do not conform to those of adult speech. For example, the formant frequencies of a 64-week-old infant's vowels "heard" as [i]'s (the vowel of "see") were actually [I]'s (the vowel of "bit"). Patricia Kuhl and her colleagues in 1992 "solved" the mystery. As we listen to speech, a perceptual magnet pulls an ill-formed formant frequency pattern toward the ideal exemplar of the language that a person is exposed to in the early months of life. In effect, our speech perception system cleans up sloppy signals.

Other studies, influenced by George (1976, 1978), attempted to use the cranial base angle supplemented by an estimate of oral cavity length to determine the tongue shapes and SVTs of extinct fossil hominids (Laitman, Heimbuch, and Crelin 1978, 1979; Laitman and Heimbuch 1982). I will return to the question of the speech capabilities of extinct hominids in light of this reappraisal of George's (1976, 1978) conclusions, which reflected both equipment limitations and the conflation of vowel perception with formant frequency patterns—an oversight shared by myself as well as other concerned parties.

Tongue Descent until Age Six to Eight Years

Unfortunately, confusion still persists (for example, Boe, Maeda, and Heim 1999; Boe et al. 2002) concerning what a vowel-like formant frequency pattern can sound like and its actual physical attributes; incorrect inferences based on George's (1976, 1978) studies persist. The initial stage of tongue displacement into the pharynx and the human flexed cranial base angle, noted in many independent studies (for example, Laitman and Crelin 1976; Lieberman et al. 2001; Nishimura et al. 2003) may at first be linked to facial retraction. However, after the cranial base angle stabilizes between ages two and three, further descent of the tongue occurs and the proportion of the human tongue (SVTv) that is in the pharynx continues to increase relative to that in the oral cavity (SVTh) until age six to eight. That proportion is then maintained as the face and jaw attain their adult sizes at about sixteen to eighteen years (Lieberman et al. 2001). There appears to be a human Bauplan, a genetic blueprint particular to humans that results in the human tongue continuing to descend into the pharynx until a 1:1 SVTh to SVTv proportion is reached. That proportion is then maintained.

The genetic mechanisms regulating these development processes are presently unknown. In a meaningful sense, we humans are monsters (Alberch 1989): some knockout gene or genes may have disrupted the normal primate developmental pattern, which otherwise would have resulted in our developing snouts. Other genetic regulatory distinctions apparently differentiate us from *Homo erectus* and Neanderthals, in whom lesser degrees of facial retraction occurred (Lieberman, McBratney, and Krovitz 2002).

Larynx versus Tongue Descent

Studies of fetal development in humans and chimpanzees have provided further insights into the significant differences between the speech-producing anatomy of apes and humans. Figure 6.8 sketches some of the relevant aspects of the basicranium, hyoid bone, and larynx of an adult human. Figure 6.9 shows an MRI image of a hu-

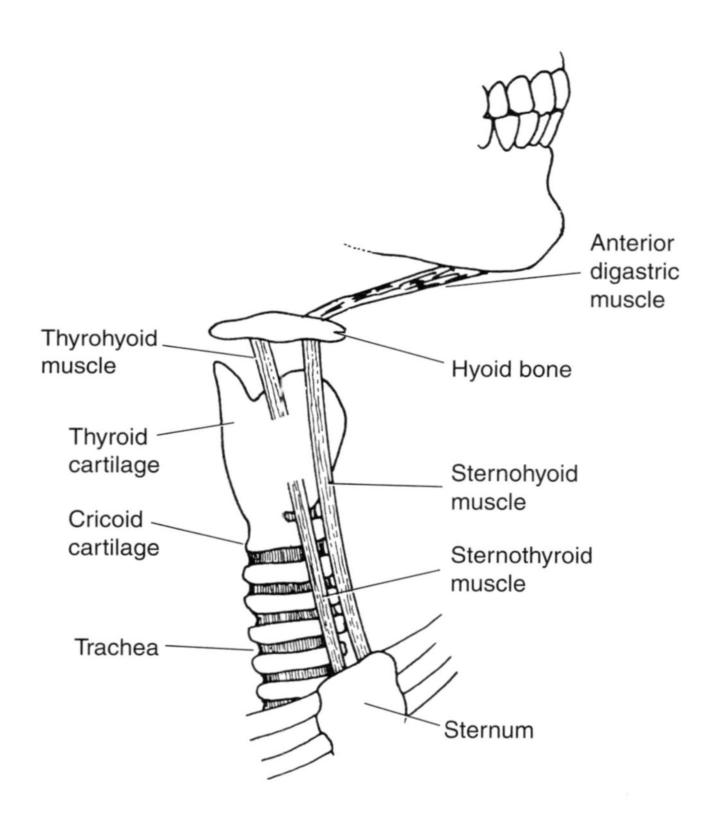

FIGURE 6.8 *The larynx in relation to the base of the mandible, hyoid bone, and sternum. The thyroid and cricoid cartilages constitute the larynx. The larynx is suspended by ligaments and muscles from the hyoid bone. In the animals studied by W. T. Fitch and his colleagues, the larynx descends either transiently during phonation or in some instances permanently because the distance between the hyoid bone and the larynx increases. However, the animals' tongues remain positioned in their mouths. In contrast, in humans the tongue descends into the pharynx, changing its shape and taking the larynx down with it.*

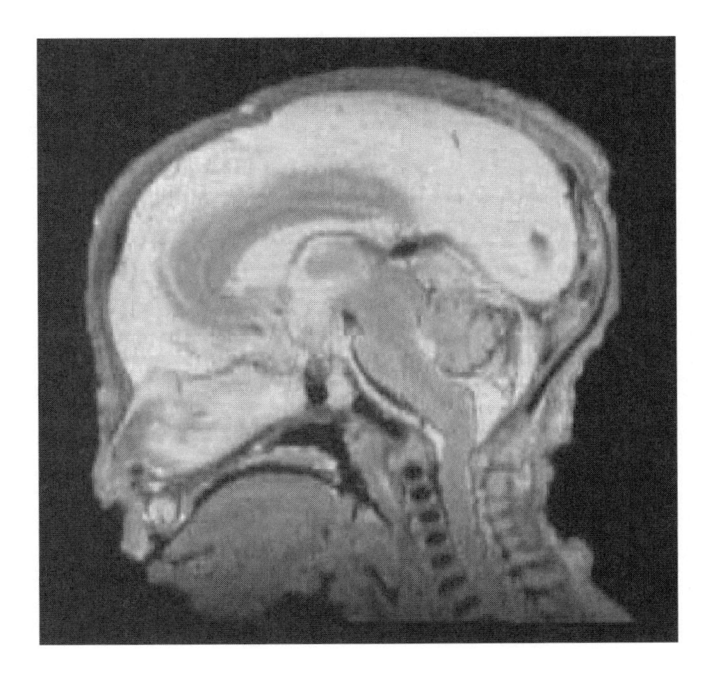

FIGURE 6.9 *An MRI midsaggital view of a human fetus at four months. The descent and reshaping of the unique human tongue starts before birth at approximately four months. (Image courtesy of N. Jeffrey)*

man fetus at approximately four months. In midterm fetal development, the basicranium (the bones that form the base of the skull) flatten, particularly at regions anterior to the spinal column (Jeffrey and Spoor 2002). This occurs while the hyoid bone descends from the basicranium, taking the tongue root down. Therefore, the larynx, which is suspended from the hyoid by cartilages and muscles, also descends (Jeffrey, in submission). The net change is an increase in the area of the oropharynx. This prenatal basicranial flattening may reflect a general primate target shared by humans and other pongids, allowing the maintenance of the airway after birth so human newborns and apes can simultaneously breathe and suckle. However, after birth the human hyoid bone continues to descend relative to the basicranium, taking the tongue root and the entire larynx down (Nishimura et al. 2003).

A similar initial phase of SVT development occurs in chimpanzees after birth, but human and chimpanzee development then diverge. Nishimura and his colleagues (2003) note that in chimpanzees further postnatal laryngeal descent occurs as the distance between the larynx and hyoid bone increases. However there is a critical difference: the tongue remains *anchored* in the chimpanzee mouth. The descent of the base of the tongue occurs only in humans. As Nishimura et al. (2003) note:

> The descent of the human larynx is achieved not only by the descent of the laryngeal skeleton relative to the hyoid bone, but also by the descent of the hyoid relative to the mandible and cranial base, even in infancy. Thus, the laryngeal skeleton descends relative to the hyoid in both chimpanzees and humans, and this descent contributes to the descent of the larynx during early infancy. By contrast, hyoid descent per se contributes to the descent of the larynx only in humans and not in chimpanzees. (Nishimura et al. 2003, p. 6932)

In other words, the distance between the hyoid bone and the cartilages that form the frame or skeleton of the larynx increases in chimpanzees, but the tongue does *not* move down into the pharynx. The lower larynx yields a longer SVT, lowering formant frequencies. However, as we shall see, a lower larynx in itself does not allow the SVT to form the abrupt and extreme changes in the cross-sectional area that are necessary to produce vowels like [i].

Larynx Descent and Signaling Size: How Big Am I?

The larynx descends in many other species. Some also can transiently increase the distance between the hyoid and larynx. Fitch and Reby (2001) show that the position of the larynx in two species of deer is quite low, and they can transiently lower their larynges still lower. However, the lower resting position as well as the momentary larynx lowering is achieved by increasing the distance between the vocal cords or folds of the larynx and the hyoid bone. The deers' tongues remain in their mouths. The thyrohyoid ligament, which connects to the thyroid cartilage of the larynx, is elastic, and the muscles that pull the larynx downward are powerfully developed in male deer, whose calls during rut involve laryngeal lowering. A similar mechanism results in lesser transient laryngeal lowering when dogs vocalize, with a slight displacement of the tongue body (Fitch 2000a). Other species have low laryngeal positions, but their tongues also are anchored in their mouths. Lions, for example, have a low laryngeal position with a long, thin tongue body that is anchored in their mouths. The lion larynx is connected to the main body of the tongue positioned in the mouth by a thin segment; the lion's tongue shape does not approach that of an adult human tongue (Weisengrubber et al. 2002). The thin tongues of lions and other big cats are incapable of forming the SVT midpoint area function discontinuities that are necessary to produce the vowels [a], [i], and [u]. They inherently cannot produce the full range of human speech sounds.

Larynx lowering in these animals seems to be the result of natural selection for auditory size enhancement. Laryngeal descent, absent tongue displacement, lowers formant frequencies but does not increase phonetic range. Fitch and his colleagues have shown that formant frequencies generally provide an "honest" indication of an animal's size (see Chapter 3). The length of the SVT correlates with height or body size in many different species (Fitch 1993, 1997, 2000b, 2000c; Fitch and Kelley 2000). Hence, a lower larynx that lengthens an animal's SVT results in lower formant frequencies, a "false" signal enhancing an individual's perceived size. Deer apparently associate lower formants with greater size and biologic fitness, which perhaps explains the lower larynges of male deer and the prevalence of calls with transiently lowered larynges during the mating season.

Similar sexual dimorphism marks human males. The distance between the hyoid bone and laryngeal skeleton increases slightly during puberty in humans (Fitch and Giedd 1999) and males generally have lower formant frequencies than females. However, the added vocal tract length does not enhance the phonetic capabilities of human males. The longer male laryngeal segment adds a tube that cannot change its shape. The range of speech sounds that devolve from SVT area function variation that men can produce is consequently no greater than that of women. The acoustics are documented in such studies as Peterson and Barney (1952) and Hillenbrand et al. (1995). Men generally have lower formant frequency patterns, but their phonetic range is no larger than that of women, if that factor is taken into account (Nearey 1979). In short, as Victor Negus first noted, the position and shape of the tongue are the key issues; the indirect significance of the low human larynx derives from its link to the descent of the tongue into the pharynx. Claims such as made by Fitch (2000b) that larynx lowering is the key factor in the evolution of speech anatomy cannot account for the evolution of the species-specific human SVT, which involves the descent of the tongue.

In another respect, humans differ vocally from many other species in conveying their size. Garments traditionally differentiate men from women; in some societies speech plays a similar role and we "improve" on nature. Some women are taller than many men and also have longer SVTs. Many women also have larger larynges with longer vocal cords than some men. Vocal cord length plays a major role in determining a person's average fundamental frequency of phonation, but acoustic analyses show that the average F0s and perceived pitch (which reflects F0) of the voices of men and women overlap (Peterson and Barney 1952; Hillenbrand et al. 1995).

Given the overlap in male versus female stature, formant frequency overlap might be expected. However, at Haskins Laboratories in the late 1960s, the late Ignatius Mattingly examined the database of formant frequency values of the Peterson and Barney (1952) study. Mattingly found almost no overlap between male and female vowels. Men whose voices had high pitches nonetheless sounded male. However, despite the overlap in stature and SVT length, male formant frequencies were consistently lower and appeared to be the primary acoustic cue signaling gender. Mattingly reasoned that adult males most likely lengthened their SVTs by protruding their lips slightly; they may have also lowered their larynges, effectively producing "male-speak" vowels that had lower formant frequencies. Women apparently employed different tactics, smiling slightly as they talked to shorten the effective length of their SVTs. Mattingly concluded that American English has subtle male and female dialects.

A subsequent study revealed that children growing up in Connecticut about fifty miles from Haskins Laboratories produced gender-specific speech at an early age. No one tutored the children; some unconscious process of imitation and "statistical" inference resulted in their modeling their vocal behavior on these prevailing male-female patterns. At age five years, boys and girls as a group do not differ in height or weight. Nor are there systematic differences in their larynges that would result in lower voice pitch for boys. However, the boys' voices were consistently recognized as belonging to boys because they lowered the formant frequencies of their vowels by protruding and rounding their lips (Sachs, Lieberman, and Erikson 1972). These articulatory maneuvers will lower formants (Chiba and Kajiyama 1941; Fant 1960). The girls, except for some who were considered tomboys, smiled and retracted their lips slightly, thereby producing higher formant frequencies. The resulting slight differences in formant frequencies conveying subtle differences in vowel quality constituted gender-specific male and female dialects.

Evolutionary Significance of Lowering the Larynx

The fact that the formant frequency dispersion, a metric developed by Fitch, shows a high degree of correlation between body size and formants in many nonhuman species demonstrates that a lower laryngeal position, in itself, does not yield increased phonetic capabilities. The formant frequency dispersion metric works only because these creatures produce the same "schwa" vowel, which is produced by an SVT that is a uniform or slightly flared tube. The relationship that holds between the formant frequencies of these SVTs is fixed; the frequency of the second formant frequency, F2 = 3 times F1, while the frequency of the third formant frequency, F3 = 5 times F1. Independent acoustic studies show these fixed formant frequency relations are generally the case for animal vocalizations, including those studied by Fitch and his colleagues (for example, Lieberman 1968; Rendall et al. in press). But the absolute values of the formant frequencies depend on the length of the SVT. Therefore, if we calculate formant frequency dispersion, F3–F1, the formant frequencies of an animal with an SVT that is twice as long as that of another animal is halved, yielding half the dispersion. But this metric only works if the *same* vowel is being compared. The distinct vowels of human speech have different relative formant frequency patterns. For example, Peterson and Barney (1952) found that the mean values for F1 and F3 of [i] for male speakers of American English are 270 and 3010 Hz, yielding a dispersion of 2740 Hz. In contrast, the mean male formant frequencies for F1 and F3 of [a] are 730 and 2440 Hz, yielding a dispersion of 1710 Hz. A body size estimate that relied on formant frequency dispersion for these two vowels would lead to different size estimates for the same person. The fact that formant frequency dispersion tracks body size in nonhuman primates and other species (Fitch 2000c) demonstrates that these creatures always produce the same vowel, produced by supralaryngeal vocal tract shapes approximating a uniform to slightly flared tube. In short, a low larynx does not signify an SVT that can produce the full range of human speech.

However, this does not negate or diminish the significance of the comparative studies of Fitch and his colleagues. Their findings establish a preadaptive basis for formant frequency variations playing a role in human speech. Coupled with the fact that formant frequencies, patterns used for communication, can be produced or perceived by a wide variety of mammals (Warden and Warner 1928; Baru 1975; Owren and Bernacki 1988; Owren 1990; Fitch 1997) and birds (Heinz, Sachs, and Sinnott 1981; Dooling 1992), their data demonstrate, beyond reasonable doubt, that many species vocally communicate critical information, thereby enhancing their biologic fitness using formant frequency patterns. Moreover, monkeys (Riede at al. 2005) and chimpanzees (Slocombe and Zuberbuhler 2005b) make use of formant frequency patterns as well as F0 patterns to convey referential information. These comparative studies established a critical link between human speech, which relies on formant frequency encoding, and the vocal communications of other species. And it may explain why we have peculiar tongues. Vocal communication that relied on formant frequency patterns would have provided an adaptive value for anatomy that enhanced the process of vocal tract normalization through the production of the supervowel [i].

Quantal Speech Sounds and the Human SVT

The selective advantages of the adult human supralaryngeal vocal tract appear to follow from the fact that it allows us to produce abrupt midpoint area function discontinuities. These extreme midpoint area function discontinuities can only be produced by a tongue that has a round posterior cross-section positioned in an SVT that has a 90° bend. Half of the tongue is positioned in the oral cavity, half in the pharynx. This SVT configuration is necessary to produce the quantal vowels [i], [u], and [a] (Stevens and House 1955; Stevens 1972; Carre, Lindblom, and MacNeilage 1995). The term "quantal" needs clarification. The term was coined by Stevens (1972) to characterize speech sounds that have two useful properties. Quantal sounds have perceptually salient acoustic properties that can be produced with a certain degree of sloppiness on the part of a speaker. Speech communication would be perfectly possible without quantal sounds, but they enhance the robustness of the communicative process.

The area function of the SVT determines the formant frequencies that play a major role in specifying the sounds of speech (see Chapter 3). Each different sound involves maneuvering the tongue, lips, and velum. The task of speech production would be simplified if it were possible to produce an invariant acoustic signal without having to use precise articulatory maneuvers. The task of speech perception also would be made simpler if the acoustic signals that were used for vocal communication were maximally distinct. These criteria are captured by Kenneth Stevens' (1972) "quantal factor." The quantal factor can be illustrated by means of the following analogy. Suppose that an elegant restaurant is preparing to open. The owner decides to employ waiters who will signal the diners' order by means of nonvocal acoustic signals. Shall he employ waiters equipped with violins or with sets of hand bells? If he wants to minimize the chance of errors in communication, he will opt for the hand bells. The hand bells would each produce a distinct acoustic signal without the waiter having to use precise manual gestures. In contrast, violins would require precise maneuvers from the waiters and would produce graded acoustic signals. The bells produce quantal signals, ones that yield distinct acoustic signals by relatively imprecise gestures.

Stevens demonstrated that the quantal vowels [i], [u], and [a] have perceptually salient acoustic correlates that can be produced while the need for precise motor control is minimized. Their perceptual salience results from the convergence of two formant frequencies. For [i] the second and third formants F2 and F3 converge at a high frequency; [a] has a low F2 and a high F1 converging at the midpoint of the frequency spectrum; while [u] has a low F1 and F2 converging at a low frequency. The acoustic properties of the supralaryngeal vocal tract yield spectral peaks when two formant frequencies converge (Fant 1956). A visual analogy will illustrate the communicative value of these spectral peaks. The effect of using quantal vowels is similar to using flags with brilliant saturated colors to communicate. Using flags of subtle pastels would be similar to using other vowels. Stevens, employing acoustical measurements of tubes with the area functions that would generate these vowels as well as computer-implemented vocal modeling, demonstrated that these distinctive formant frequency patterns can be generated by means of relatively imprecise articulatory gestures.

Figure 6.10 presents a stylized model of the cross-sectional area of the SVT for the vowel [a]. The discussion that follows para-

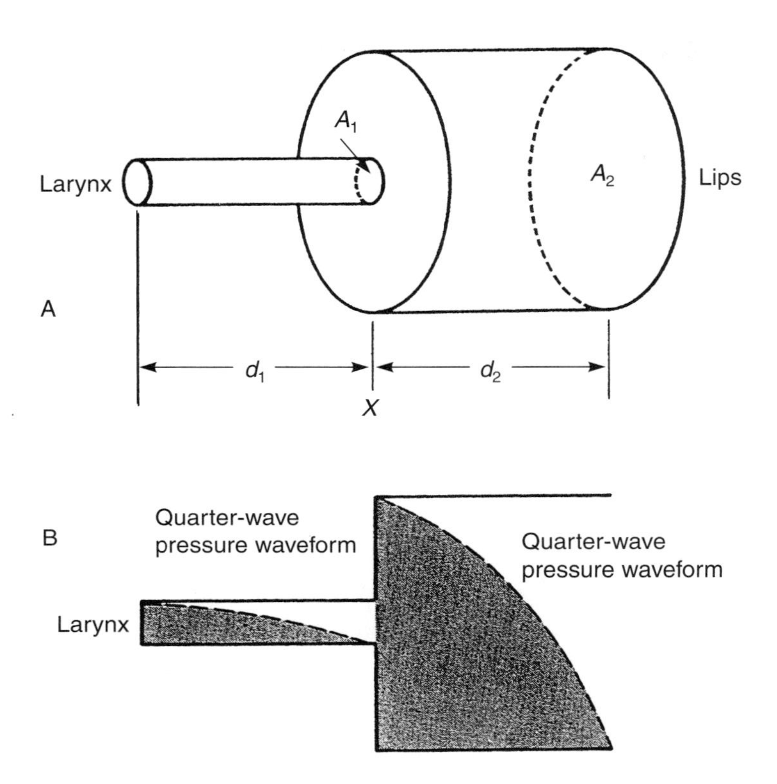

FIGURE 6.10 *A stylized "two-tube" model of the SVT shape for the quantal vowel [a]. The ten-to-one midpoint area function discontinuity results in F1 and F2 converging. The lower sketch illustrates the first-order approximation for calculating F1 and F2. Because the two tubes have equal lengths, they would yield identical formant frequencies. In real life, coupling effects between the two tubes result in F1 and F2 not being equal but instead converging.*

phrases Stevens's 1972 analysis. Note that the shape of the supralaryngeal vocal tract for the vowel [a] approximates a two-tube resonator. The cross-sectional area, (A1) of the posterior portion of the SVT, which corresponds to the pharynx, is constricted. The cross-sectional area (A2) of the anterior, oral cavity is ten times as large as A1. To a first approximation, the first two formant frequencies can be calculated as simple quarter-wavelength resonances of these two tubes. The physical reasoning behind this approximation is not difficult to follow. At the closed end of the back tube, the air pressure that best matches the closed end is a pressure maximum. The air pressure that best matches the open end of the tube at point X is zero. This follows from the fact that the cross-sectional area A2 is ten times larger than A1. The opening of the unconstricted tube is ten times larger than that of the constricted tube. A 10:1 difference in cross-sectional area is enormous. The effect on air pressure can be visualized by imagining a crowd exiting a three-foot-wide passage and emerging into a thirty-foot-wide passage. In the narrow passage, the crowd would jostle against each other; but once they reached the larger passageway, they would spread out and never touch each other. If you think of the people as particles, pressure would build up as the people collide in the smaller passageway. The collision of the gas molecules that generated the air pressure waveform in the constricted tube is thus minimized at point X. The cross-sectional area pressure waveform in the unconstricted tube is also a quarter-wave pattern because the oral tube is nine-tenths closed at point X. The two pressure waveforms are sketched in Figure 6.10.

The quarter-wave resonance model is a first approximation of the behavior of the vocal tract for the vowel [a] (the vowel of the word "ma"). It does, however, point out the salient speech-producing physiologic feature of formant frequencies of the vowel [a]. The change in sectional area, point X, occurs at the midpoint of the SVT. F1 and F2, the first and second formant frequencies, therefore, are equal. If we perturb the position of point X from the midpoint, we would not expect these two formant frequencies to change very abruptly. For example, if we move point X 1 cm forward or backward, we could generate the same first and second formant frequencies. The front tube could be longer and would generate the lower resonance F1 if point X were moved 1 cm backward. If point X were instead moved 1 cm forward, the back tube would generate the lower first formant. The first formant frequency would be identical for these two situations. It is immaterial whether the front or the back cavity generates the first formant frequency; all that matters is that the same frequency is generated. The second formant frequency would also be similar in these two cases. It would be generated by the shorter tube. The first and second formant frequencies for the vowel [a] thus would not change very much so long as point X was perturbed about the midpoint of the SVT. An increase in the length of the front, oral cavity necessarily results in a decrease in the length of the back, pharyngeal cavity, and the two cavities "trade off" in generating the first and second formant frequencies.

The quarter-wave model for the vowel [a] is a first approximation because, in actuality, coupling occurs between the front and back tubes, thereby limiting the convergence of F1 and F2. In Figure 6.11

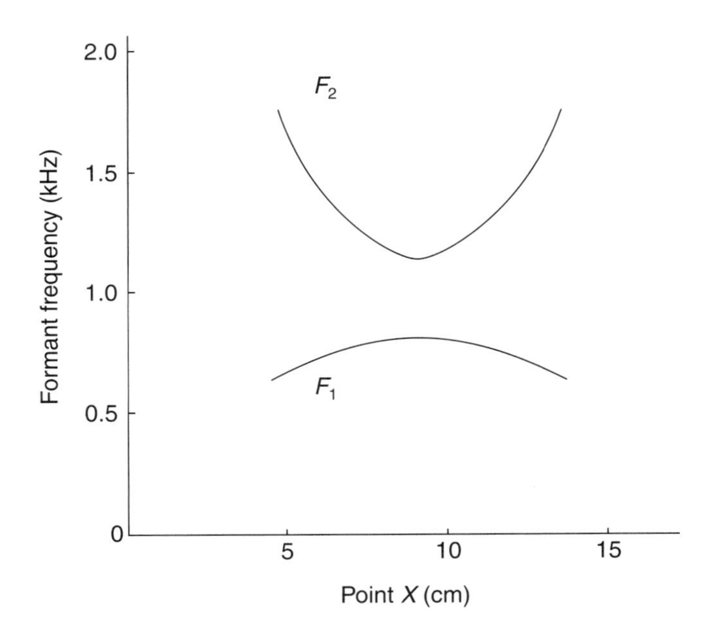

FIGURE 6.11 *The actual first formant frequency F1 and second formant frequency F2 for a two-tube model of the vowel [a] for different positions of the area function discontinuity (shorter or longer back and front tubes with a combined length equal to 16 cm). The formant frequencies were measured by Stevens (1972) for different positions of the area function discontinuity by changing the lengths of the tubes. Note the "quantal" formant frequency convergence and stability at the midpoint.*

calculated values for F1 and F2 are plotted for various positions of point X about the midpoint of a supralaryngeal vocal tract 17 cm long. These calculations were made using a computer-implemented model of the SVT (Henke 1966). The computer program calculates the formant frequencies of the supralaryngeal vocal tract for specified area functions. Note that the first and second formant frequencies converge for X = 8.5 cm, the midpoint of the SVT. There is a range of about 2 cm in the middle of the curve in which the second formant varies over only 50 Hz and the first formant changes even less. Within this region the two formants are close together. The transfer function for [a] thus has a major spectral peak, analogous to a saturated color. In contrast, for the 2 cm range from X = 11 to 13 cm, the second formant frequency changes by about 0.4 kHz and the central spectral peak would be absent. Similar effects occur for the vowels [i] where F2 and F3 converge because of the area function discontinuity at the midpoint of the SVT. Stevens (1972) verified these phenomena by first placing an acoustic energy source at the laryngeal end of tubes with these dimensions and recording the signals produced in a soundproof anechoic chamber (which is a room treated to eliminate sound reflections). Stevens then measured the formant frequencies that were produced as he changed the position of the area function discontinuity.

Figure 6.12 sketches midsagittal views of an adult SVT for the quantal vowels [i], [a], and [u], the corresponding area functions, and the resulting formant frequency patterns. Note the peaks in the frequency spectrum that follow from the convergence of two formant frequencies. The discontinuity at the approximate midpoint of the supralaryngeal airway allows speakers to be imprecise and still generate a vowel sound that has the quantal properties identified by Stevens (1972). Radiographic (Perkell and Nelson 1982) and microbeam X-ray studies that track tongue movements (Beckman et al. 1995) confirm Stevens' theory. Beckman and her colleagues showed that as Stevens claimed, speakers can be sloppy and still produce [i]'s that have stable formant frequencies. The exact position of the speaker's tongue with respect to the midpoint constric-

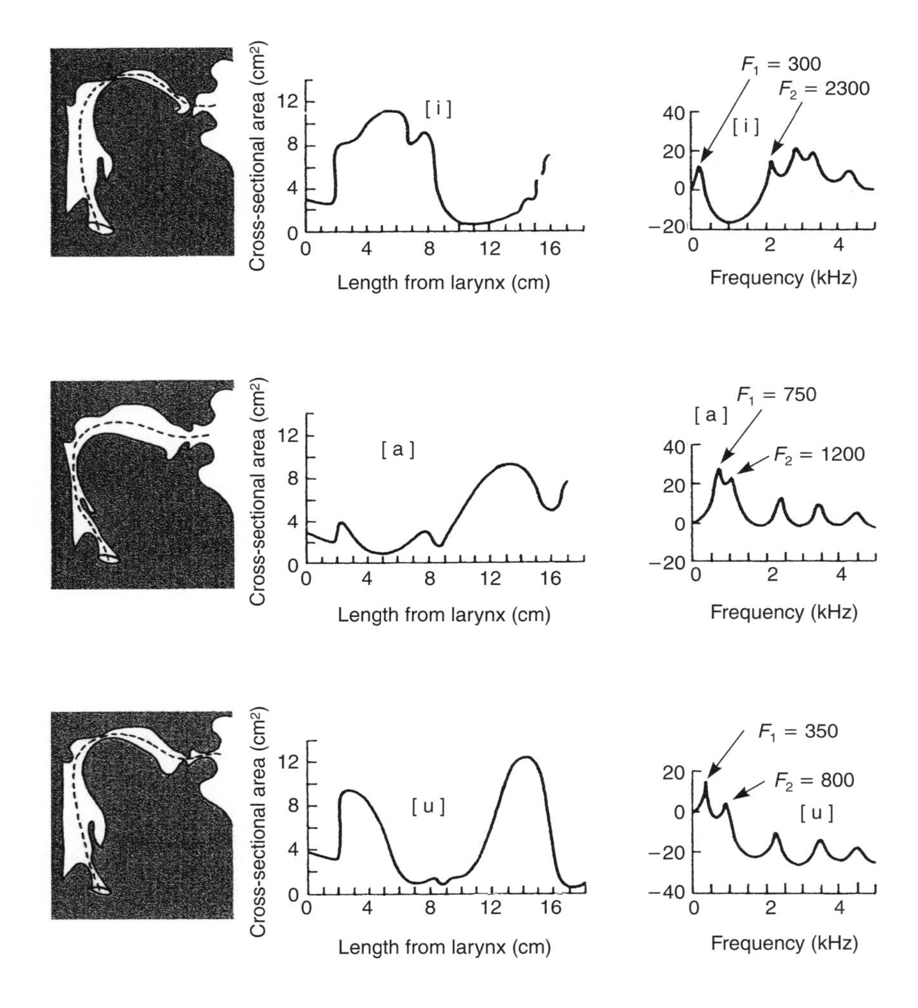

FIGURE 6.12 *Midsaggital SVT sections for the vowels [i], [a], and [u], the corresponding cross-sectional area functions plotted in terms of the distance from the larynx, and the resulting formant frequency functions.*

tion for [i] does not have to be precisely specified. The species-specific morphology of the human SVT makes it possible to produce the quantal [i] as well as [u] and [a]. Carre, Lindblom, and MacNeilage (1995), using a radically different procedure, independently reached similar conclusions. Their computer model of the SVT "grew" a pharynx that was equal in length to its oral cavity

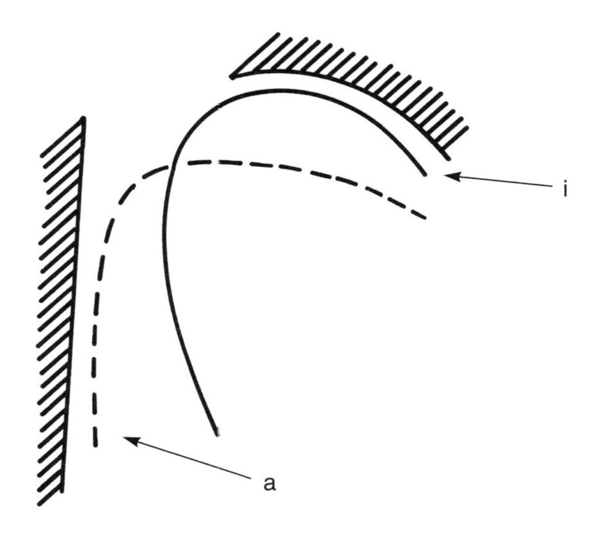

FIGURE 6.13 *Simple movements of the undistorted human tongue with respect to the hard palate and rear pharyngeal wall produce the SVT shapes necessary to produce the vowels [i] and [a]. Humans essentially have a "two-tube" SVT that has an inherent discontinuity built in at the juncture of SVTh and SVTv. The posterior semicircular tongue contours were documented by Nearey (1979).*

when directed at producing the full range of human vowels, delimited by [i], [u], and [a].

Figure 6.13 illustrates how these abrupt area function changes are produced by simply moving an undeformed human tongue about in the right angle space defined by the oral cavity and pharynx. Radiographic studies from the last seventy years (Russell 1928; Carmody 1937; Chiba and Kajiyama 1941; Perkell 1969; Ladefoged et al. 1972; Nearey 1979; Hiiemae et al. 2002) and MRI studies (Baer et al. 1991; Stone and Lundberg 1996; Story, Titze, and Hoffman 1996) show that the tongue body has a circular midsagittal posterior shape and is almost undeformed when we produce vowels. Producing an [i] involves the tongue's extrinsic muscles moving the tongue upward and forward. An [a] can be produced by simply moving the tongue back and down. Peter Ladefoged's research group at UCLA in the 1970s came to that conclusion through statistical analyses of tongue movements captured in cineradiographic studies. The computer model developed by Story and Titze (1998) employs these two "orthogonal modes," which reflect the forces exerted by the extrinsic muscles that move the tongue. The human tongue and those of virtually all mammals is a hydrostat (Stone and Lundberg 1996). Although muscular, it cannot be squeezed into a smaller volume by the extrinsic muscles of the tongue that propel it up or down, forward or backward, as we produce different vowels. The intrinsic muscles of the tongue are sometimes bunched up when speakers produce an [i] (Fujimura and Kakita 1979), but inspection of the tongue contours in the studies I have highlighted confirms Nearey's (1979) observation that this is not necessary. These quantitative studies consistently show that the human tongue is simply moved about as an almost undeformed body with appropriate lip maneuvers, sometimes raising the larynx (for an [i]) or lowering it (for a [u]).

The vowel [i] is an optimal signal for estimating the length of a speaker's SVT because it is produced at the limit of the SVT's capacity for producing area function discontinuities. The anatomic prerequisites that would enable the Knights of Ni to say [i] are:

- A tongue that has an almost round posterior contour, forming the lower surface of the oral cavity and the anterior surface of the pharynx. This provides an oral cavity and pharynx that have about the same length.
- A right-angle bend at the junction of the mouth and pharynx.
- Muscles that move the tongue within the space formed by the roof of the mouth and posterior pharyngeal wall.

These features appear in the course of human postnatal development as our faces restructure and our tongues migrate down into the pharynx.

Stevens (1972) demonstrated similar quantal effects for the formant frequency patterns of the consonants that occur most often in the world's languages (Greenberg 1963; Maddieson 1984). The

English stop consonants [b], [d], and [g] and their "long-lag" VOT cousins [p], [t], and [k] have quantal properties. Spectral peaks occur owing to the convergence of formant frequencies, and slight errors in tongue or lip placement do not dramatically alter the formant frequency pattern. These consonants again are acquired with greater facility by children than other speech sounds (Olmsted 1971). The conclusion that may be reached is that the mutations that restructured the human SVT were retained to enhance speech communication at the expense of vegetative deficiencies.

Easy Talking

In short, the unique shape of the human SVT allows us to simplify the production of the sounds of speech. The human SVT also is well adapted to enable us to compensate for extreme perturbations, such as would occur if you talked with a straw or tube in your mouth. Savariaux, Perrier, and Orliaguet (1995) used midsagittal X-ray images to study the vocal tract shapes of French speakers attempting to produce the lip-rounded vowel [u] (as in the word "boo") with a tube placed between their lips. This normally requires protruding and narrowing one's lips. Although the tube precluded lip-rounding, one speaker was able to approximate the vowel's first and second formant frequencies (F1 and F2) by an alternative SVT maneuver, moving his tongue farther back and closer to the roof of his mouth. Other speakers were less successful: some simply were unable to approximate the F1 F2 pattern that largely conveys this quantal vowel. However, all of these speakers' attempts to produce a [u] preserved the round shape of the tongue noted by Nearey (1979). They simply moved their tongues to an alternative position, taking advantage of the right-angle bend in the human SVT to simultaneously narrow the constriction formed by moving the tongue body closer to the roof of the mouth and increasing the size of the pharynx. Story (2004), using a computer model based on MRI studies (Story, Titze, and Hoffman 1996), confirmed that this compensatory maneuver can be achieved by the extrinsic muscles of the SVT moving the tongue body up or down or backward and forward. The salient aspects of the computer model used by Story are:

(1) it is based on actual SVT area functions determined by MRI imaging, and (2) the maneuvers of the tongue that it executes are constrained by the muscles and geometry of the human SVT (Story and Titze 1998). The X-ray study (Savariaux, Perrier, and Orliaguet 1995) and computer-modeling study based on MRI area data and physiologic constraints (Story 2004) complement each other. They confirm the findings for "normal" speech: the tongue moves about as an almost-undeformed body in the right-angle space of the SVT to produce the full range of vowels. Even under extreme perturbations, the adult human SVT can in some instances produce appropriate formant frequencies without changing the shape of the tongue.

Speech Capabilities of Nonhuman Tongues

Early acoustic analyses of the vocalizations of apes and monkeys suggested that these mammals were incapable of producing the full range of human speech sounds even though they clearly produced the formant transitions (Lieberman 1968) that later studies (Riede et al. 2005; Slocombe and Zuberbuhler 2005b) show convey referential information. However, it could have been the case that the range of vowels that the nonhuman primates were observed to produce reflected the limits of neural control, a possibility supported by observations of apes in their natural habitat (Goodall 1986). The inherent phonetic limitations of the supralaryngeal airways of nonhuman primates were first assessed by means of computer-modeling studies in which the supralaryngeal airways of rhesus monkeys (Lieberman, Klatt, and Wilson 1969), chimpanzees, and newborn human beings (Lieberman, Crelin, and Klatt 1972) were systematically perturbed through their possible range of area-function variation. The range of articulatory gestures for the newborn human SVT modeled was based on that evident in the Truby, Bosma, and Lind (1965) cineradiographic study of newborn infant crying and swallowing. These studies all show that the primate standard-plan supralaryngeal airway cannot produce the quantal vowels [i], [u], and [a].

This finding was not unexpected in the first of these studies, which modeled the range of area functions that the supralaryngeal airway of a rhesus macaque could produce (Lieberman, Klatt, and Wilson 1969). The unperturbed monkey SVT area function was established by introducing liquid silicone rubber into the monkey's upper airway immediately after its death (from another cause). The silicone rubber solidified, yielding the airway's "negative" space—its shape. The Stevens and House (1955) model indicated that a tongue with a round contour and a pharynx was necessary to produce the full range of human speech sounds that was confirmed by the 1969 computer modeling of the monkey airway. Independent computer-modeling studies subsequently replicated this finding: they show the need for an abrupt area-function discontinuity at the midpoint of the SVT (Stevens 1972; Beckman et al. 1995; Carre, Lindblom, and MacNeilage 1995). The phonetic limitations of the nonhuman primate SVT follow from the fact that the 10:1 area-function midpoint variations at the SVT's midpoint are necessary to produce quantal vowels and cannot be produced in an SVT with a relatively thin tongue positioned almost entirely within the oral cavity. These area-function variations can be generated at the midpoint of the human SVT by shifting the curved "round" human tongue in the right-angle space defined by the spinal column and palate. The nonhuman vocal tract is essentially a single-tube system in which a thin tongue defines the floor of the mouth. There is no natural discontinuity in the system at which an abrupt area-function change can occur. Nor are there any muscles that could produce an extreme midpoint constriction (Lieberman and Crelin 1971).[2]

Figure 6.14 makes this point clear. The nonhuman tongue could be raised to form a constriction toward the front of the mouth in an attempt to form an [i]. However, the soft tissues of the nonhuman tongue cannot then abruptly deform, shifting downward to form the wide rear posterior tube that is necessary to produce the formant frequencies of the [i]. Similar constraints prevent [u] and [a] production. A constriction necessarily must gradually give way to an unconstricted area in the single tube airway because the tongue

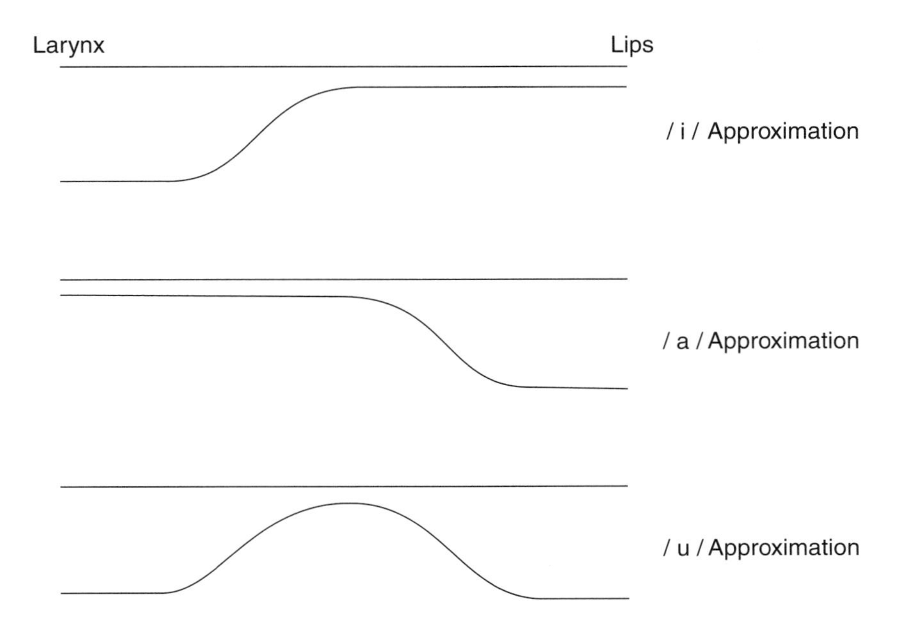

FIGURE 6.14 *The midpoint area function discontinuities necessary to produce the quantal vowels [a], [i], and [u] cannot be produced by nonhuman SVTs because the tongue is largely positioned within the oral cavity. The sketch illustrates the limits imposed by deforming the long, relatively thin nonhuman tongue in a "one-tube" SVT.*

cannot itself sustain a singular, abrupt discontinuity. You cannot form an abrupt constriction at the midpoint of a tongue positioned almost entirely within the oral cavity with an abrupt transition to the wide open area functions necessary to produce quantal vowels. Cineradiographic studies of newborn infants while they vocalize show that they do not produce the area functions that would be necessary to form quantal vowels because their long, relatively thin tongues are almost entirely positioned in their mouths (Truby, Bosma, and Lind 1965).

Phonetic Capabilities of Extinct Hominids

Here again, we must proceed with caution because we do not fully understand the morphologic bases of the development of the human SVT and know almost nothing concerning the genetic mechanisms that control its development. Neither do we have a time machine or recordings of the speech of extinct hominids (at one conference on the evolution of language, a paper was presented that claimed that early hominids' speech was impressed on clay tablets and thus could be recovered). The most contentious debate surrounds the speech-producing abilities of the Neanderthals. Barring the discovery of a frozen Neanderthal preserved in a glacier, this issue may never be fully resolved. All that remain are bones and skulls—the relevant soft tissue is gone and so we are reduced to theories that attempt to infer the relations that existed between skeletal structure and the complex soft tissue that constituted the SVT—tongues, lips, larynges, and so forth. However, the findings of recent independent studies show that Victor Negus was basically correct. These studies, moreover, replicate Edmund S. Crelin's reconstruction of the La Chapelle Neanderthal SVT reported in Lieberman and Crelin (1971).

Tongue length, oral cavity length, neck length, and hyoid position (which is tied to the human tongue) all enter into reconstructing an SVT. The proportion of the tongue that rests in the oral cavity can be determined if the total tongue length and the length of the oral cavity of an extinct hominid can be established. Impossible SVT shapes that would place the larynx in the fossil's chest can be ruled out if the neck length of an extinct hominid can be determined from inspection of surviving cervical vertebrae. The reconstruction of the SVT of an extinct hominid becomes more certain if the probable position of the hyoid bone can be determined. Negus and Crelin did not have the quantitative database that now exists. The findings of a comparative study of apes and living humans has shown that a fixed relationship holds between body mass and tongue volume, thereby providing a means for estimating tongue lengths in extinct hominids (Clegg 2001). McCarthy and Strait (2005) derived oral cavity lengths for hominid fossils from their skulls. McCarthy and colleagues (in submission) have derived estimates of neck lengths for these fossils from measurements of cervical vertebrae and have determined hyoid placement with greater precision from comparative X-ray studies of living primates.

Insights from the Vocal Communications of Living Primates

The comparative method provides some guidance. Although chimpanzees are not living examples of early hominids, they provide some insights on the earliest stages of the evolutionary processes that shaped human anatomy. Chimpanzees cannot produce quantal speech sounds, but they have the anatomic capability to produce most of the phonetic distinctions that convey words, including the formant frequency patterns that convey most of the sounds of speech. Nonhuman primates also produce a wide range of F0 patterns. There is no reason to suppose that early extinct hominids had any anatomic deficiencies that would prevent them from producing similar F0 or formant frequency patterns. Moreover, the studies I have reviewed show that nonhuman primates make use of formant frequency patterns to signal their size and referential information, signaling the presence of predators and certain social situations (attacks) by means of both F0 and formant frequency patterns (for example, Cheyney and Seyfarth 1980, 1990; Hauser 1996; Zuberbuhler 2002; Riede et al. 2005; Slocombe and Zuberbuhler 2005a, b).

At least one primate species appears to use a very different mechanism from humans to generate alarm calls that have different formant frequencies. Diana monkeys produce alarm calls for eagles and leopards that have different F0 characteristics and different formant frequency patterns that most likely reflect the contributions of laryngeal air sacs, which are absent in humans. Riede and his colleagues (2005) attribute one of these formant frequency patterns to an SVT shape similar to that used by humans to produce the vowel [a], but this cannot be the case because the monkeys' tongues (positioned in their mouths) inherently are incapable of forming the requisite abrupt area function shape. Moreover, even granting the possibility that they could form their SVT into the human [a] shape, their relatively short SVT length would preclude the generation of the low-formant frequencies observed. The details are in Lieberman (in press b). However, what is extremely significant about these monkeys' vocal signals is that formant frequency patterns are apparently so useful for vocal communication that alternative mechanisms have independently evolved in different primate species. Therefore, barring compelling evidence to the contrary, we may conclude that fundamental frequency and formant frequency patterns could have served as mediums for language in the earliest phases of hominid evolution.

Anatomic Limits and the Antiquity of Speech

One reasonable inference that may be drawn from the studies of human development and apes is that extinct hominids who had very long mouths had nonhuman tongues positioned in their mouths and larynges that could lock into their nose while drinking fluids. The skeletal remains of archaic hominids such as the Australopithecines pictured in Figure 6.15 resemble present-day apes insofar as they have long, prognathic snouts (Wood 1992) as well as body dimensions that differ from those of members of the genus *Homo* (Wood and Collard 1999). As is the case for present-day apes, there is a long space behind the palate into which a high laryngeal position can form a seal with the nasal passage, isolating the food and breathing airways. There has been little dispute of Edmund S. Crelin's reconstruction of the SVT presented in Lieberman (1975). The long prognathous mouths of Australopithecines (as well as earlier fossil ape-like hominids) would have contained their tongues; they would not have been capable of producing quantal vowels.

Similar considerations may apply to *Homo erectus* hominids. Their faces are not prognathous and they lack snouts; but taken as a whole, their faces project, unlike *Homo sapiens,* whose faces are retracted (Lieberman, Ross, and Ravosa 2000). The archaeological record suggests that these hominids did not possess the level of technology of later hominids, although they and most likely Australopithecines (Roche et al. 1999) made stone tools (Toth and Schick 1993; Susman 1994). Fire-hardened spears associated with *Homo*

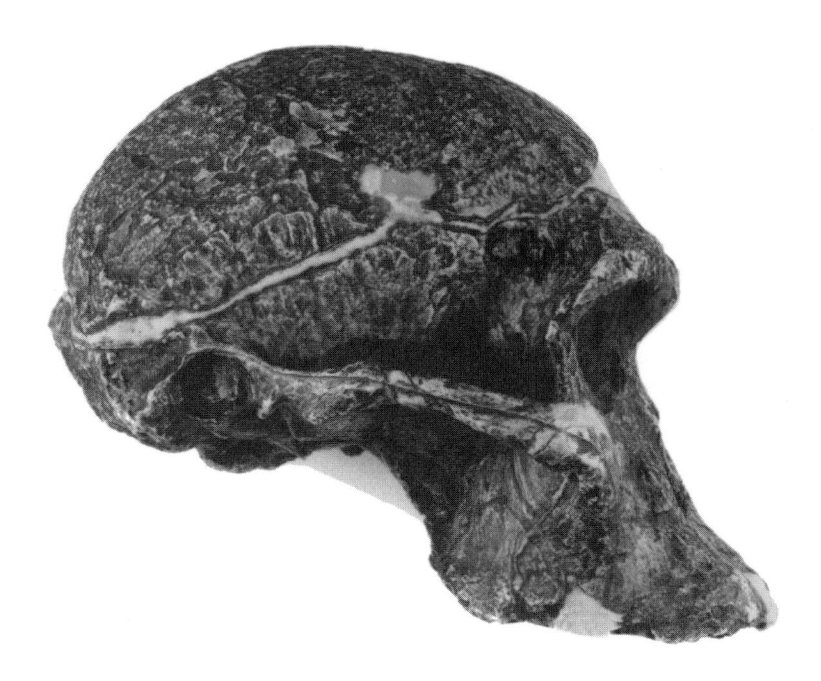

FIGURE 6.15 *Lateral view of a cast of Australopithecus africanus recovered at Sterkfontein, South Africa.*

erectus have been found (Thieme 1997). Erectus-grade hominids also migrated out of Africa (Templeton 2002), later reaching China (Zhu et al. 2001). It is unlikely that quantal vowels could have been produced by these archaic hominids. Nonetheless, it is highly probable, almost a certainty, that speech served as a medium of communication in the erectus-grade hominids who are ancestral to anatomically modern *Homo sapiens* as well as in our Neanderthal cousins. Natural selection can only act on overt behaviors. Given the increased possibility of choking on food associated with the modern human tongue and supralaryngeal vocal tract, speech most likely was in place *before* the restructuring of the human face took place. Absent speech, there would have been no selective advantages for the retention of the series of deviations from the primate standard-plan tongue and vocal tract. Hence the roots of human speech capabilities and a level of neural control that would have enabled some degree of voluntary speech must have a time depth extending well beyond the 200,000-year period associated with modern humans.

Neanderthal Speech

The same relationship between the selective advantages of the human SVT for speech and increased possibility of choking applies with equal force to the Neanderthals who possessed large brains (as large as ours). The conclusion of the Lieberman and Crelin (1971) paper "On the speech of Neanderthal man" was not that these hominids lacked speech and language. We did *not,* as did some studies, particularly those of Boe and his colleagues (1999, 2002), claim that Neanderthals were speechless. Our exact words were:

> He [Neanderthal] was not as well equipped for language as modern man. His phonetic ability was, however, more advanced than those of present day nonhuman primates and his brain may have been sufficiently well developed for him to have established a language based on the speech signals at his command. The general level of Neanderthal culture is such that this limited phonetic ability was probably used and that some form of language existed. Neanderthal man thus represents an intermediate stage in the evolution of language. This indicates that the evolution of language was gradual, that it was not an abrupt phenomenon. The reason that human linguistic ability appears to be so distinct and unique is that the intermediate stages in its evolution are represented by extinct species (Lieberman and Crelin 1971, p. 221).

Over the years much ink has been spilled in a continuing debate about the validity of Edmund S. Crelin's 1971 reconstruction of the SVT of the La Chapelle-aux-Saints Neanderthal fossil and the speech capabilities of Neanderthals. Unfortunately, many of the questions raised have little to do with the data or conclusions reported in that study or, for that matter, in later publications (Lieberman 1984). Because we are now in the age of electronic publication and electronic archives, access to the original study may be difficult. Therefore I begin with a brief overview of the study and note what appears to be a limit on Neanderthal speech capabilities in the light of present knowledge. The detailed discussion that follows will, hopefully, lead to more informed debate and further studies to fill in voids in our understanding of the evolution of human speech.

Crelin's reconstruction was an analogy based on the similarities between the skull and mandible of the La Chapelle-aux-Saints Neanderthal fossil and those of newborn humans. Crelin (1969) had expert knowledge of the detailed anatomy of human newborns and had published the first detailed atlas of newborn human anatomy. Although the complete basicranium of the fossil is not complete, enough material was present to make comparisons with data in the Marcellin Boule (1911–1913) initial study (Boule, 1911–1913) of the La Chapelle-aux-Saints fossil. The morphologic affinities included skeletal features supporting the muscles that move the tongue, such as the pterygoid process of the sphenoid bone and the angulation of the basilar portion of the occipital bone. The total length of the basicranium and the distance between the end of the palate and the foramen magnum (the "big hole" into which the spinal column inserts) were similar in newborns and in the fossil. In human newborns the larynx moves upward to lock into the space between the palate and foramen magnum.

Crelin observed that it would be impossible for an adult human larynx to fit into the shorter palate-to-foramen magnum space of the adult human skull. Figure 6.16 shows some of the skeletal features that led to the conclusion that the Neanderthal fossil did not have an adultlike human SVT. Crelin made use of a plaster copy of the La Chapelle-aux-Saints fossil, which was checked against Boule's reconstruction in Paris by P. Lieberman.

Could a Neanderthal Skull Support a Modern Vocal Tract?

Before going into the details of Crelin's reconstruction, let's place some reasonable limits on the possible position of the tongue in a Neanderthal. These limits follow from the "hard evidence" of the

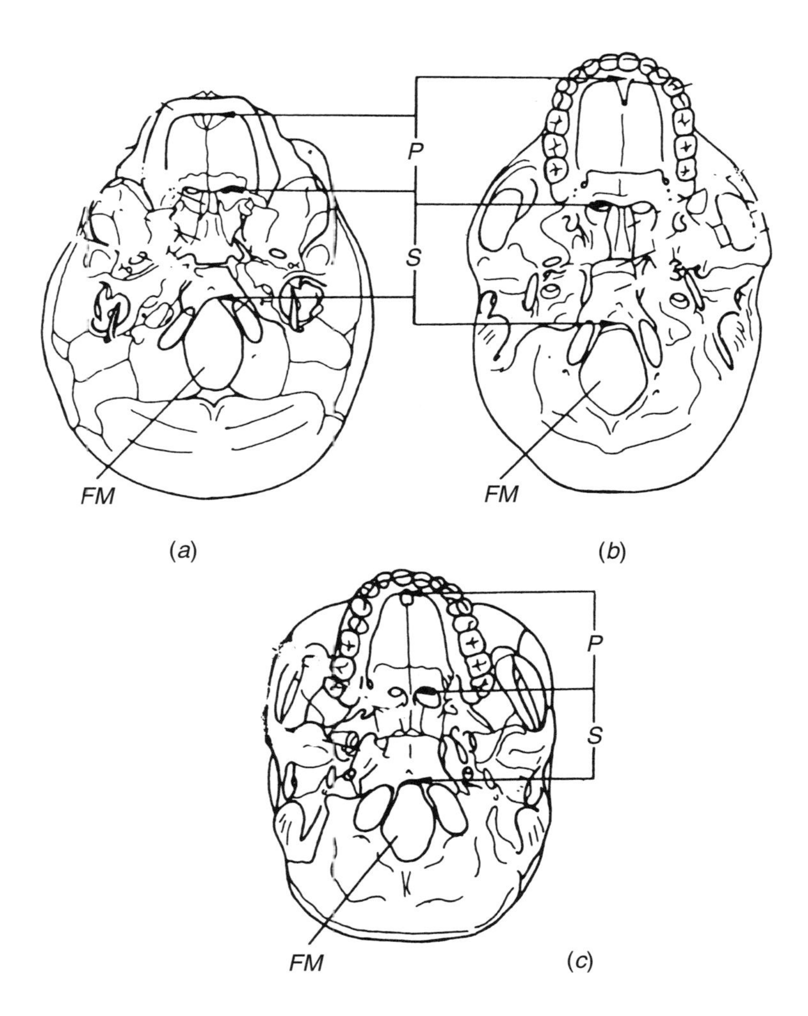

FIGURE 6.16 *Views of the bottoms of the skulls of a human newborn (a), the La Chapelle-aux-Saints Neanderthal adult (b), and an adult human (c). The skulls, which are quite different in size, have been drawn to the same scale. "FM" indicates the foramen magnum into which the spinal column inserts. The distances P + S provide indices of the length of the oral cavity. (After Lieberman and Crelin 1971)*

skeletal features sketched in Figure 6.16. Neanderthals could not have had a fully human SVT. The "hard" skeletal evidence that demonstrates that Neanderthals did not have a normal adultlike SVT is the long span of the oral cavity between its anterior end, marked by the prosthion (approximately at the front incisor teeth) and the anterior margin of the foramen magnum.

Because papers such as Boe et al. (2002) and Boe, Maeda, and Heim (1999) continue to claim that the La Chapelle-aux-Saints Neanderthal fossil had an adult human SVT, let us go through the steps of a reconstruction that would yield a human SVT. We must keep in mind the length of the oral cavity, the horizontal part of the supralaryngeal vocal tract (SVTh).

The first step entails placing the La Chapelle skull on a "normal" vertebral column. This follows the observation of Strauss and Cave (1957) and yields normal upright posture.

Step 2 involves placing a human tongue on the fossil. As noted above, the curved human tongue body forms both the floor of the oral cavity and the anterior wall of the pharynx. Let us place an adult human tongue from the independent radiographic study of Ladefoged et al. (1972), scaled to the minimal size that would allow the Neanderthal to swallow food on the Chapelle-aux-Saints skull. Studies of swallowing in humans and other species show that the Neanderthal's tongue would have had to be large enough to propel food along the length of his oral cavity to enable him to eat (Palmer et al. 2002; Hiiemae et al. 2002).

Step 3 involves attaching the hyoid bone and complete larynx from the cineradiographic study of Perkell (1969) to the Neanderthal. Human necks accommodate an SVTv that is equal in length to SVTh, but the "fit" is quite close. The cartilages of the larynx just fit into the human neck. Fink and Demarest (1978, p. 25), in their detailed study of the larynx, place the lower margin of the cricoid cartilage of the larynx during expiration at the lowest cervical vertebra (CV7) of the human neck. Their cineradiographic data shows that the cricoid transiently descends below CV7 during inspiration. Although Negus (1949) places the lower margin of the cricoid at

CV6, Perkell's (1969) cineradiographic study locates the vocal cords at the lower margin of C6, placing the cricoid cartilage lower. In short, the larynx just fits into the human neck. Heim (1976) who studied the La Ferrassie 1 Neanderthal fossil, whose cervical vertebrae were preserved, states that the length of the Neanderthal neck was no longer than that of a modern adult human, although the sketches on page 315 of his study show that it is shorter than the neck of the single "Frenchman" illustrated. Spoor also concludes that the La Ferrassie neck was shorter than that of an average adult human (personal communication based on his ongoing study of the La Ferrassie 1 fossil).

As noted above, SVT length correlates with height in many animals (Fitch 2000c) and in humans (Fitch 1993), though to a lesser degree because body shape and height differ in human populations adapted to different climates. However, neck length appears to be longer for taller people. Mahajan and Bharucha (1994), who studied 2724 children in India, derived an age-independent linear regression that correlated neck length with height, neck length = 10 cm + (0.035 × height) cm. Different scaling factors probably hold for human populations adapted for life in cold climates; but the point to keep in mind is that, as children mature, their larynges most likely just fit into their necks. This may explain the fact that the tongue does not descend into the pharynx, thereby achieving SVTh = SVTv, until after the age of six years.

A Most Improbable Primate

Figure 6.17 shows the Neanderthal skull and the putative human-like vocal tract. Note that the larynx is positioned below the seventh cervical vertebra somewhat below the level of the sternum. The larynx's low position results from the 12.66 cm Neanderthal basicranial span between the incisors (the usual skeletal feature marking this point is the prosthion) and the anterior margin of the foramen magnum, the basion. This skeletal length (prosthion to basion) appears to have been subjected to intense selective pressure during the course of human evolution because it has low variabil-

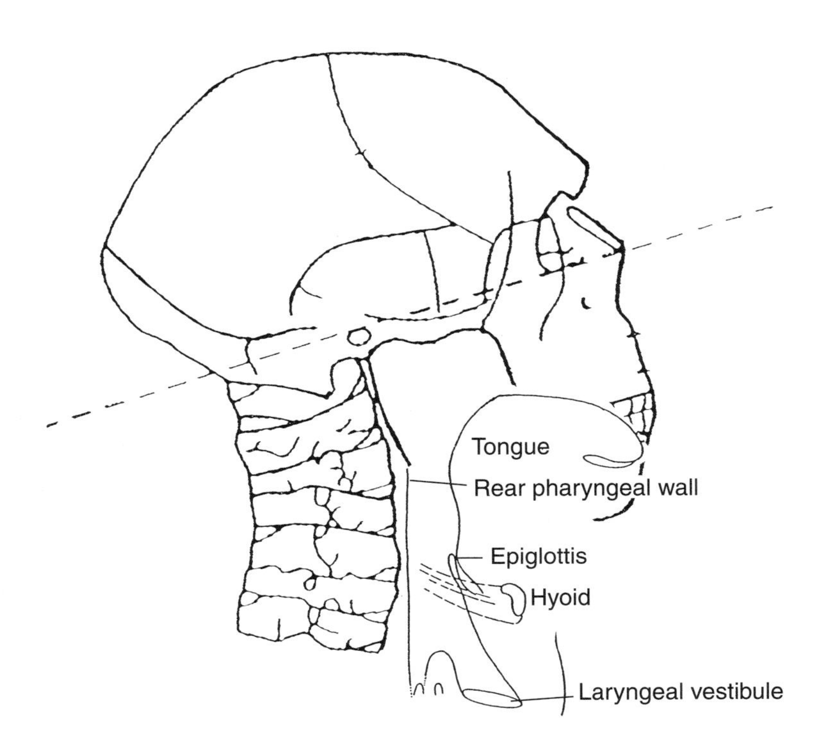

FIGURE 6.17 *Why Neanderthal hominids could not have had a human SVT. If an SVT with equal SVTh-to-SVTv lengths is placed on a Neanderthal skull, the larynx will fall below the cervical vertebrae at the level of the sternum bone in the Neanderthal chest. This follows from the long Neanderthal oral cavity (defining SVTh) and the relatively short Neanderthal neck. Swallowing would have been impeded because the sternum would have blocked the larynx from moving upward and forward.*

ity across modern human populations. Howells (1989) studied the skulls of 2504 adult males and females drawn from groups distributed around the world. The mean length for males is 100.46 mm with a standard deviation (SD) of 4.6 mm. The longest male mean length for a robust Melanesian group is 107.04 mm, SD 4.76 mm; the shortest length is 93.75 mm, SD 5.7 mm for a European group (Howells 1989, pp. 10, 125, and 141). The 126.6 mm Neanderthal length is one element of the Neanderthal skeletal complex that Howells's statistical analyses show fall outside the human range of variation. The recent findings of McCarthy and his colleagues (in submission) are consistent with Howells's data.

If, as an estimate, we use this dimension to scale the Neanderthal SVT, it would be approximately 126/101 times longer than that of an adult male human, if his SWVt and SVTh had equal lengths. SVT length is correlated with height in humans, though to a lesser extent than in rhesus monkeys (Fitch 1993); and the SVT lengths of the tall (6 foot (183 mm) and 5 foot 9 inch (175 mm)) adult males studied by Baer et al. (1991) were 17 cm long. The formant frequencies of their [i] vowels, which are determined by SVT length, are somewhat lower than those measured by Hillenbrand et al. (1995) for forty-five adult men, which suggests that 17 cm is most likely the longest "normal" SVT length for adult human males. (A comprehensive study that includes disparate human groups, similar to Howells's (1989) monograph on features of the skull, is not available.) The length of the La Chapelle Neanderthal SVT thus would be about 21.3 cm long, given its basion-to-prosthion length (1.25 × 17 cm). This would result in an SVTv length that was 20 mm longer than an adult human male's, placing the lower margin of the cricoid below D1, the first thoracic vertebrae of the La Ferrassie I Neanderthal, whose length is 17 mm (Heim 1976). We have again tilted the exercise toward the benefit of the Neanderthal, because the SVTvs of adult males are somewhat longer that those of females (Fitch and Giedd 1999). A longer male SVTv would place the larynx even lower.

In short, the human structures of the larynx just fit into the human neck. If a Neanderthal had a functionally human SVT, its long SVTh would have to be matched by an equally long SVTv, placing the cricoid cartilage of the larynx below the sternum. This would interfere with the laryngeal maneuvers that are necessary to swallow food (Negus 1949, p. 176). The hyoid moves upward and forward about 13 mm, opening the esophagus and placing the larynx into a position in which food will not fall into it while the person swallows (Ishida, Palmer, and Hiiemae 2002). A larynx placed in the neck can execute these maneuvers, moving upward and forward. A larynx placed below the sternum would be blocked by this bone from executing these movements. Because the swallowing pattern generator—the movements that are involved in swallowing—are similar in humans and apes (Palmer et al. 2002), no human or ape descended from our common ancestor has a larynx in its chest.

The relevance of the length of the Neanderthal oral cavity and oropharynx to the reconstruction of a Neanderthal SVT has been discussed before (c.f. Lieberman 1984, pp. 290–296; 1982). Any claim that a Neanderthal had a human SVT entails fitting the Neanderthal mouth with the tongue of a human adult. Human tongues have equally long oral and pharyngeal sections and the larynx would end up placed in the Neanderthal's chest. Similar constraints apply to the SVTs of Australopithecine and Erectus hominid fossils.

The 1971 Neanderthal Reconstruction and Modeling Techniques

Some details of Edmund S. Crelin's 1971 Neanderthal SVT reconstruction and the computer modeling of its probable phonetic capabilities are necessary to assess subsequent conflicting claims, counterclaims, and the speech capabilities of different SVT configurations. Crelin noted a number of skeletal features of the Neanderthal basicranium and mandible that appeared to be similar to features that, in newborn humans, support the muscles and ligaments of the supralaryngeal vocal tract. For example, the pterygoid process of the sphenoid bone is relatively short, and its vertical lamina is more inclined away from the vertical plane than is the case in adult human skulls. The medial pterygoid plate, which is one of the points where the inferior pharyngeal constrictor muscle is attached, is also similar in the modern newborn and in fossil Neanderthal skulls. This muscle plays a part in swallowing (Bosma 1975) and in speech production (Bell-Berti 1971). It is active in the production of nonnasal sounds, during which it helps seal the nasal cavity, and in the production of sounds like the vowel [a], during which it pulls the tongue body back.

Other skeletal features relating to the soft tissue of the SVT are similar in the newborn and Neanderthal skulls. The distance between the vomor bone and the synchondrosis of the sphenoid and occipital bones is relatively long in newborns. In contrast, the vomar often overlaps the synchondrosis in adults (Takagi 1964). This may be a reflex of the rearward shift of the face during human development. The relatively longer distance in the newborn contributes to the long space between the back of the palate and the anterior border of the foramen magnum into which the newborn infant's larynx can rest when it locks into the nasal airway.

On the basis of the total complex of anatomic similarities, Crelin and I concluded that the Neanderthal SVT was similar to a human newborn's and would have imposed similar phonetic constraints. Grosmangin (1979) in an independent study reached similar conclusions. When the range of vowel formant frequency patterns that Crelin's SVT could produce was computer-modeled, it was evident that Neanderthal speech would not have been as efficient a means of vocal communication. It would not have been able to produce the supervowel [i] and vocal tract normalization consequently would not have been as robust as is the case for human speech. Nor would Neanderthal speech have been as resistant to articulatory errors (Stevens 1972; Beckman et al. 1995). However, Neanderthal phonetic capabilities would have not precluded speech communication because a wide range of formant frequency patterns conveying vowels and consonants could have been produced to which distinctions in duration and diphthongization could have been added, as is the case for human speech (Hillenbrand et al. 1995).

Figure 6.18 displays a diagram of Crelin's reconstruction. Although Crelin concluded that the Neanderthal SVT resembled an exceedingly large newborn human's, he positioned the reconstructed Neanderthal larynx below the position that would be typical of a human newborn's to tilt its phonetic capabilities toward that of adult humans. The tongue, like that of a human newborn's, is positioned almost entirely within the mouth. The pharynx is positioned behind the larynx. The total length of Crelin's Neanderthal SVT, including the laryngeal portion and lips, was 19 cm. When it was

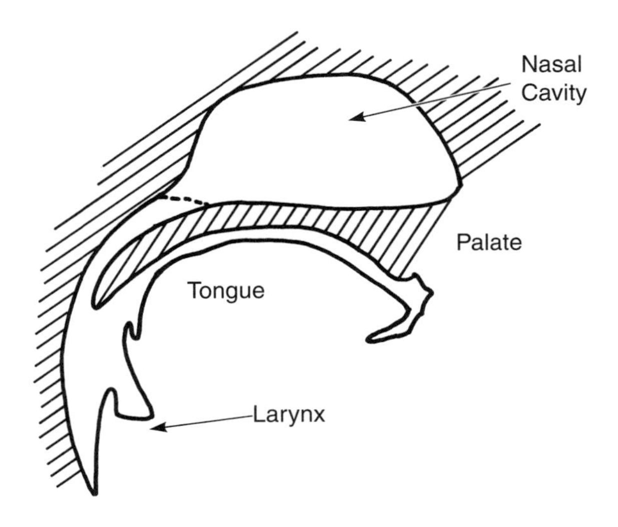

FIGURE 6.18 *Edmund S. Crelin's 1971 reconstruction of the La Chapelle-aux-Saints Neanderthal SVT. The tongue occupies proportionately more of the oral cavity than is the case for modern humans. The SVTh-to-SVTv proportions are approximately 1.3 to 1, similar to those deemed most likely by McCarthy et al. (forthcoming).*

computer-modeled, it failed to produce the formant frequency patterns of the quantal vowels—like the newborn human SVT—although acrobatic tongue maneuvers were programmed to tilt the modeling toward human capabilities.

Computer modeling of the reconstructed Neanderthal SVT was conducted using Henke's (1966) computer model, which represents the supralaryngeal vocal tract by means of a series of contiguous cylindrical sections, each of fixed area. Each section can be described by a characteristic impedance and a complex propagation constant, both quantities for uniform cylindrical tubes. Junctions between sections satisfy the constraints of continuity of pressure and conservation of volume. In this fashion, the computer program calculated the three lowest formant frequencies of the vocal tract filter system that specify the acoustic properties of a vowel (Chiba and Kajiyama 1958; Fant 1960; Nearey 1979; Stevens 1998). Tongue movements based on the anatomic constraints imposed by having the tongue positioned almost entirely within the mouth were then modeled. It is not possible to achieve abrupt changes in SVT cross-sectional area with a long, relatively thin tongue positioned in the mouth. The cineradiographic study of newborn tongue movements during cry of Truby, Bosma, and Lind (1965) served as a guide because the Neanderthal reconstruction was modeled on the newborn SVT. For example, frame 61 in Figure 1 (p. 75 in that study) of a newborn cry, which showed extreme constriction of the back of the oral cavity and maximum opening of the mouth, served as a guide for our [a] approximations. Figure 6.19 shows the area functions that were modeled.

The Neanderthal SVT was constricted to 0.30 cm^2. This is not apparent in the sketch in Figure 6.19. However, in response to later inquiries on this point, a nonlinear ordinate scale was used that showed the minimal constriction in the chimpanzee SVT modeled in Lieberman, Crelin, and Klatt (1972), which had the same minimal constriction as the modeled Neanderthal SVTs (a smaller constriction at that time was thought to have produced turbulent noise at the constriction, destroying vowel quality). Boe and his colleagues (1999, 2002) have claimed that the constriction modeled in our studies was 1.0 cm. That is not the case. Boe and his colleagues also state that constriction size is an exceedingly critical factor in correctly modeling the phonetic range of an SVT. However, they appear to overstate that issue; the MRI study of Baer et al. (1991) reports constrictions of 0.69 to 0.50 cm^2 for two adult speakers when they produced the quantal vowels [i], [u], and [a]. Story, Titze, and Hoffman (1996), in contrast, report a minimal constriction of 0.10 cm^2 for their male subject. The shift in formant frequencies that might be expected in the formant frequencies of the quantal vowel [i] from these different constriction areas appears to be minimal. The data of the Beckman et al. (1995) study of the production of the vowel [i] predicts a shift of less than 50 Hz for F2, the second formant frequency of this vowel for these different degrees of constriction, which may explain the observed variations in constriction area evident in MRI studies of adult human speakers. Beckman and

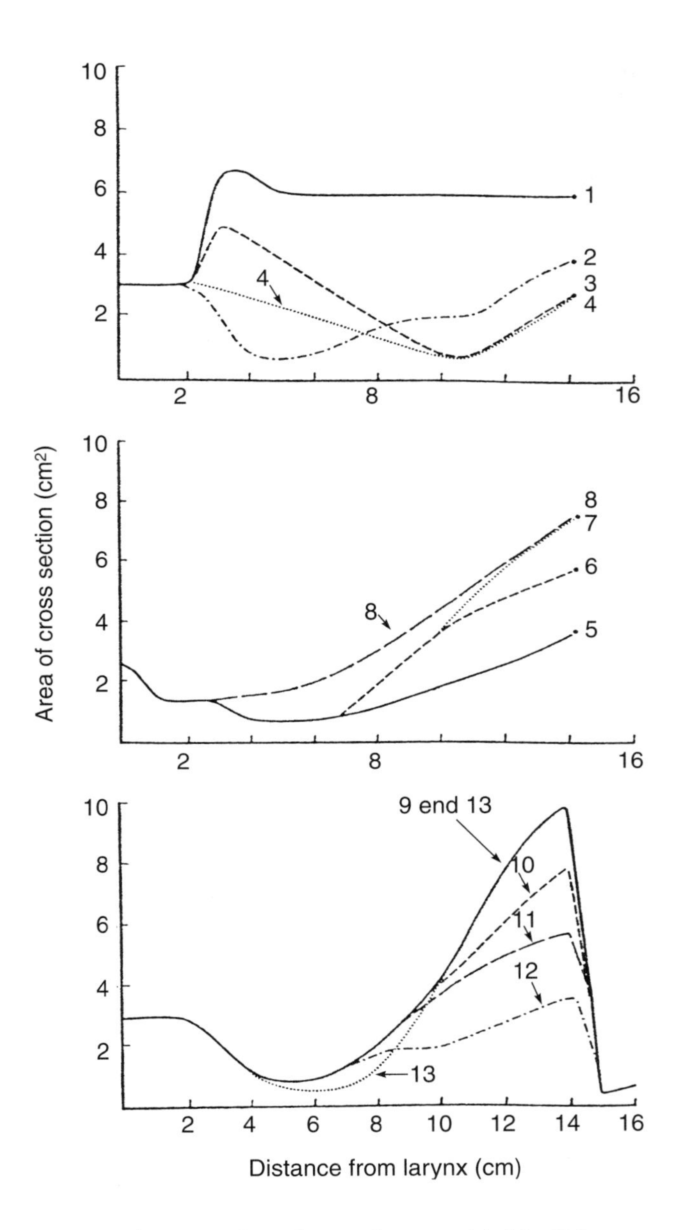

FIGURE 6.19 *The range of vocal tract shapes modeled in Lieberman and Crelin (1971) was guided by the tongue shapes in the Truby, Bosma, and Lind (1965) cineradiographic study of newborn infant vocalization and swallowing and the degree to which the tongue could plausibly be deformed. Taking these factors into consideration, the SVT area functions in Figure 6.19 aimed at producing the best approximations to the quantal vowels [i], [a], and [u], respectively.*

her colleagues concur with previous assessments (Perkell and Nelson 1982; Fujimura and Kakita 1979) regarding the relative insensitivity of the quantal vowels [i], [u], and [a] to variations in tongue placement; they conclude that [i] is the most stable quantal vowel. The vowel [u] shows the next smallest variability and these two vowels along with [a] are the most common in the vowel inventories of the world's languages (Greenberg 1963; Maddieson 1984).

To guard against underestimating Neanderthal speech capabilities, we forced the computer-modeled SVT in 1971 toward adult human capabilities by means of some rather acrobatic approximations to the vowels [i], [u], and [a]. After the area functions shown in Figure 6.19 were modeled, the calculated formant frequencies were "normalized" to the 17 cm length noted by Fant (1960). The reconstructed Neanderthal SVT was 19 cm long and would have produced formant frequency patterns beyond the outer limits of the human range, a point to which I shall return. Figure 6.20 shows the normalized Neanderthal formant frequencies that were plotted against the first and second formant frequency "loops" of the vowels of English by Peterson and Barney (1952). The labeled closed loops enclose the formant frequencies that accounted for 90 percent of the samples produced by the human speakers in each vowel category.

It was not necessary to simulate the sounds of all languages with the computer-implemented Neanderthal vocal tract because the main point was whether the Neanderthal SVT could produce the full range of human speech. The absence of Neanderthal data points in the vowel loops for [u], [i], and [a], which delimit the human vowel space, shows that the Neanderthal phonetic repertoire is inherently limited. This occurred even when we forced the reconstructed Neanderthal vocal tract toward humanlike configurations by means of improbable contortions such as functions 3, 9, and 13 in Figure 6.19. Despite these gyrations, the modeled Neanderthal SVT could not produce the supervowel [i], which yields the optimal vocal tract length-calibrating signal (Nearey 1979). The computer simulation was also used to generate consonantal vocal tract func-

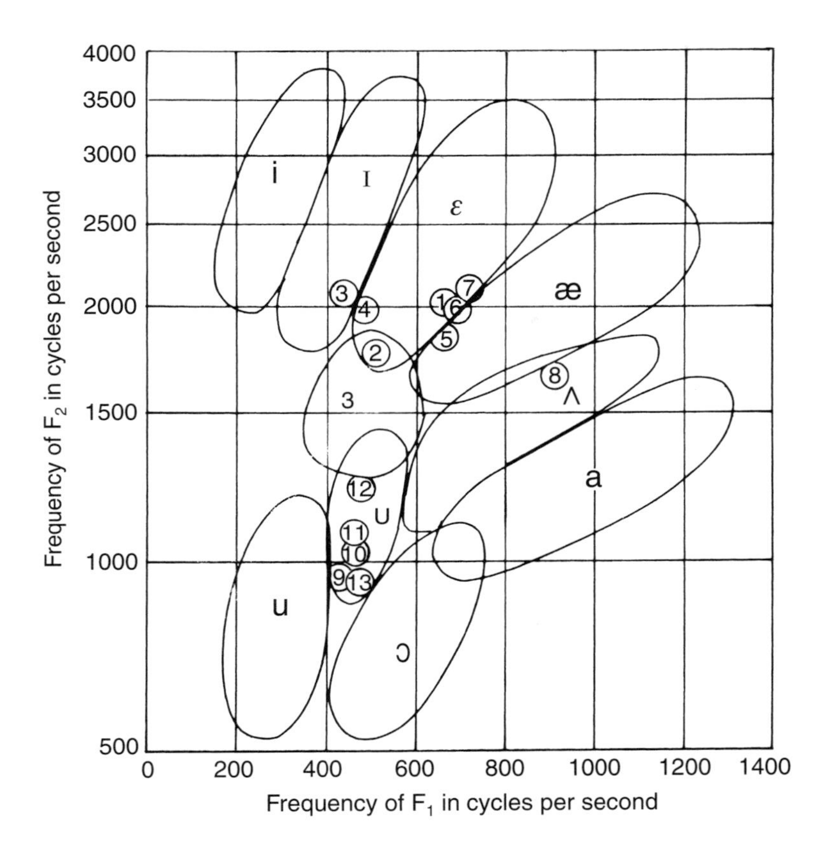

FIGURE 6.20 *The hypothetical Neanderthal SVT area functions shown in Figure 6.19 were modeled using Henke's (1966) computer-implemented algorithm. The loops marked with phonetic vowel symbols are those observed by Peterson and Barney (1952). They encompass 90 percent of the tokens within each vowel class. The range of vowels that the Neanderthal SVT could have produced appears to be somewhat greater in McCarthy et al. (forthcoming), which takes into account the findings of MRI studies subsequent to 1971 of human SVT area functions for quantal vowels and the greater range of F1 and F2 values for vowels measured by Hillenbrand et al. (1995). A Neanderthal SVT most likely could have produced [i] vowels at the verge of being fully quantal. The 1.3 to 1 SVTh-to-SVTv Neanderthal SVT proportions would have precluded their being able to produce quantal vowels that were as insensitive to articulatory imprecision as those produced by human SVTs. Human SVTs having 1:1 SVTh-to-SVTv proportions are not evident in the present fossil record until 50,000 years ago.*

tions; the Neanderthal SVT was limited to labial and dental consonants like [b] and [d].[3]

The Neanderthal Debate

Continuing controversy has surrounded our assessment of Neanderthal phonetic capabilities. This is not surprising, given the general heated level of debate surrounding Neanderthals. Many skeletal studies (for example, Howells 1976, 1989; Lieberman 1995; Ponce de Leon and Zollikofer 2001) suggest that Neanderthals were a different species than modern humans. Moreover, genetic studies of DNA recovered from Neanderthal remains (Krings et al. 1997; Ovchinnikov et al. 2000) indicate that Neanderthal hominids were a species that diverged from modern humans about 500,000 years ago. Archaeological studies have found Neanderthal remains associated with artifacts that are similar to those of contemporary human populations (for example, d'Errico et al. 1998), although others disagree (Hublin et al. 1996); but being a separate species does not necessarily predicate having different levels of cognitive ability (for example, different species of fish). A number of studies have claimed that the SVTs of Neanderthals were virtually identical to those of adult humans (DuBrul 1977; Arensburg et al. 1990; Houghton 1993; Schepartz 1993; Boe, Maeda, and Heim 1999; Boe et al. 2002).[4] However, as noted above, this would place their larynges in their chests.

The recent study by Robert McCarthy and his colleagues may perhaps bring closure to the Neanderthal speech debate. McCarthy and colleagues (in submission) take account of the set of skeletal features that place limits on what kind of SVT can be reconstructed on the fossil remains of an extinct hominid. The Lieberman and Crelin (1971) study to a great extent relied on correlations between skeletal morphology and the soft tissue that forms the SVT. Other studies such as those of Laitman and his colleagues (Laitman et al. 1978; Laitman, Heimbuch, and Crelin 1979; Laitman and Heimbuch 1982) also relied on correlations between larynx position and skeletal features such as the cranial base angle and palatal length. In contrast, McCarthy et al. (in submission) take into account skeletal features that place limits on where the cartilages of the larynx could be for SVTs having different SVTh/SVTv proportions. A SVT that has a larynx positioned below the neck in the thorax would make it impossible to swallow. The probable length of the tongue, which Clegg (2001) shows is closely related to body size in apes and humans, is also taken into account, as is the position of the hyoid bone relative to the mandible.

McCarthy and his colleagues obtained measurements of mandibular length and height and the vertical height of the cervical spine from a sample of 11 chimpanzees, three Neanderthals, *Homo ergaster* (the fossil identified as WT 15000), the Skhul V fossil dated to approximately 90,000 years before the present (BP) that is generally considered to be an early specimen of anatomically modern *Homo sapiens,* seven human skulls from the Upper Paleolithic dated to about 50,000 years BP, and direct measurements of a large database of recent modern humans. *Homo ergaster* generally is believed to be the species ancestral to both Neanderthals and humans (e.g., Wood and Collard 1999). The upper and lower positions of the hyoid bone with respect to the mandible also were determined with greater precision for humans and rhesus monkeys. The probable tongue lengths of these hominids were taken into account following Clegg's (2001) procedures, which show that tongue mass in apes and humans tracks body mass. These parameters place limits on reconstructing the SVTs of fossil hominids.

McCarthy and his colleagues tested their procedures and measurement base by seeing whether they could predict the actual SVTs of living chimpanzees; the results suggest that the procedures are valid because they correctly show that SVTh, the horizontal portion of the chimpanzee SVT, is 2.5 to 3 times longer than SVTv, the vertical portion of the chimpanzee SVT. The reconstructed SVT proportions for the specimen of *Homo ergaster* were within the range for chimpanzees. The neck lengths of Neanderthals were at the lower limit for those of modern humans, but their oral cavities were longer. A modern human SVT having 1:1 SVTh/SVTv proportions would place the Neanderthal larynx within the chest at vertebra T2, precluding swallowing. If Neanderthals had larynges positioned at the same level in the neck as modern humans and had tongues scaled to their bodies, their SVTs would have 1.3:1 SVTh/SVTv proportions, similar to the Lieberman and Crelin (1971) reconstruction. This would allow them to swallow but would preclude their producing quantal vowels. If a human SVT capable of fully producing quantal speech sounds were placed on the *Homo ergaster* fossil, the larynx also would have been positioned in the chest, placing the larynx at vertebrae T2 or lower. This again would have precluded swallowing.

Surprisingly, Skhul V most likely would have had a similar problem if his SVT were fully human because his neck length is at the lower limit for modern humans and his face is long. The most likely SVT reconstruction for this early anatomically modern human yields a SVTh that is longer than its SVTv, resulting in speech production capabilities at the outer margin for producing quantal vowels. Quantal vowels could be produced, but with less tolerance for articulatory imprecision. Neck length and skull measures indicate that fully human 1:1 SVT proportions first occurred in the fossil sample studied in the Upper Paleolithic, 40,000–50,000 BP. However, the fossil record of anatomically modern humans in the period between Skhul V and 50,000 years BP is incomplete.

Speech and Speciation

The limitations of Neanderthal speech could have served as a genetic isolating mechanism. Dialect differences serve as genetic isolating mechanisms for living human populations. Barbujani and Sokal (1990, 1991) have shown that, in European populations, individuals generally mate with a person speaking a similar language or dialect. Even minute distinctions in dialect exert an effect on gene transfer (Barbujani 1991). Skeptics can point to many instances in which men and women mate and have surviving heirs, even though they do not speak the same language or dialect. Examples that readily come to mind range from royal marriages arranged for reasons of state to rape by invading armies or militias. However, the Barbujani and Sokal data show that a general, more harmonious pattern prevails over these perturbations: marriage usually takes place between couples who can communicate—and who share a particular culture marked by a particular dialect.

If we were to follow Darwin's principle of uniformitarianism—the principle that humans behaved as they do today in the epoch in which both Neanderthal and human populations existed in Europe and Asia—the lower Neanderthal male formant frequencies would have signaled that the speaker was not "one of us," a male specimen of *Homo sapiens.* Neanderthal speech could have served as an isolating mechanism. Distinctions in speech might have been one factor isolating humans from Neanderthals throughout a long period in which behavioral disparities relevant to fitness, such as different techniques for hunting (Lieberman and Shea 1994), resulted in the demise of the Neanderthals (Zubrow 1990).

Neanderthals Don't Matter

In the end, the question of Neanderthal speech capabilities does not bear on the significance of the human tongue and SVT. The chimpanzee SVT is a reasonable approximation of the start point for the evolution of human morphology; comparative studies of human and chimpanzee SVTs and developmental studies of humans reveal adaptation that enhances speech communication at the cost of increased morbidity from choking on food while swallowing. If it were the case that Neanderthals possessed a fully human SVT, the date at which this attribute occurred would simply shift back in time to the common ancestor of Neanderthals and anatomically modern *Homo sapiens* without changing that conclusion.

Evolutionary Significance of the Human Tongue

To sum up, the anatomy involved in talking has a long evolutionary history. Virtually all of the phonetic contrasts that differentiate words are employed to some degree by other species. The fundamental frequency (F0) variations that convey distinctions in meaning in most human languages could potentially be produced by the larynges of many primates. Similar patterns are used by many primate species to convey referential and emotional information. Formant frequencies that, when encoded, yield the high information transfer rate of human speech, are used by many species in their vocal communications. The unique human supralaryngeal airway results from the descent of the tongue into the pharynx. It appears to be the product of natural selection that enhanced the robustness of the human speech "code." Facial retraction, which characterizes the human face, may have fortuitously triggered the initial phase of the descent of the human tongue. However, the subsequent pattern of tongue descent to achieve a 1:1 SVTh to SVTv proportion and then hold that proportion suggests that selection occurred for speech. The uncertainty concerning the precise character of Neanderthal speech capabilities or, for that matter, other archaic hominids follows from lacunae in our understanding this second phase of tongue restructuring. However, it does not detract from the messages of this chapter. First, given the clear end points, the vocal tracts and corresponding basicranial morphology of chimpanzees and those of modern adult-like humans, it is clear that humans possess vocal tract anatomy adapted to enhance vocal communication. Second, vocal communication has a long evolutionary history and must have been present in some form in the species ancestral to modern *Homo sapiens;* otherwise, there would have been no selective advantage for the retention of whatever mutations were responsible for the morphology of the human tongue and supralaryngeal vocal tract.

It is curious to view the continuing controversy concerning Neanderthal speech. If Neanderthals had had brains similar to ours, they would have talked using sounds that were not as resistant to misinterpretation as those of human speech. The brain is the key to speech and language. Apes possess the anatomic prerequisites for speech, albeit with probable higher perceptual error rates; but they appear to lack the ability to freely produce different words—the "reiterative" ability discussed in the previous chapters. As I noted in 1985, studies of the neural bases of motor control focusing on cortical-striatal-cortical circuits may furnish the key to human cognitive and linguistic ability (Lieberman 1985). The probable date at which the FOXP2 gene reached its human form, 100,000 years in the past, points to the coevolution of the anatomy and neural mechanisms that make human speech possible—the gift of tongue that sets us apart from other species.

Take-Home Messages

- We share much of the anatomy involved in the production of human speech with other species—the evolution of the anatomic bases of vocal communication can be traced back to anurans. Many mammalian species, including humans, have retained laryngeal morphology adapted for phonation at the expense of respiratory efficiency.
- Both gradual changes and abrupt discontinuities at branch points deriving from ecological changes and chance genetic events appear to have entered into the evolution of vocal anatomy.
- The human tongue and supralaryngeal vocal tract (SVT) are species-specific. The human tongue appears to be adapted to produce sounds such as the supervowel [i] (the vowel of the word "see") that enhance the process of speech perception at the expense of difficulties in swallowing. However, speech communication is possible absent this sound, which enhances the process of vocal tract normalization. Vocal tract normalization makes use of a "primitive" mechanism whereby formant frequencies signal a speaker's size.
- A low larynx, in itself, is not an index of an SVT that can produce the full range of human speech. The descent of the human tongue into the pharynx also yields the low position of the human larynx. Although other species have a low larynx, their

tongues are anchored in their mouths and they cannot produce the range of SVT shapes necessary to produce quantal vowels. Acoustic analyses show that their vocalizations are generally limited to the "schwa" vowel that results from a uniform or slightly flared tube, except for some species who appear to have evolved specialized, "derived" anatomy that produces additional formant frequency patterns.

- Comparative studies of the ontogenetic development of humans and other primates suggest that facial retraction played a part in the evolution of the human SVT. However, factors that are presently unclear appear to have entered into that process.
- Early hominids such as the Australopithecines most likely lacked tongues and SVTs that could produce the full range of human speech sounds, including the supervowel [i]. We must conclude that language employing some form of speech must have existed in some form of *Homo erectus* ancestral to modern human beings; otherwise, there would have been no selective advantage for the retention of the peculiar human tongue and SVT in humans. In short, given the fact that the only apparent contribution of the human SVT is to slightly enhance the process of speech perception, speech must have been present long before the appearance of anatomically modern *Homo sapiens.*
- Speech may have served as a genetic isolating mechanism in the era in which humans and Neanderthals shared the planet.

CHAPTER 7

Linguistic Issues

IN THIS CHAPTER I present several modest examples to suggest that the study of language might profit by taking account of the general principles, methods, and findings of evolutionary biology. My intent here is not to deride or attack theoretical linguists, although in their insularity they could rightly serve as subjects for an essay by Jonathan Swift if he were still with us. Given the defensiveness that results from hermetic isolation, I also do not expect that any paradigm shift in linguistic research will occur. Nonetheless I shall present some concrete examples, many of which have been noted before by other scholars, that point the way toward integrating linguistic research within the broader framework of biology. I do not intend to—indeed, I cannot—present a comprehensive analysis of how to bring the principles and findings of biology to the attention of linguists. That is an enterprise that should command the attention of a new generation of speech scientists and linguists.

Darwin versus Occam

One theme that has been present throughout this book is that evolution is a miserly tinkerer. Structures and systems that were adapted for one purpose are modified at lowest "cost" to take on new tasks. If we take account of this basic principle of evolutionary biology, we may be able to better understand some linguistic phenomena. There is no reason to expect that biological "solutions" to the demands of life will be elegant or logical. However, linguistic studies continually invoke Occam's razor, a concept that places priority on the simplest, most economical solution to a problem.

The concept of "simplicity" is itself difficult to define. If simplicity were a meaningful criterion for the design of ships, would a ship powered by diesel engines or a steam turbine be a better way to cross the Atlantic Ocean? How about the "Flying Cloud" with its towering sails versus Leif Erikson's Viking boat? How can simplicity be calculated? Is it simply a matter of some vague concept of parsimony? But it is clear that the design of biologic organisms is neither simple nor parsimonious in any meaningful sense. The logic of evolution is, as Ernst Mayr (1982) demonstrated, proximate and historical. Consequently, when applied to biology, Occam's razor is a dull instrument. Nonetheless this relic of medieval philosophy structures linguistic theories in areas of research ranging from pragmatics to phonology.

Grice (1989, pp. 26–31), for example, claimed that speakers communicate the minimal amount of material necessary to convey a message. His theory boils down to the advice that a defense lawyer might give to his mobster client: "Say as little as possible." However, although some readers may prefer short, simple sentences, are George W. Bush's speeches more cogent than Winston Churchill's? Was Hemingway a "better" writer than Conrad? Nonetheless, linguists—for example, Chomsky and Halle (1968) in discussing the sound pattern of English—often make use of "simplicity" metrics, even in phonologic studies that ultimately relate to observable events such as the movements of a person's tongue or formant frequencies. In practice, simplicity generally reduces to a maxim that might read, "God loves to conserve ink."

Breathing Prosody and Grammar

I will start by focusing on phonology because the predictions of any phonologic theory must result in observable phenomena. Let us begin by considering two seemingly unrelated phenomena—sentences and breathing. One of the most basic aspects of life is breathing, and one of the central concepts in linguistic theory is the sentence. The complex biologic system discussed here that provides life-sustaining oxygen to our bodies does not represent the most economical solution to this problem. The "design" of the human respiratory system, which structures the linguistic uses of sentence prosody, the acoustic events that play a large role in segmenting the flow of speech into sentences, is the result of evolutionary tinkering over millions of years. Its resulting structure is a design nightmare that eludes any analysis motivated by Occam. The manner in which humans make use of this existing anatomic system and its physiologic constraints for linguistic ends are neither elegant nor simple.

One point of agreement among all theories of grammar is that a sentence is the minimal unit of speech for a complete semantic interpretation (Trager and Smith 1951; Chomsky 1957). A sentence expresses a complete thought. The primary function of orthographic punctuation is to indicate the scope of a sentence. A simple experiment will test this claim—take a printed text and move the final period and initial capitalization of a sentence one word to the right. The meaning of the string of words between the capitalized word and the period will usually be incomprehensible. You will probably also find that reading the words with normal intonation or prosody, which is the melody of speech, has become difficult because you cannot breathe at the right time.

Anatomy and Physiology of Respiration

The sentence-delimiting intonation pattern that appears to characterize virtually all known human languages (Greenberg 1963) derives from the interplay of linguistic intent with phylogenetically primitive anatomic and physiologic constraints. As we talk, we override motor activity directed toward the maintenance of adequate oxygen levels in the bloodstream and brain that evolved over hundreds of millions of years. The primary function of the respiratory system is to pump air into and out of the lungs to provide oxygen to the bloodstream, and the anatomy of the respiratory system reflects its

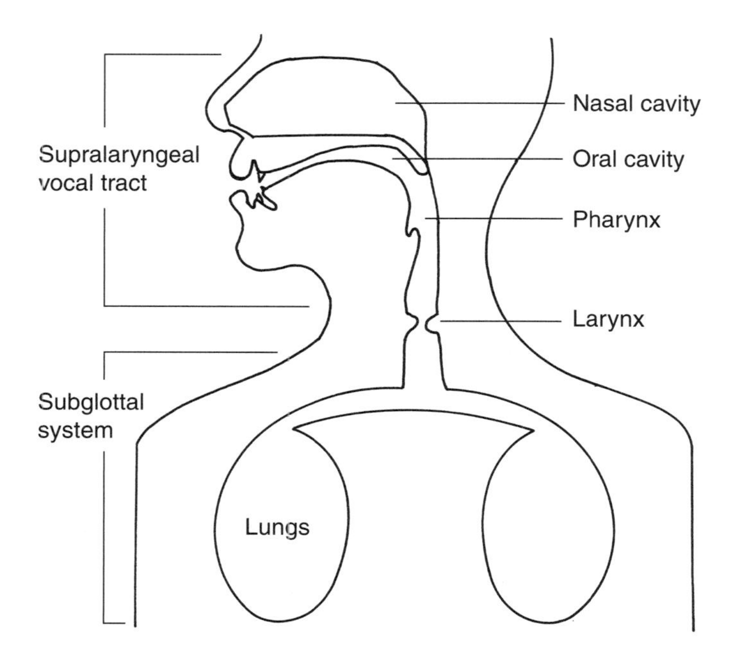

FIGURE 7.1 *The three physiologic systems involved in producing speech.*

evolutionary history. The schematic diagram in Figure 7.1 shows the lungs, the larynx, and the upper supralaryngeal airways.

Two rubber balloons suspended in a sealed box that can change its volume constitute a simple, but reasonable, mechanical model for the lungs. The balloons in the sealed box connect to the atmosphere via a tube (a pseudotrachea) that can be closed by a valve (a pseudolarynx). The mammalian respiratory system most likely is the product of a tinker working on the swim bladder of a fish. Primitive fish, such as sharks, do not have swim bladders and they must swim or sink. Advanced fish can pump air that their gills extract into two elastic internal bladders. As a fish pumps air into its swim bladders, they expand and the volume of the fish increases. Releasing air decreases the fish's volume. In this manner a fish can change its size to displace water, adjusting its buoyancy to float at a specific depth. A blimp floats in the air in the same manner by matching its weight to that of the air it displaces. A fish equipped with swim bladders can slowly swim or effortlessly hover. We and all terrestrial animals probably have two lungs for the same reason that water wings come in pairs. Two water wings make it less likely to spin around and around as we swim.

The resulting system was "designed" by the cartoonist Rube Goldberg, not Occam. Figure 7.2 shows the basic mechanical arrangement for pumping air into and out of the lungs. To keep matters simpler, the sketch shows only one lung sac. The lung sac is modeled by the elastic balloon. The two side walls of the model can

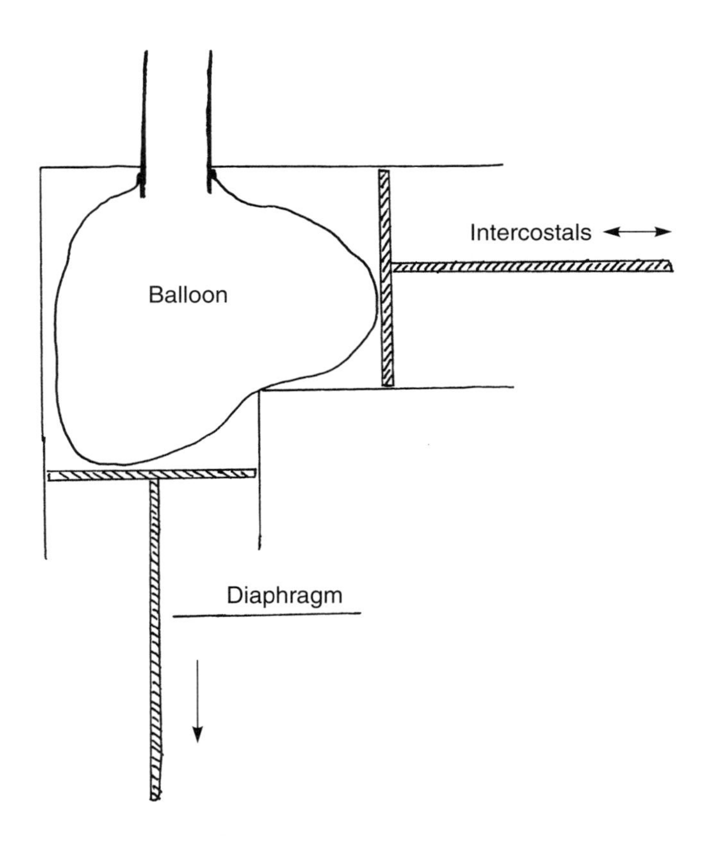

FIGURE 7.2 *Schematic diagram of one lung. The two pistons can change the volume of the lung balloon through the action of the intercostal muscles or the diaphragm muscle. Air and recoil energy are stored in the elastic lung balloon.*

move inward or outward as the rib cage is moved by the intercostal muscles. The volume inside the box formed by the rib walls, abdomen, and diaphragm thus can increase or decrease as the intercostal muscles move the rib walls outward or inward. The abdominal muscles can also change the box's volume, as can the diaphragm, which in this diagram moves in the vertical plane. Note that the lung-balloon is open to the outside and is not attached to the moving walls of the box. The space between the lung and the box, the pleural space, is sealed.

Inspiration involves filling the lung with air. It takes place in this model and in your respiratory system in an indirect manner. The walls of the chest can move outward, or the diaphragm can move downward. Either, or both, maneuvers will increase the volume of the box. The pressure within the sealed pleural volume, Pp, will fall as the volume of the pleural space, Vp, increases because Vp(Pp) = K, where K is a constant. Because the inside of the lung balloon is open to the outside air, the balloon will expand as the pleural air pressure falls, since the higher atmospheric air pressure pushes against the inside wall of the lung balloon. As the box expands, the lung balloon will stretch and expand. And as the lung balloon expands, it stores air and also stores energy in its "elastic recoil force." Think of a rubber balloon: when you blow it up, you are storing energy as you stretch its elastic walls. If you let go of the balloon's open end, that energy stored will force the air out of the balloon.

The lungs of human beings, after age three months, operate in a manner similar to the balloon-and-box model. The external intercostal and part of the internal intercostal muscles of the chest can move outward to inflate the lungs—which they can do because the ribs are slanted downward from the spinal column after the third month of life. Langlois and his colleagues (1980) showed that the near-perpendicular orientation of the ribs to the spine in early infancy makes it impossible to use these muscles to inflate the lungs. The diaphragm also can inflate the lungs by moving downward. In short, the energy that drives quiet inspiration is applied through the intercostal muscles and diaphragm. These muscles also store energy for expiration in the elastic recoil force of the lungs. During expiration, the elastic recoil forces air out of the lungs. Part of the internal intercostal muscles and the abdominal muscles can also act to force air out of the lungs.

Neural Mechanisms Regulating Respiration

Maintaining proper oxygen levels in the bloodstream and brain is clearly important, and layered regulatory mechanisms monitor breathing in humans and other animals to ensure that the respiratory system meets the physiologic demands of both normal and strenuous activities. A readable account is presented in Arend Bouhuys's 1974 book, *Breathing.* One layer of control involves mechanical stretch receptors in the lung tissue that send signals back via the vagus, or tenth cranial nerve to the brain. These stretch receptors monitor the degree of inflation of the lungs and activate a control system that limits the depth (that is, the magnitude) of inspiration. Herring and Breuer described this feedback control system over 150 years ago.

There are two additional layers of feedback control that make use of chemoreceptors that monitor the levels of dissolved carbon dioxide and oxygen as well as the degree of acidity or alkalinity (pH) of blood and cerebrospinal fluid. These feedback mechanisms are basic in that they sustain the ventilatory conditions that are necessary to support life. The layering most likely reflects evolutionary tinkering that was retained to maintain redundancy in the life support system. The two layers of chemoreceptor-actuated feedback are central and peripheral with respect to the brain. The central chemoreceptors are located near the ventrolateral surface of the medulla, or spinal bulb of the brain. The medulla is continuous with the spinal cord and is one of the most primitive parts of the human brain. They monitor the CO_2 and pH of both the cerebrospinal fluid and the blood that perfuses the medulla. Peripheral chemoreceptors are located in two places: near the bifurcation of the common carotid artery in the neck and in the aortic bodies. The aorta is the main artery that carries oxygenated blood from the heart. The peripheral chemoreceptors monitor pH and oxygen in the arterial blood (Dejours 1963; Bouhuys 1974).

The central and peripheral chemoreceptor feedback systems both overlap and complement each other. The peripheral system acts rapidly to make small changes in respiration. The central system operates slowly, but it can effect large changes in respiration. When healthy people breathe low concentrations of carbon dioxide in air (3–7 percent), their breathing rate, the depth of their breathing, and the volume of air that passes through their respiratory system per minute all increase. The chemoreceptors are quite sensitive. For example, they initiate increased respiratory activity when you breathe in a closed room with a number of other people, because the oxygen content of the stale room air is lower than it should be. The chemoreceptor feedback systems can operate rapidly; when you are breathing stale air, a single breath of pure oxygen will lead to a temporary reduction of ventilation. What's pertinent here is that we override all of these regulatory systems when we talk, sing, or play wind instruments. Linguistic-cognitive intent takes over.

Breathing while Talking—Prosody and the "Reality" of Sentences

As soon as we start to talk, we depart from the control pattern that prevails during quiet respiration. Speech production usually results in airflow rates that are too low to meet basic vegetative constraints during strenuous activities. As Figure 7.3 shows, though speakers in some cases adopt patterns of respiration that maintain optimum air transfer with flow rates compatible with intelligible speech, they usually give priority to the lower flow rates that are necessary for speech production and override the regulatory mechanisms. That explains why it is difficult to talk while you are running.

The primary departure from quiet breathing during speech is the variable duration of the expirations. The duration of each expiration is usually keyed to a linguistic construct—the length of the sentence that you *intend* to say. It has been evident for many years that the pattern of breathing during speech is quite different from that typical of quiet respiration with respect to the relative duration of

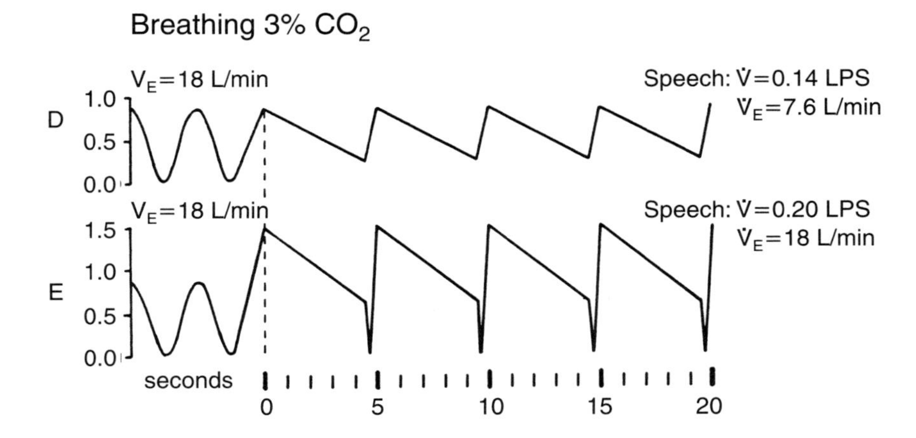

FIGURE 7.3 *Lung volume as a function of time for subjects breathing at an average flow rate,* V_E*, of 18 liters per minute (L/min) during quiet respiration, which is plotted to the left of the vertical dashed line. Note that the inspiratory and expiratory phases of respiration have about the same duration during quiet respiration. The plots to the right of the dotted vertical line show lung volume during speech for two flow rates: D at 7.6 L/min and E at 18 L/min. Note the long linear fall in lung volume during expiration, which is typical of respiratory activity during speech. (After Bouhuys 1974)*

the inspiratory and expiratory phases of respiration, the control of air pressure, and the muscles involved (Draper, Ladefoged, and Whitteridge 1960; Lieberman 1967; Klatt, Stevens, and Mead 1968; Mead, Bouhuys, and Proctor 1968; M. Lieberman and Lieberman 1973; Bouhuys 1974). During quiet respiration, the durations of the expiratory and inspiratory phases—when you take air into your lungs and exhale—are almost equal.

In contrast, when you talk, a short inspiration is followed by an expiration whose length is usually, though not universally, keyed to the length of a sentence or a phrase. Figure 7.4 shows this pattern. The duration of an expiration most often segments the flow of speech into sentence-like units during spontaneous speech (M. Lieberman and Lieberman 1973; Sandner 1981; Tseng 1981; Landahl 1982; Lieberman et al. 1984), but it is also apparent when speakers read material (Lieberman 1967; Atkinson 1973; Maeda

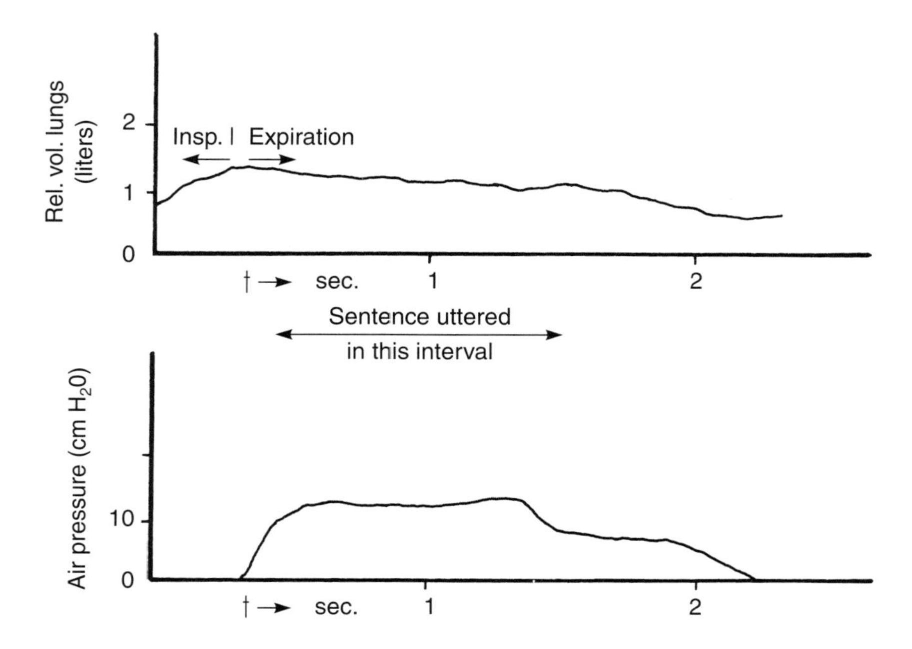

FIGURE 7.4 *Lung volume and air pressure in the lungs during the production of a sentence. Note the long length of the expiratory phase relative to inspiration. Note also the air pressure function, which does not gradually decrease as it does during quiet respiration as it follows the decrease in lung volume.*

1976). It also is evident that a speaker can, if she or he is prompted, further subdivide a sentence by segmenting smaller syntactic units with expirations (Armstrong and Ward 1926; Lieberman 1967; Atkinson 1973; Cooper and Sorenson 1977, 1981; Tseng 1981). Moreover, speakers can indicate that the sentence is not over by means of prosodic signals at the end of expiration (Armstrong and Ward 1926; Pike 1945; Lieberman 1967; Tseng 1981; Lieberman et al. 1984).

One point needs emphasis because it appears to have been overlooked by many linguists: the respiratory maneuvers that a speaker uses during normal discourse reveal both the psychological "reality" of a sentence and the planning involved in discourse. The data show that a speaker has in his or her mind blocked out the general framework of a sentence before uttering a single sound. Before a long sentence, a speaker generally draws more air into her or his lungs.

Therefore, there is more air and a greater elastic recoil force present at the outset of a long sentence than for a short sentence. Humans, being human, occasionally make mistakes and do not take enough air into their lungs before they start a long sentence. They "run out of air" when this happens. Small children often do this when they excitedly enumerate the animals at the zoo or the good things to eat at a party. Adults obviously also make "production errors"; but even in these instances, they tend to stop at points, as Armstrong and Ward (1926) note, at which the words of the sentence indicate that the sentence probably is not over. Indirect markers of breathing furnished by the amplitude, F0, and spectrum of the speech signal (the balance of energy at higher harmonics of F0) provide cues that signal that the sentence is not over.

The same pattern of "cognitive" respiration, wherein the duration of an expiration is determined by a higher-level structure, holds for respiration during wind instrument playing and during singing (Bouhuys 1974). The duration of the expiratory phase of respiration is keyed to the length of a musical phrase. The duration of expiration is not merely keyed to the constraints of vegetative breathing—that is, maintaining particular oxygen and carbon dioxide levels in the bloodstream. Instead it is influenced by seemingly abstract cognitive events such as the duration of a sentence, the duration of a major syntactic constituent (a relative clause), or the duration of a musical phrase. The intuitive relation between breathing and cognitive structure has been incorporated into a theory of poetic structure and practice. The American poet Charles Olson regarded breath control as an inherent structural element in poetry: "The line comes (I swear it) from the breath, from the breathing of the man who writes it, at the moment that he writes . . . for only he, the man who writes, can declare at every moment, the line, its metric and its ending—where its breathings shall come to termination" (1959, p. 2). Figure 7.5 shows the breathing pattern of a member of Olson's circle of poets while he was reading. Olson did not devise a notation to transcribe the appropriate breathing pattern. He instead depended on the poetic oral tradition that was partially transferred to orthography some few thousand years ago. The points at which expiration

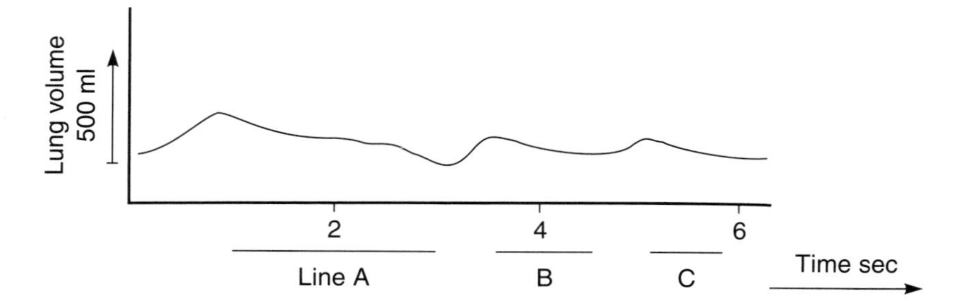

FIGURE 7.5 *Lung volume plot for a poet reading one of Charles Olson's poems. The individual lines of this part of the poem, A, B, and C, are each produced on an individual expiration. Note that the magnitude of inspiration is greater for the line that has a longer duration. The speaker has taken a deeper breath before he started to talk. (After Lieberman and Lieberman 1966)*

ends in the readings of Olson's poems do not always coincide with sentences, sentence-like intervals, or major elements of the phrase structure. The poem achieves a rhythmic structure through the interplay of the breathing pattern and the meaning derived from the strings of words. Words and breathing points that are syntactically anomalous achieve emphasis, and we become aware of the poet's breath. The Olson poems work because they violate the usually close relation between syntax and respiration (Lieberman and Lieberman 1973).

Speakers also often pause while they are talking, thereby interrupting the pattern of respiration when they stop to think. Goldman-Eisler (1958) related thinking time to pauses. When the pauses of individual air traffic controllers were measured as they spoke in a workload study (a simulation of communications with pilots landing simulated airplanes), pause durations sharply decreased when they were unable to think about potential hazards (Prinzo, Lieberman, and Pickett 1998). The pattern of respiration also changes for speakers under emotional stress (Lieberman and Michaels 1962; Protopappas and Lieberman 1997; Chung 2000). We really do not know how these phenomena intersect with the linguistic aspects of respiratory control. Indeed, information is lacking on how we balance the conflicting requirements of airflow and speech when we run or walk uphill.

Sentence Prosody and the Breath-Group

A breath-group is a phonologic construct that attempts to take account of biologic constraints reflecting the mark of evolution (Lieberman 1967, 1975, 1984).[1] The breath-group theory derives from earlier proposals (Jones 1932; Armstrong and Ward 1926; Stetson 1951) insofar as it notes that the flow of speech is often segmented into sentences, or other major syntactic units, by grouping words in a single expiration. The basic, "normal" breath-group reflects the biologic constraints imposed by breathing. The prosody of a sentence—the acoustic cues that delimit the end of a sentence—usually follow from articulatory maneuvers that represent the state of minimal departure from quiet respiration. Human speech is generally produced on the expiratory phase of respiration because we can generate a steady subglottal air pressure function throughout most of the expiration. However, this entails planning and executing a complex pattern of inspiratory and expiratory muscle activity. As we shall see, this requires anticipating the approximate length of the sentence before we utter a single word.

Most human languages for which phonetic data exist signal the end of a simple declarative sentence with a breath-group having these prosodic acoustic cues at its very end—a fundamental frequency (F0) contour that abruptly falls, with a concomitant fall in amplitude and the acoustic energy present in the higher harmonics of F0. These cues derive from the vegetative constraints of respiration. A positive air pressure must be generated in the lungs during expiration in order to force air out of the lungs. During inspiration, a negative air pressure must be generated in the lungs in order to pull air into them. If speech and phonation are prolonged until the end of an expiration, there has to be an abrupt terminal fall in the subglottal air pressure function. This is an absolute biologic constraint: anyone talking must open his or her larynx and must develop negative air pressure in the lungs in order to breathe. The larynx also must be open during inspiration. As noted in Chapter 3, it takes at least 60 msec to open the larynx from the constricted setting necessary for phonation. This means that the larynx is opening during the terminal portion of expiration. As the larynx opens, the mass of the arytenoid cartilages enters the phonatory cycle. The combination of increased mass and a falling subglottal air pressure results in the vocal cords moving more slowly, producing a falling F0 contour with less acoustic energy at higher harmonics of F0. Therefore, absent additional tensioning of the vocal cords (which would in itself raise F0), the fundamental frequency of phonation must fall at the end of the expiration, together with the amplitude and energy present at higher harmonics of F0.

In short, sentence prosody—the fundamental frequency of phonation, amplitude, and energy present at higher harmonics of F0—will abruptly decrease at the end of a breath-group unless the speaker executes a set of additional muscular maneuvers (tensioning the vocal cords) that are not necessary for breathing. The salient acoustic features of the breath-group that signal the end of a sentence are the terminal falls in the fundamental frequency of phonation and the amplitude of the speech signal. There is also a concomitant decrease in the energy of higher harmonics of F0. And a gap, a silent interval, necessarily follows during inspiration when the vocal cords are opened from their closed phonatory position to present an unobstructed air passage. These acoustic parameters follow from the articulatory maneuvers that are necessary to sustain life by breathing. It is absolutely necessary to change both the laryngeal opening and the alveolar air pressure in the lungs in going from the expiratory to the inspiratory phase of respiration. The vocal cords are detensioned as the lateral cricoarytenoid and vocalis muscles relax. The mean opening of the glottis increases, reducing both the aerostatic and aerodynamic forces generated by the airflow from the lungs. The vibrating mass of the vocal cords also increases as the posterior section of the glottis opens, putting the arytenoid cartilages into the vocal cord mass in motion. All of these factors lower the fundamental frequency, amplitude, and energy in higher harmonics of phonation (Van den Berg 1958; Atkinson 1973). Figure 7.6 illustrates the resulting F0 contour.

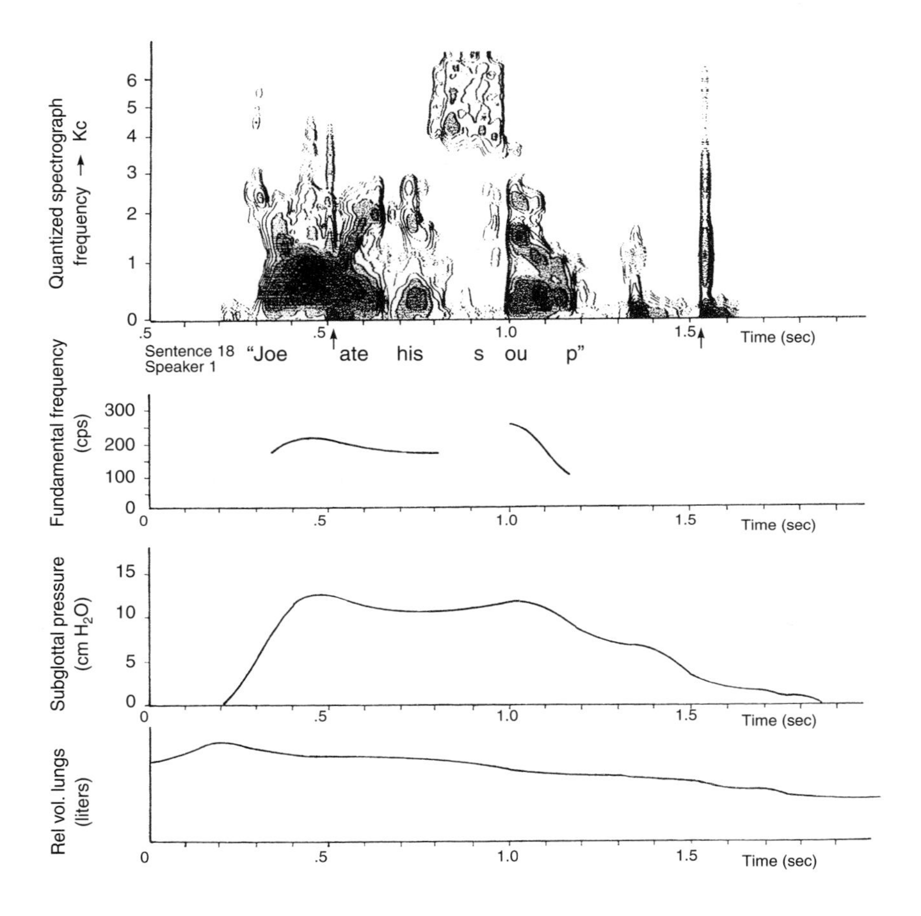

FIGURE 7.6 *Traditional auditory-based phonetic studies (e.g., Armstrong and Ward 1926; Jones 1932; Pike 1945; Trager and Smith 1951) and instrumental analyses (e.g., Stetson 1951; Lieberman 1967) identify a basic "intonation" contour or "tune" that delimits the scope of a sentence. The salient property of this intonation contour, termed the "normal breath-group," is a rapid "terminal" fall in F0 at the end of the expiration on which a sentence is spoken. The upper plot shows a spectrographic analysis of the sentence "Joe ate his soup." The plot below shows the F0 contour. The plot below the F0 contour shows the air pressure generated by the lungs that impinged on the vocal cords of the larynx. Note that it is almost steady throughout the sentence, falling at its very end. The lowest plot of lung volume shows the speaker's lungs inflating before he speaks and the gradual fall in lung volume as air is released to provide the driving force for speech production. (After Lieberman 1967)*

Perceiving Prosodic Cues

Adult listeners are able to segment the words of an unknown synthetic language using statistical procedures after a short familiarization to a continuous speech stream. Eight-month-old infants also are able to perform this act (Saffran, Aslin, and Newport 2001). When gap-like pauses are introduced, human listeners also can extract structure equivalent to the syntax of "natural" languages (Pena et al. 2002). Pena and her colleagues, who hold to the view that knowledge of syntax is innate, view breath-group pauses as subliminal and somehow conclude that these structural regularities are perceived by means of the listener's innate knowledge of syntax. But the perception of intonational cues is no more subliminal than the acoustic correlates of vowels or consonants. A somewhat more plausible explanation is that the adult listeners are responding to pauses in the synthetic language as they would to the breath-group pauses in their native natural language.[2]

In short, the state of minimum control for breathing yields the pattern of fundamental frequency and amplitude that defines the normal breath-group. It reflects the evolutionary history of the respiratory system, breathing, vocalization, and speech. Adult speakers of English appear to supplement these acoustic cues by increasing the duration of the last word or syllable of a sentence (Klatt 1976). This cue appears to be learned by children who are raised in an English-speaking environment. They gradually learn to shorten the duration of words that do not occur in a breath-group final position (Kubaska and Keating 1981).

Declination theories have been proposed that claim that F0 always gradually falls throughout the expiration (Pierrehumbert 1979; Liberman 1978; Maeda 1976), but this is not the case, especially for spontaneous speech (Lieberman et al. 1984). Although the F0 may vary in different ways in the nonterminal parts of the breath-group, it must fall at the end of an expiration unless steps are taken to counter the underlying constraints imposed by the necessity to breathe—the primary function of the respiratory system. It is possible to counteract a fall in subglottal air pressure by tensing laryngeal muscles (Muller 1848; Van den Berg 1958; Ohala 1970; Atkinson 1973). In many languages this forms the basis for a phonetic—in other words, linguistic—opposition with the normal breath-group, and some sentences can end with an opposing breath-group or tune (Armstrong and Ward 1926) that has a rising or level fundamental frequency of phonation. This is often the case for English yes-no questions.

Controlling Lung Air Pressure

Because the fundamental frequency and amplitude of phonation are functions of the subglottal air pressure (Van den Berg 1958; Lieberman 1967, 1968b; Atkinson 1973; Ohala 1970), it is necessary to stabilize the subglottal air pressure to avoid uncontrolled variations in perceived pitch and amplitude. Newborn human infants do not do this in their birth cries (Lieberman et al. 1972; Truby, Bosma, and Lind 1965), and the subglottal air pressure frequently rises to levels that are too high to sustain phonation during newborn cry; the vocal cords blow open and phonation is interrupted by breathy noise excitation of supralaryngeal airways (Truby, Bosma, and Lind 1965). Most children soon are able to control subglottal air pressure function throughout the duration of the breath-group. Whether the ability to control the articulatory patterns that maintain a steady nonterminal subglottal air pressure function is an innate characteristic deriving from our mammalian ancestry, or whether it is learned, is still an open question. However, infants do start to imitate the intonation of their caretakers between the sixth week and the third month of life (Lewis 1936; Lieberman 1984; Sandner 1981; Kuhl and Meltzoff 1996). Patients with profound bilateral damage to the basal ganglia also have problems in coordinating breathing activity with tongue, lip, and laryngeal maneuvers during speech production (Pickett et al. 1998). This suggests the involvement of the striatal component of the circuits regulating other aspects of speech production.

Surprisingly, we lack evidence from comparative studies that might suggest whether the complex muscle commands necessary to stabilize subglottal air pressure are learned or genetically transmitted. Lenneberg, in one of the first studies of the biologic bases of language, claimed that human beings are "endowed with special physiological adaptions which allow us to sustain speech driven by expired air" (1967, p. 81). Lenneberg is correct insofar as humans have this ability, but virtually all terrestrial animals vocalize on the expiratory phase of inspiration. This follows from the conservative nature of evolution. As is the case for human speech, animal vocalizations take place on the expiratory phase of respiration because the elastic recoil force of the lungs provides most of the power for phonation.

However, this fact presents some control problems. The air pressure in a speaker's lungs gradually falls during expiration in the absence of speech. This gradual decrease of air pressure follows from the gradual decrease of the elastic recoil force of the lungs during expiration (Mead, Bouhuys, and Proctor 1968). The elastic recoil force is greatest when the lung is stretched to hold more air at the start of the expiration. As air flows out, lung volume decreases; the lung is stretched less and less, so the elastic recoil force gradually decreases. The situation again is similar to the elastic recoil force generated in a rubber balloon: it is greatest when the balloon is blown up and decreases as the balloon empties.

The situation is quite different when a person talks. The speaker maintains a reasonably steady air pressure throughout a long expiration until its end, when the air pressure abruptly falls. Figure 7.7 shows lung volume and air pressure during the production of a sentence. Electromyographic recordings of the activity of the intercostal and abdominal muscles show that human speakers maintain a steady subglottal air pressure throughout an expiration by setting up a preprogrammed pattern of muscular activity. They first use their inspiratory intercostal muscles to oppose the force developed by the elastic recoil force of the lungs and then gradually bring their expiratory intercostal and abdominal muscles into play to supplement the elastic recoil force as it gradually falls (Draper, Ladefoged,

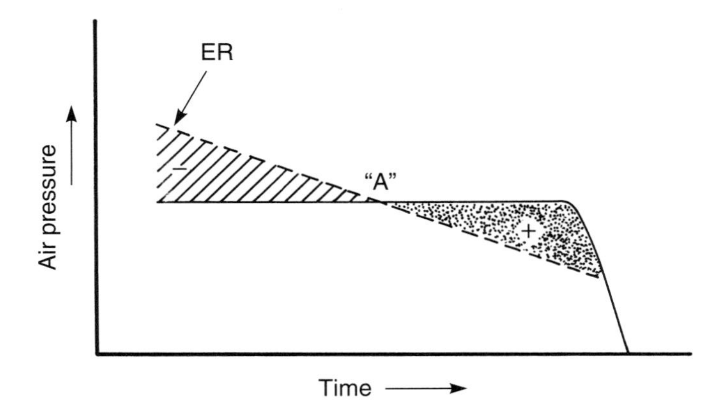

FIGURE 7.7 *Regulation of alveolar air pressure during speech. The elastic recoil (ER) of the lungs generates the linear falling air pressure indicated by the interrupted line. At the start of a long expiration, this air pressure exceeds the level suitable for phonation. The speaker opposes the elastic recoil, generating a force indicated by the diagonally lined area. When the elastic recoil force reaches point A, the speaker starts to supplement it with the force indicated in the stippled area. The net result is the steady air pressure indicated by the solid line throughout the nonterminal portion of the breath-group.*

and Whitteridge 1960; Mead, Bouhuys, and Proctor 1968). The diaphragm is almost inactive in this process, probably because it lacks proprioceptive muscle spindles that would send monitoring signals back to the nervous system.

The diagram in Figure 7.7 illustrates the control problem. The vertical axis plots lung volume and air pressure, and the horizontal axis, time. The interrupted line shows the linear air pressure function that would result from the gradual decrease of the elastic recoil force as the volume of air in the lungs gradually falls during expiration. (Note the almost linear decrease of lung volume during speech production in the diagram.) The solid line shows the level subglottal air pressure that a speaker usually produces in the nonterminal portion of an expiration (Lieberman 1967; Ohala 1970; Atkinson 1972).[3] The elastic recoil function intersects this air pressure line at point A. The speaker initially produces the steady air pressure function by opposing the elastic recoil force by pulling outward on the rib cage with the inspiratory intercostal muscles until point A, as the diagonally lined area indicates. After point A is reached, the speaker must supplement the elastic recoil force by compressing the lungs with the expiratory intercostal and abdominal muscles. The stippled area indicates the gradual increase of expiratory muscle activity that supplements the falling elastic recoil force.

As I pointed out, the connection between anatomy, motor control, and the regulation of subglottal air pressure rests on the fact that the system that we use could not work unless our ribs slanted downward from our spine. The geometry of the downward slant and the insertion of the intercostal muscles into the ribs allows the inspiratory intercostals to expand the chest and oppose the elastic recoil force to effect a steady subglottal air pressure at the start of an expiration. Newborn humans, whose ribs are almost perpendicular to the spine, cannot do this. They do not start to use the control pattern that adults use for expiration during speech until about the third month of life, when their ribs have assumed the adult-like configuration. Subglottal air pressure cannot readily be measured in newborn infants, nor are such data available for monkeys. The ribs of adult chimpanzees and gorillas slant downward, so they too have the anatomy necessary to regulate subglottal air pressure during expiration. Preliminary acoustic analyses of the vocalizations of chimpanzees in the Gombe Stream Reserve suggest that they regulate subglottal air pressure during expiration. Some of the chimpanzee vocalizations involve long call sequences. It would be virtually impossible for human beings to achieve similar patterns unless they regulated the subglottal air pressure throughout the utterance. However, we still do not know how chimpanzees produce these calls; further research on this issue clearly would yield insights into the evolutionary bases of human language.

Mothering and the Breath-Group

The evolutionary antecedent of the human breath-group may go back in time to the first mammals. Mammals evolved from the therapsids, mammal-like reptiles who lived long before the dinosaurs, between 230 and 180 million years ago. Therapsid fossils have been found on every continent (including Antarctica) because the world then consisted of one large continent, the great land mass of Pangaea. MacLean (1986) pointed out two changes in behavior marking the evolutionary transition from reptiles to mammals that most likely account for the evolution of the separation cry and the ultimate adaptation of this cry for linguistic purposes: (1) mammals nurse their infants and (2) infant mammals have to maintain contact with their mothers. Therefore infant mammals produce separation or isolation calls (the terms are used interchangeably) when they are separated from their mothers, who respond to these vocalizations.

Figure 7.8 shows the similarities between the separation calls of a squirrel monkey, a macaque monkey, and a human. The normal crying pattern of human infants conforms to the general primate form (Lieberman et al. 1972; Newman 1985; Hauser and Fowler 1991). The pitch of the infant's voice first rises, stays almost level, and then falls. We retain this vocalization pattern, although we have adapted it for language, using it to segment the flow of speech into sentences.

This derived mammalian behavior appears to devolve on the evolution of "matched" anatomic and neural systems—an attribute that I believe generally characterizes the biologic bases of speech production and other aspects of human linguistic and cognitive ability. Several mammalian specializations, apart from mammary glands, work with the brain to achieve successful nursing. If vocal communication is to be effective in maintaining the mother-infant bond, the mothers obviously have to be able to hear their infants' separation calls. Therefore, unlike reptiles, mammals have a middle ear, which allows them to hear quieter sounds. The middle ear has two small bones, the malleus and incus (hammer and anvil), which evolved from two small bones of the joint that hinges the jaw to the skull. These bones, which were still part of the jaw hinge in therapsids, act as a sort of mechanical amplifier. The first true mammals

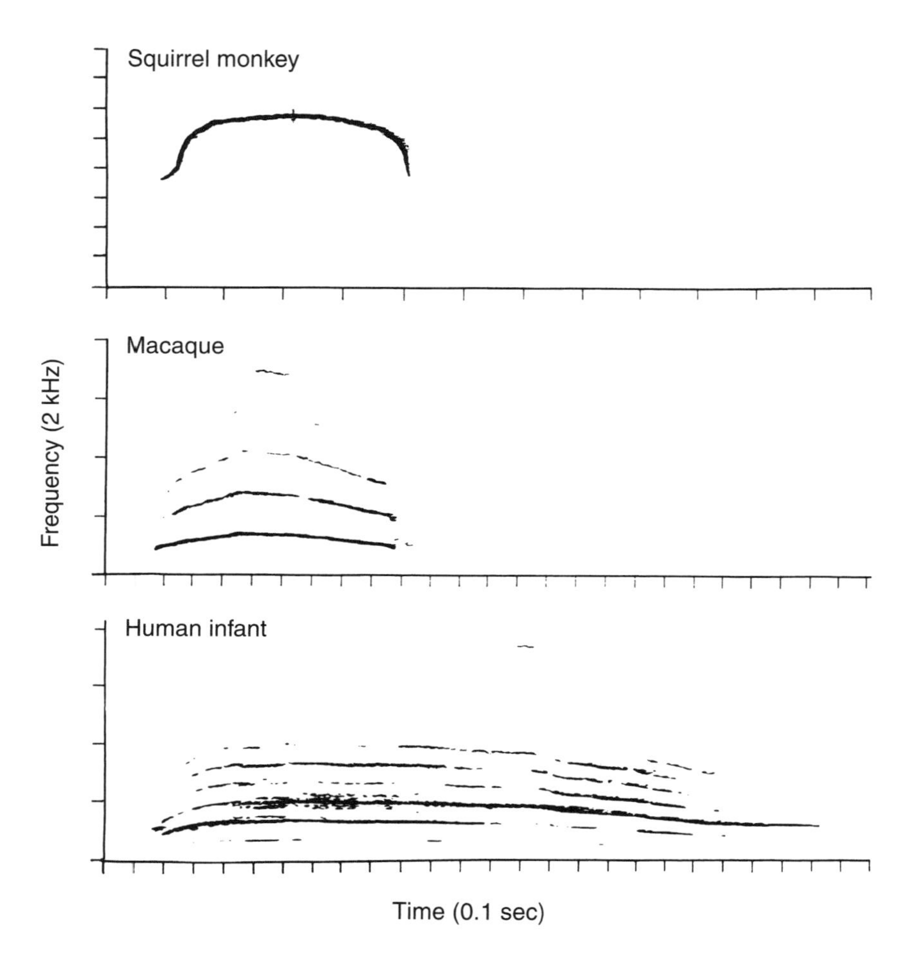

FIGURE 7.8 *Spectrograms showing the fundamental frequency contours (F0) for the separation calls of a squirrel monkey, a macaque monkey, and a human infant. The horizontal axes plot time. The vertical axes show F0 and some of its harmonics. (The harmonics occur at multiples of F0—for example, 2F0, 3F0, 4F0, and so on. The F0 pattern is similar for these primates over a complete expiration, despite different average F0 ranges and durations. (After MacLean and Newman 1988)*

can be identified in the fossil record because they had middle ears. These bones suggest that therapsids also had the neural substrate that is implicated in mother-infant interactions in living species—the anterior cingulate cortex. Lesions in the cingulate cortex can extinguish the isolation cry and infant care (Stamm 1955; Slotnick 1967; Newman and Maclean 1982; Newman 1985) as well as result in mutism (Cummings 1993).

As I pointed out in Chapter 4, the anterior cingulate cortex is activated in circuits that direct attention to some ongoing activity as well as in circuits regulating phonation. Studies of the neural circuits that regulate human speech show that the anterior cingulate cortex is implicated in laryngeal control (Cummings 1993). Neurophysiologic studies show that the anterior cingulate cortex also plays a part in regulating maternal behavior triggered by the separation cry as well as the production of this cry. Experiments that destroy part of the brain in rats show that lesions in the cingulate cortex result in marked deficits in maternal behavior (Stamm 1955; Slotnick 1967). The lesioned rats neglected nest building and nursing and retrieving pups; only 12 percent of their pups survived. The separation calls of primates also are regulated by circuits that include cingulate cortex. A series of experiments that MacLean started in 1945 shows that stimulation of the anterior cingulate cortex elicits the separation call in the monkey. The almost routine activation of the anterior cingulate cortex in fMRI studies, which is associated with attention, may have its evolutionary antecedent in a mother's need to pay attention to her infants.

The evolution of the anatomic and neurologic substrate serving the mammalian isolation call clearly increased biologic fitness. The neural substrate involved in generating and responding to the isolation call is most likely guided by an innately specified grammar. We know that cortical-striatal-cortical circuits are involved (Cummings 1993) and that Parkinson's disease interferes with the linguistic uses of respiration (Illes et al. 1988) as well as vegetative respiratory activity. However, we lack data from the comparative studies that might cast some light on how humans came to make use of this same system for linguistic ends. Proposals for studies of the long-distance calls of chimpanzees (noted by Goodall 1986) have in past years been rejected by advisory panels dominated by linguists, who take the absolute uniqueness of human language as a given. Perhaps it is time for a change.

Segmental Phonologic Features

But although prosodic cues clearly play a part in spoken language, the focus of most phonologic studies is on phonemes—the vowels and consonants of the words transcribed in writing systems such as English. Though English orthography is not as close to a phonetic transcription of speech as French, phonemes roughly correspond to the segmental letters of the alphabet that differentiate words.

In linguistic theories such as that proposed by Roman Jakobson, one of the most productive linguists of the past century, phonemes are themselves fractionated into "features"—atomic units that, when combined, define a phoneme. For example, the vowel [a] is composed of a cluster of features such as grave, compact, tense, and so on. The features each have acoustic correlates that, according to the theory, allow a listener to perceive the feature as well as a set of articulatory gestures that are used to produce the acoustic correlates that convey the feature (Jakobson, Fant, and Halle 1952). For example, the acoustic correlate of the feature "grave" is a low-frequency energy concentration resulting from formant frequency patterns at the low end of the acoustic spectrum; the acoustic correlate of the feature "compact" is a central energy concentration that followed from formant frequencies convergence. The acoustic correlate of the feature "voiced" is phonation, produced by laryngeal maneuvers. The features all have binary values. A phoneme thus could be either + grave or − grave.

Phonologic rules in Jakobson's linguistic studies attempted to describe sound changes such as those documented in the historical record by means of binary changes in the features that specified a phoneme. A sound change that involved a single feature like voicing, such as "vater" to "father" (the + voiced "v" became − voiced), was judged to be more probable than one involving several features. Although the binary aspect of this theory can be disputed, Jakobson attempted to take into account biologic constraints and the fruits of quantitative acoustic speech analysis. His theory was subject to empirical test and many of Jakobson's specific hypotheses have not held up. However, that is not the central problem with the feature systems commonly used in current phonologic studies. Unfortunately, linguistic research strayed from Jakobson's path and my intent here is essentially to reintroduce linguists to Jakobson and his approach, which took into account biologic principles and empirical facts. I do not attempt to present a comprehensive set of segmental phonologic features; to do so would call for extensive research similar in nature to that documented by Kenneth Stevens (1998), which was directed toward resolving linguistic issues.[4]

A few examples may suffice to point out how linguists could advance phonologic enquiries if they took account of even two of the issues that were discussed in some detail in previous chapters—vowel production and voice-onset time.

Vowels

Linguists have long been aware of the fact that languages change through time. We don't speak Middle English, nor do contemporary children necessarily preserve the syntax and sounds of the dialect that their parents and grandparents speak (Labov, Yaeger, and Steiner 1972). One of the goals of linguistic science has been to describe and account for these changes. Chomsky and Halle (1968), in one of the founding studies of generative phonology, *The Sound Pattern of English,* attempted to describe historical changes in the vowel system of English. However, the "features" that they used bear little relation to the manner in which vowels are produced or perceived, and they do not provide insights on how the vowel system shifts from one generation to the next over the course of time.

The system that is now commonly used by linguists to describe how vowels are produced derives from Melville Bell, Alexander Graham Bell's father. In the final stages of editing, Chomsky and Halle (1968) abandoned the feature system proposed by Roman Jakobson and reverted to Bell's traditional theory, thereby setting the course for subsequent phonologic studies.

In the middle years of the nineteenth century, Bell devised a system aimed at teaching deaf people to talk. Bell believed that the position and shape of the tongue in a person's mouth was the primary determinant of vowel quality. He provided deaf persons with a set of instructions that he believed would enable them to speak, such as "move your tongue down and back in your mouth to produce the vowel [a]." Bell, of course, had neither radiographs nor MRIs available to confirm these hypothetical tongue positions. Theoretical linguists continue to use these tongue features to the present day in phonologic studies. Binary features such as − low, + high, and + back generally are used to specify tongue position. The binary notation that is necessary to code the vowels of languages such as English, which supposedly have vowels with intermediate "heights," also makes use of feature combinations specifying them, such as being both + high and + low. These combinations have an arbitrary quality (a "high" tongue position seemingly cannot be "low"). But these supposedly binary systems make it possible to code intermediate positions of the tongue such as those that supposedly represent the lower position of the tongue in a word such as "bet" compared to its position in "bit."

Unfortunately, the tongue positions that are central to the Melville Bell articulatory vowel feature system have no basis in reality. As I pointed out in Chapters 3 and 6, this has been evident in a series of independent studies that date back to 1928. Over the course of more than seventy years, the tongue shapes and positions that are necessary to produce the vowels of English and a number of other languages have been established in independent studies that, as time progressed, used still radiographs (X-rays), cineradiography, and MRIs (Russell 1928; Carmody 1937; Perkell 1969; Ladefoged et al. 1972; Nearey 1979; Perkell and Nelson 1982; Baer et al. 1991;

Story, Titze, and Hoffman 1996; Hiiemae et al. 2003). Reviewing these findings shows that the shape of the tongue is almost identical for all vowel sounds and is moved about as an almost-undeformed body. Some speakers produce all of the vowels of English (the most intensively studied language) with their tongues in the *same* position, except for the quantal vowels [i], [u], and [a]. These speakers use lip maneuvers (protrusion and constriction) and adjustments in larynx height to generate the formant frequencies that specify different vowels. Moreover, different speakers employ different tongue positions when they produce the same vowel.

As Nearey (1979) showed, the traditional vowel height system does not even describe the relative differences in tongue height that supposedly distinguish close vowel pairs such as [I] versus [e]. The vowel [I] supposedly is always produced with a higher tongue position than [e], but some speakers produce [e] with a higher tongue position. Some speech researchers (for example, Ladefoged et al. 1972; Nearey 1979; Lindblom 1988) have concluded that gradual shifts along acoustic dimensions account for sound changes over time and space. The few linguists who have actually studied ongoing changes in the manner in which words are pronounced across generations, such as William Labov, concur (Labov, Yaeger, and Steiner 1972). The binary tongue features of mainstream linguistic research lead to implausible categorical jumps along nonexistent specifications of tongue position.

Melville Bell was wrong, but he cannot be faulted because neither X-rays nor MRIs that would have allowed him to test his theory existed in his lifetime. But the present situation is different. We must conclude that the hermetic character of linguistics accounts for the fact that Bell's tongue features are retained. The cost is not trivial. The convoluted manipulations of binary tongue features in Chomsky and Halle (1968) obscure the probable mechanism involved in the historic sound patterns of English and other languages. As is the case for the present-day vowel sound changes documented by Labov, these vowels shifted along acoustic dimensions, generally toward the highly valued quantal vowels [i], [u], and [a].

Stop Consonants

Other facts drawn from research on the production and perception of speech could advance phonology. Many phonologists continue to attempt to characterize the "manner" distinction that differentiates a stop consonant such as [b] from [p] by means of the binary feature "tenseness." However, the voice-onset time dimension discussed earlier that actually differentiates these consonants is inherently a three-part distinction. Studies dating back to 1964 (Lisker and Abramson) show that in many languages other than the "universal" language of English, words can be specified by stop consonants that differ in whether the "release" occurs before the onset of phonation, near the onset of phonation (English [b], [d], or [g]), or after the onset of phonation (English [p], [t], or [k]).

English makes use of only two of these categories and a traditionalist might argue that the distinction between a [b] and a [p] is that the [p] is "tense" whereas [b] is "lax." However, there is nothing tenser about a [p] than a [b]; the tension of the muscles that open a person's lips for these consonants is similar. The tension of the vocal cords also appears to be similar (Atkinson 1973). The only reason for the feature tenseness would appear to be Occam's dull razor—simplicity. The long vowels of English such as [i] and [æ] (the vowel of the word "bat") can be differentiated from the short vowels by the label "tenseness," which never has been shown to correspond to a speaker's using increased muscle tension to differentiate these vowels. If duration were used to differentiate the long versus short vowels of English, and VOT to differentiate stop consonants, two features would be needed in place of one—the putative feature "tenseness." But the guiding light of many formal linguists is vaguely defined "simplicity" or "parsimony"—in practice, using the minimal number of features to specify all of the sounds of speech.

But parsimony is for misers, not well-designed communications systems. Using distinctions in duration as a cue to differentiate vowels that have adjacent formant frequency patterns, such as [i] and [I], is a clear case of language making use of multiple, redundant acoustic cues to convey meaningful information (Lieberman and Blumstein 1988; Lindblom 1988). The appropriate phonologic solution is analogous to that employed by real-world electronic communications systems—multiple redundant information transfer gets the message through.

If we take perception as well as articulation into account, other useful insights on the nature of the sound pattern of language fall into place. The traditional linguistic binary features used to characterize the "place" distinctions that differentiate a consonant, such as [d] from [g], do not appear to have an articulatory basis. There are no apparent anatomic constraints on where the tongue blade can form a constriction along the roof of the mouth (the hard palate). Studies of swallowing show that our tongues effortlessly propel food along the entire length of the palate. The factor that structures the consonant sound patterns of many languages seems to be Stevens's (1972) quantal factor. Stevens showed that, at certain locations along the roof of the mouth, a constriction will result in two formant frequencies converging. These formant frequency convergences yield readily perceived peaks in the frequency spectrum, analogous to saturated colors. But unfortunately, virtually all phonologists hold to binary all-or-nothing features coding hypothetical articulatory maneuvers.

Modularity

One of the dubious contributions of linguistic research to cognitive science has been the concept of "modularity." According to Fodor's (1983) influential book, the functional organization of the mind-brain involves a set of modules that instantiate a self-contained process. Modular models (for example, Levelt 1989) propose that, in perceiving speech, we start with a module that first converts the acoustic signal to a set of discrete phonetic features. Another module accepts the output of this first module and discerns phonemes, a third module recognizes words formed by sequences of phonemes, and so on. The input from one module is not available to the next. Studies such as Utman, Blumstein, and Sullivan (2001) show that this is not the case. Semantic word priming occurs when a person who is asked to signal whether a word is an actual English word when preceded by a semantically related word. For example, subjects consistently respond faster to the word "dog" when it is preceded by the word "cat." Minute alterations to the voice-onset time (VOT) of the word "cat" diminish the effect. According to modular theory, listeners should have no knowledge of the acoustic parameter VOT at this point—they should have grouped the phonemes into the word "cat" and should have no access to the VOT parameters that specify the initial consonant of the word "cat."

However, stronger evidence that does not entail computer-implemented modification of VOT or other acoustic parameters refutes modular theories for both speech perception and "higher" levels of language processing, such as word recognition and the syntactic processing that leads to sentence comprehension. In the 1960s, it was apparent that the speech produced in normal conversational situations was underspecified. Pollack and Pickett (1963) found that 200-msec segments cut out of the stream of speech were generally unintelligible, even when recordings were made under optimal conditions. The message would suddenly "pop up" as the duration of the excised segment was increased. In some cases the apparent message was incorrect and the listeners "heard" words that the speakers never had intended and had not uttered. The apparent explanation was that listeners formed a running hypothesis on the basis of impoverished acoustic cues; when the hypothesis was consistent with an informed guess based on the listeners' expectations, they "heard" the message, reconstructing nonexistent phonetic detail from their expectations.

Pitt and Samuel (1995), Samuel (1996, 1997, 2001), and Shockey (2003) have replicated and refined the Pollack and Pickett (1963) study. Similar effects occur for the lexical word tones of Chinese (Tseng 1981), which can be "heard" though they are not produced when Chinese speakers converse. Speakers seem to be selectively sloppy when they believe that a listener can guess the word from its context or prior knowledge (Lieberman 1963b). In an exhaustive study, Johnson (in submission) has shown that the effect is general. When people talk, their speech output is underspecified. What we "hear" is in part determined by what we think we are hearing.

One obvious conclusion is that perceptually based phonetic transcriptions of a foreign language or unfamiliar dialect should be supplemented by objective acoustic analyses. Apart from the effects of limits on auditory-based transcriptions such as the perceptual magnet (Kuhl et al. 1992), which tilts the categorization of speech signals into the frame of one's own linguistic background (see Chapter 3), the phonetician's knowledge of a message's context affects the transcription. Objective acoustic analysis should be part of the linguist's tool kit. Modular theories cannot account for how we understand the meaning of conversational speech. Human listeners clearly do not employ a series of sequential modular processors. Perhaps it is time to consider theories that invoke distributed, parallel processing, which appears to better approximate the "computations" that occur in biological brains.

Universal Grammar?

I recall a session at the Boston University Language Conference at which the audience, who adhered to the theories of Noam Chomsky, became electrified. An apostate returned to the fold; she declared that she had been in error–she too now "believed in Universal Grammar." Her words were appropriate–she might well have declared that she believed in the Holy Trinity, or the power emanating from the image of the Tibetan god Mahakala. According to Piatelli-Palmarini (1989), Universal Grammar is an established, indisputable truth. At a meeting in Florence, Italy, on the genetic bases of language change, he declared that there is a worldwide "religion" whose creed is Chomskian grammar. But Universal Grammar is a biologic claim that must be consistent with the body of knowledge that has steadily grown since Charles Darwin's time. The plausibility of UG, therefore, rests on the findings of biology.

As I pointed out at the beginning of this book, Chomsky's central claim is that human syntactic ability derives from an innate organ of the human brain—the Universal Grammar, in which all of the syntactic rules and principles of every human language that currently exists, has ever existed, or will ever exist is genetically coded. The UG supposedly makes it possible for a child to effortlessly acquire a language without formal instruction from the impoverished speech that she or he hears. The basic argument by Chomsky and his supporters is that the speech of the adults and older children that a young child hears is sloppy and often ungrammatical. Yet all children, they claim, attain knowledge of the one true grammar of the language to which they are exposed.

No objective data support this claim, though this deficiency appears to be irrelevant to Chomsky's adherents, whose clinching argument is the productivity of language. People cannot attain linguistic ability by simply imitating the words or sentences they hear. No person ever hears the infinite number of possible sentences that she or he has the capacity to produce. This argument has been repeatedly presented—for example, Chomsky (1986), Bickerton (1990), Pinker (1994, 1998), Jackendoff (1994), and Calvin and Bickerton (2000). The principles and parameters coded in the Universal Grammar are supposedly triggered to yield the correct grammar of a particular language when a child is exposed to normal discourse. If this theory were true, humans would have a genetically specified, detailed program for language similar to that of duck hatchlings that "acquire" calls after limited exposure to tape recordings of duck calls (Gottlieb 1975).

Chomsky is not entirely off the mark. It would be folly to claim that innate capacities do not play a role in shaping human language. Human children clearly are predisposed to acquire language. Infants learn to respond to the sounds of their native language early in life (Eimas et al. 1971; Eimas 1974; Grieser and Kuhl 1989). The brains of three-month-olds respond to speech signals in much the same manner as those of adults (Dehaene-Lambertz, Dehaene, and Hertz-Pannier 2002). Children respond differentially to the vowel sounds of their native language by age six months (Kuhl et al.

1992). Young children start to communicate with a mixed array of gestures, eye contact, and sounds as they gradually move towards the adult model (Greenfield 1991). Any child younger than six years will acquire any language on earth with native fluency by simply living with the language (Johnson and Newport 1989). Deaf children will attempt to establish language-like communication by means of gestures (Goldin-Meadow 1993).

However, the question is the specificity of the innate biologic substrate involved in acquiring language. Children learn all manner of culturally specific behavior, including language, by means of such processes as imitation or associative learning (for example, Meltzoff and Moore 1977, 1983; Lieberman 1984; Tomasello and Farrar 1986; Gopnik and Meltzoff 1985, 1987; Donald 1991; Greenfield 1991; Karmiloff and Karmiloff-Smith 2001; Saffran, Aslin, and Newport 2001; Tomasello 2004a, b). Indeed some of the sound contrasts that human children acquire early in life, such as the VOT distinctions that differentiate stop sounds (Eimas et al. 1971; Eimas 1974), are also acquired by rodents (Kuhl 1978, 1981, 1988), birds (Kluender, Diehl, and Killeen 1987), and computer-implemented neural nets that instantiate associative learning (Seebach et al. 1994). Other neural nets can acquire fragments of syntax (Elman et al. 1997). Capacities that are equal to or exceed those of birds, rodents, monkeys, or simple computer networks surely are present in human infants and children.

In short, it is clear that some innate capacities must exist that allow humans to effortlessly acquire the complex syntactic processes and sound patterns of human language. But two questions must be addressed: are the details already built into the human brain, and what constraints exist on the form of possible grammars and the nature of any language-specific mechanisms? The seductively simple solution offered by Chomsky is that virtually all of the rules of syntax are specified in the innate UG. When a child is exposed to a fragment of a particular language, the UG triggers a detailed representation of the language's syntax.

Some aspects of life clearly devolve from innate neural mechanisms. In the minutes after birth, breathing must be established. There is little time in which the newborn infant could learn to breathe; the motor control patterns are genetically specified and there is a normal progression of respiratory activity in the hours after birth. However, even in this act, variation occurs and medical intervention is necessary to save the lives of infants who exhibit life-threatening deviations from the normal pattern (Bouhuys 1974; Langlois, Baken, and Wilder 1980). Kagan, Reznick, and Snidman (1988) present evidence for a genetic component to shyness. However, as Ernst Mayr (1982) has forcefully pointed out, population thinking, which takes account of the constant presence of variation, is the model for biologic inquiry. Variation is the feedstock for selection and Darwin's *On the Origin of Species* starts by discussing variation. Barring the biologic bases of language being unique—the gift of some god—biologic variation rules out any form of Universal Grammar (UG). As I pointed out in 1984, some "normal" individuals would have markedly different capacities for acquiring a particular language if a UG existed. Genetic variation is a given and some individuals would lack a particular genetically transmitted "parameter," "markedness condition," or aspect of "optimality" (the terms keep shifting as generative grammar jumps to a "new" theory). Color blindness does have a genetic basis. A color-blind child does not see colors. Imagine a color-deficient child being asked to master an orthographic system that used colors to transcribe words. The same child could learn to read using the letters of the alphabet. If a UG really existed in the human brain, a similar situation would occur. A particular "parameter-deficient" child might be able to acquire Italian but not English, if these languages differed with respect to the missing fragment of the UG's genetic code. But that doesn't happen—place a child in any group talking and acting in a normal manner and the child will acquire the group's language.

Moreover, an innate UG would have to result from the same biologic processes that yield any of the properties of the human brain or body. But we do not have to speculate on how natural selection can shape the biologic endowment of human beings. Biologic fact can be used to test the premise that a detailed, genetically transmitted UG exists. Natural selection acts on population isolates to adapt them to their environment. For example, yaks are buffalo-like animals that have adapted to living on the high Tibetan plateau. Yaks have respiratory systems that differ from those of lowland buffaloes and that allow them to effortlessly carry heavy loads over long distances at altitudes exceeding 3500 meters above sea level. Yaks have no trouble crossing 6000-meter mountain passes. I once briefly rode a horse climbing a 5000-meter Himalayan pass; it became so breathless that I dismounted. No student of evolutionary biology would argue with the premise that natural selection acting over many generations of yaks favored the survival of individual yaks and their progeny that had respiratory systems that functioned efficiently at these extreme altitudes.

Similar selective pressures and adaptation mark the people who share the Tibetan plateau with yaks (Moore, Niermeyer, and Zamudio 1998). As I have noted in this chapter and earlier, breathing is regulated by innate mechanisms; human infants do not have to learn to breathe. A set of chemoreceptors immediately comes into play after birth to regulate the amount of air that an infant must breathe in order to transfer sufficient oxygen into the bloodstream. Increased physical activity requires more air. However, respiratory physiologists studying large numbers of people living at low altitudes have found a great deal of variation in respiratory efficiency. Some persons must inhale almost 10 liters of air to extract the same amount of oxygen as other individuals inhaling one liter (Bouhuys 1974). At low altitudes, in oxygen-rich air, these variations have little effect on the individual's biologic fitness, defined as the survival of that individual and his or her children. However, the low oxygen content of the air at altitudes exceeding 3500 meters places a premium on efficient lung-to-blood oxygen transfer.

Tibetans, like all human beings, derive from people who lived in Africa at comparatively low altitudes. In the period (estimated at 40,000 years) in which Tibetans have lived at extreme altitude, Darwinian natural selection has acted on the pool of respiratory variation in the original Tibetan population isolate. Tibetans as a group have exceedingly efficient oxygen transfer capabilities; they can transfer the same amount of oxygen into their bloodstream from the thin air at extreme altitudes with far less effort than most lowlanders. A period of 40,000 years appears to be sufficient for natural selection to enhance innate characteristics; similar selection involving mitochondrial DNA has enhanced cold tolerance in arctic populations (Wallace 2004). Therefore, Tibetans have very efficient innate oxygen extraction capabilities (Moore, Niermeyer, and Zamudio 1998).

The innate nature of this capability is evident in Tibetans who have been born and raised at low altitudes. For example, on Mount Everest, Sherpa guides whose ancestors left Tibet more than 300 years ago and who were born and spent their youths at comparatively low altitudes in India have little difficulty breathing compared to most of their European, Japanese, Korean, and other lowland clients. Tibetans are able to survive in oxygen-rich low latitudes; they simply don't have to transfer as much air through their lungs as many other people. Perhaps Tibetans will someday establish new long-distance track records in the Olympic Games.

Other adaptations to extreme altitude differentiate Tibetans from lowlanders. The birth weight of infants born at high altitudes to Tibetan women is about 500 grams greater than that of infants of Han Chinese women who have recently emigrated to Tibet. Tibetan women are genetically different from these Chinese women; they deliver more oxygen to their developing fetuses by means of increased placental blood flow (Moore et al. 2001). In other words, natural selection has acted on Tibetans, who form a population isolate, filtering out variations to enhance both breathing efficiency and oxygen delivery to the fetus during pregnancy.

How does this biologic evidence bear on UG? During this same 40,000-year period, Tibetans have been speaking Tibetan; and as any English-speaking person attempting to learn that language knows, it has remarkably different syntactic rules. If we indeed acquire our native language from detailed, genetically transmitted knowledge of its syntax, then natural selection surely should have yielded an optimal Tibetan UG by filtering out the principles and parameters that apply to Indo-European languages. These superfluous Indo-European entries in the UG would only complicate the process of acquiring Tibetan. The Tibetan UG, if it existed, would facilitate Tibetan children's acquiring Tibetan but would impede their acquiring Hindi or English. But no evidence supports this scenario. Tens of thousands of Tibetan children have been born and raised in India since the Chinese invasion of Tibet. These Tibetan children have no more difficulty acquiring Hindi or English than the indigenous population of India. Similar facility in acquiring Swiss-German or English applies to children of Tibetan ancestry raised in Switzerland or the United States. In Cambridge, Massachusetts, near Central Square, a few miles from Chomsky's office at MIT, we can find children whose ancestors lived for 40,000 years in Tibet and spoke the Tibetan language, and these children are acquiring American English as readily as are the descendants of the voyagers on the *Mayflower.*

Similar considerations preclude a specialized UG for the Australians who reached that isolated continent about 50,000 years ago. Although native Australian languages such as Walpuri differ radically from English, there is no evidence for natural selection that yielded an optimal Walpuri UG (Bavin and Shopen 1985). It is most unlikely that any detailed innate store of syntax exists in the brains of Tibetans, Australians, or anyone else. In fact, there is no evidence at all to support the claim that normal children have any difficulty learning a language that is not spoken by their parents, as long as they start early.

The so-called language gene (Pinker 1994), FOXP2, is a regulatory gene that affects neural circuits that regulate motor control, mood, cognition, and language. It also regulates the expression of the lungs and other parts of the body. It is not a language gene in the Chomskian sense–a unique human linguistic capability divorced from other aspects of our behavior and that of other species. Undoubtedly, other genetic distinctions exist that account for the singular properties of the human brain that confer cognitive and linguistic ability, and it is possible that these properties place constraints on the form of human language. But these constraints are unlikely to be language-specific; to explore these properties rigorously would require a range of data beyond the purely linguistic phenomena focused on by traditional linguistic studies.

The Nicaraguan Sign Language Story

The highly publicized claim that deaf Nicaraguan children spontaneously created a new complex sign language, Nicaraguan Sign Language (NSL), has been used as evidence for an innate Universal Grammar (Kegl, Senghas, and Coppola 1999). American Sign Language (ASL) and other manual sign languages make use of complex syntax and morphology to convey meaning (Stokoe 1978). Kegl and her colleagues, who appear to hold to Chomsky's world view, claim that deaf children in Nicaragua—without benefit of instruction from any individual who knew ASL or any other manual sign language, without any material on ASL, and without contact with adults communicating by means of sign language—spontaneously developed a sign language with a morphology and syntax equivalent to ASL. To Kegl, Pinker (1994), and many linguists, the spontaneous generation of NSL is proof of an innate UG. Otherwise, how could untutored children, raised in the absence of ASL or any other sign language, acquire a complex manual sign language? The answer appears to be that the events that led to these children acquiring NSL have been misrepresented.

ASL experts such as William Stokoe (unpublished letter to *The New York Times* on the Gaulladet University website) have pointed out some of the similarities that exist between NSL and ASL. Because more than 100 different sign languages are used throughout the world, we might wonder why these similarities between NSL and ASL exist. However, the question is moot. Contrary to Kegl's claims, education for deaf children was provided in Nicaraguan government schools, and ASL was not unknown in Nicaragua. Thomas Gibson, an American Peace Corp volunteer, provided ASL manuals and instruction in 1979. Nicaraguans were in contact with deaf educators in neighboring Costa Rica between 1974 and 1979. Polich (2006) documents the facts. NSL at first lacked complexity, and many of the gestures used by hearing Nicaraguans were incorporated into NSL over a period of time. The claim that NSL spontaneously generated appears to have the same validity as the story, cited by Charles Dickens in *Bleak House,* of an Italian countess spontaneously bursting into flame in Verona.

Motor Control and Language

This takes us to one of the major points of this book. The Broca-Wernicke language-organ theory is wrong. Cortical-striatal-cortical circuits clearly regulate many aspects of human and animal behavior, including walking, manual motor control, speech production, syntax, shifting cognitive sets, mood, and personality. Independent neurophysiologic studies show that these neural circuits include the basal ganglia, which support anatomically segregated, independent neuronal populations. These neuronal populations project to different cortical targets that respectively regulate different aspects of motor control and cognition, including sentence comprehension. The basal ganglia constitute a sequencing engine that can link routine actions or interrupt an ongoing motor act or cognitive process. The sequencing engine performs in the same way, whether it operates on the motor pattern-generators that are the elements reiterated to form motor acts, or on the cognitive pattern-generators.

Studies of motor control have informed neuroscientists about the nature of motor learning: how animals, including humans, acquire motor pattern-generators. At all levels of the neural motor control system, including the putamen, supplementary motor cortex, premotor cortex, and primary motor cortex, neuronal populations coding motor acts are formed as an individual learns through practice to execute a particular motor act. Whether it is a monkey moving a baton or a human shifting the gears of a stick-shift car, playing the piano, or executing the stylized sword cuts of a Japanese Samurai, automatization occurs. As noted in eighteenth-century Samurai training manuals, the practitioner carefully and with much concentration slowly carries out the motor act. At some point, neuronal populations form that allow the motor act to be carried out precisely and rapidly. The process clearly does not involve the release of innate information. Detailed discussions of the manner in which children acquire language, such as those of Karmiloff and Karmiloff-Smith (2001) and Tomasello (2004a), leave little doubt that a child learns the particular aspects of syntax as well as the words that characterize their native language.

Linguists cannot, without any evidence to support their claim, state that the basic operations of the neural substrates for language and motor control differ profoundly, given similar circuitry incorporating common neuroanatomic structures. The data clearly show that most aspects of motor control that we wish to associate with cognitive behavior are learned. But, as I noted at the beginning of this book, no syntactic linguistic universals have been documented. Noam Chomsky turned the world upside down. Words appear to be conserved across related languages and time but the particular rules of syntax vary; they clearly are not biologically predetermined.

Recursion and Reiteration

In Chapters 1 and 5, I stated that recursion is a restricted, theory-specific form of the process of reiteration, which allows us to reorder or repeat a finite set of elements to form an infinite number of sentences, musical compositions, dances, and other creative acts. According to Hauser, Chomsky, and Fitch (2002), recursion is restricted to language and language alone. They state that "[in] natural language go beyond purely local structure by including a capacity for recursive embedding phrases within phrases which can lead to statistical regularities that are separated by an arbitrary number or words or phrases." Just what does this mean?

In practice, recursion is the process that forms "complex" sentences such as *The boy who kissed the girl fell down.* The relative clause *who fell down* hypothetically results from words that could form the simpler sentence *The man fell down* that were embedded in a syntactic structure that could yield the sentence *The man saw the ship,* in the absence of the embedded string of words (see Figure 7.9). In Chomsky's 1957 theory, the relative clause is the surface manifesta-

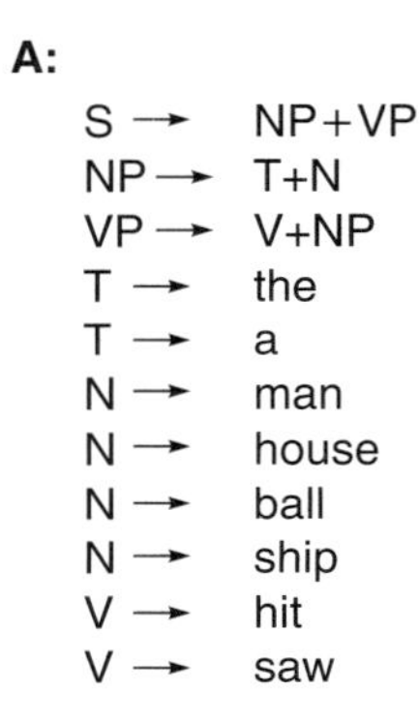

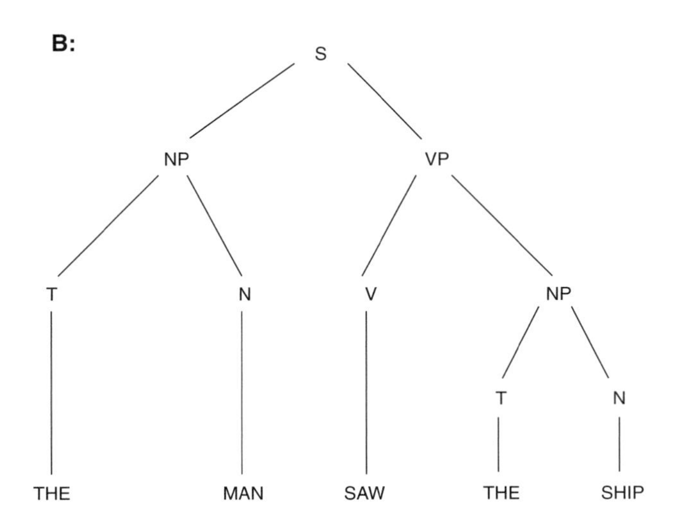

FIGURE 7.9 *The phrase structure "rules" noted in 7.9A, when applied sequentially, can generate the tree diagram for the sentence, "The man saw the ship."*

tion of a deep structure that presents semantically relevant information and a starting point for mental processes that convey the sentence's meaning. A set of phrase structure rules describes the syntactic relations that hold between the words of the deep structure. Phrase structure rules are a component of virtually all formal grammars whether they strictly differentiate between syntactic and semantic processes or differentiate between deep structures that convey semantic information and surface structures tailored to optimize communication.

A phrase structure grammar of English could have the rules in Figure 7.9A that could generate the syntactic structure represented in the tree diagram Figure 7.9B. Recursion is introduced into syntax by phrase structure rules such as:

VP → V + NP + S

This rule, in somewhat different notation, can be found in Chomsky (1975, p. 75). This rule could generate the diagram in Figure 7.10, with the embedded sentence node "S" that yields the string of words *the man fell down.* Subsequent, hypothetical syntactic operations then rewrite the words *the man* of the embedded sentence as *who,* forming the relative clause of the final surface structure sentence. However, the presence of an embedded S = Sentence in a deep structure is itself a hypothetical construct, as is the concept of a deep structure. Different versions of generative linguistic theory, including Chomsky's own theories subsequent to 1957, have dispensed with the deep-to-surface distinction; but recursive processes have been retained and mark the current minimalist theory (Chomsky 1995). The relative clauses as well as prepositional and adverbial clauses that occur in complex sentences are still derived from embedded sentences.

It is an open question as to whether Chomsky's theories have any biologic validity. Can we really be sure that the relative, prepositional, and adverbial clauses that occur in complex sentence are derived from embedded sentences? Competent linguists before

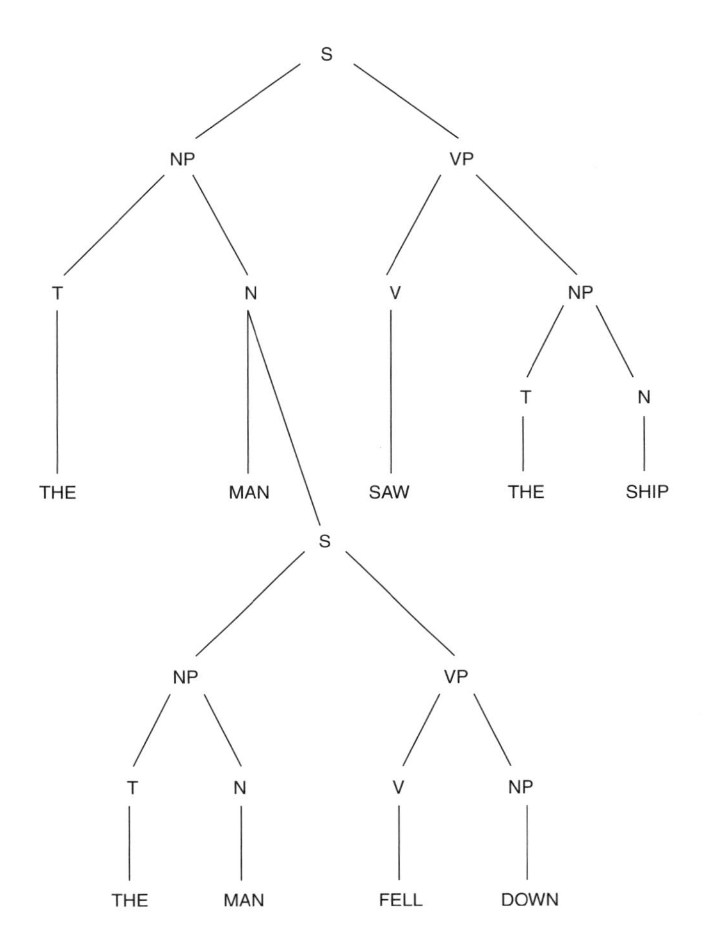

FIGURE 7.10 *A hypothetical linguistic tree diagram for the sentence, "The man who fell down saw the ship." In the generative grammars proposed by Chomsky and his adherents, a sentence containing a relative clause hypothetically results from an S node (a sentence) being recursively inserted within the phrase structure tree of a carrier sentence. In this example, the relative clause "who fell down" (that one actually hears or reads) is the product of syntactic processes that rewrite the words "The MAN" (under the S node dominated by the N node in the carrier sentence) to yield the word "who." Whether the hypothetical syntactic operations that introduce S nodes and then rewrite the resulting nodes and words have any foundation in mind or brain is an open question.*

Chomsky came onto the scene, such as Dwight Bolinger and Kenneth Pike, described the semantic distinctions conveyed by embedded relative, prepositional, adverbial clauses in complex sentences without the formal apparatus introduced by Chomsky. They noted that a relative, prepositional, or adverbial clause could be embedded within the framework of a "simpler" sentence. In short, recursion as defined by Hauser, Chomsky and Fitch (2002) follows from hypothetical syntactic structures, hypothetical insertions of sentence and phrase nodes, and hypothetical theory-specific rewriting operations on these insertions to form the sentences that we actually hear or read.

In contrast, the behavioral and electrophysiologic studies reviewed in this book provide a biologic basis for reiteration. Reiteration, which is a repeating and reordering motor or cognitive pattern, accounts for complex sentences that hypothetically derive from recursion as well as a wider range of seemingly unrelated phenomena. The music of Bach and Mozart, for example, makes use of repeated musical phrases, slightly different in form. In a traditional square dance, repeated elements are reordered and repeated to form a complex series of novel patterns that could go on indefinitely if the conventions of linguistic theoreticians were adopted. That also applies to other forms of dancing, such as tap dancing. Imagine the plot that could be written about a malefactor condemned to tap dance to eternity.

Competence and Performance

This brings to mind one aspect of the competence-performance distinction noted in many linguistic studies. If by competence, we mean the study of the biologic capacity of a species to acquire some aspect of language, we are on firm ground. The ape language studies of the Gardners, Savage-Rumbaugh, and others show that apes have the capacity to acquire some aspects of human language. However, if the object of linguistic research is to understand human capacities, we can rule out definitions of competence and resulting syntactic theories that entail being able to produce infinitely long sentences and phrases separated by an infinitely long sequence of words or phrases. One of my favorite Monty Python television skits is an interview of an aged man telling a TV interviewer the name of a great German Baroque composer: "Johann Gambolputty de von Ausfern-schplenden-schlitter-crass crenbon-spelltinkle-grumblemeyer-gutenabend . . . "—an apparently endless string of titles, but dies before he reaches the end.

On the Nature of Linguistic Processes

As Marcel Mesulam pointed out in 1990, the brain is not an amorphous blob in which distributed computations take place throughout the total system. Particular structures, whether subcortical structures or regions of the cortex, appear to perform particular local operations. However, as numerous studies show, these operations in themselves are only parts of a neural system that regulates an observable aspect of behavior which can contribute to biologic fitness. Thus the subcortical basal ganglia perform similar local operations when a person switches categories on the Wisconsin Card Sorting Test, talks, or walks. Similar operations mark the basal ganglia and cortical targets when monkeys or mice perform motor acts.

If we were to analyze the operations of a grinder, we would assume it worked in fundamentally the same way, regardless of the substance being ground. Studies of the operations carried out by the structures that support the neural circuits regulating motor control show that they do not involve serial algorithmic processes. Cortical and striatal activity can be directly monitored during studies of motor control in species other than humans. However, Alexander, DeLong, and Crutcher in 1992 concluded that the neural bases of motor control in biologic brains are not affected by a set of algorithmic processes that sequentially follow one another, each process going to completion before the next one starts. Their conclusion is shared by virtually every subsequent study, for example, Graybiel (1995, 1997, 1998), Graybiel et al. (1994), Aldridge et al. (1993), and Meyer-Luehmann et al. (2002). Marsden and Obeso (1994), in their seminal paper on basal ganglia function derived from observations of neurosurgery and neurodegenerative diseases affecting human subjects, reached a similar conclusion.

As we have seen, basal ganglia structures projecting to the cortex play a similar sequencing and set-shifting role in cognition (Monchi et al. 2001; Scott et al. 2002) and in sentence comprehension (Lieberman et al. 1990, 1992; Grossman et al. 1991, 1992, 1993; Hochstadt 2004). But as Marsden and Obeso (1994) pointed out, the processes that are sequenced appear to be implemented in parallel operations. For example, neurons firing in the putamen that project to neurons in the supplementary motor area (SMA) do not cease to fire when the SMA neurons fire as we execute a motor act, as would be expected if the process was simply a sequence performed by discrete structures (Cunnington et al. 1995).

There is no reason to assume that neural physiology shifts to the serial algorithmic processes that, with little success, have marked linguistic studies for the past 50 years. Indeed, theoretical linguists know that they have failed at even being able to describe the grammar of any present human language using these algorithmic processes. As a reminder—to close the circle started in the first chapter of this book—Ray Jackendoff, one of Chomsky's advocates, notes, "Thousands of linguists throughout the world have been trying for decades to figure out the principles behind the grammatical patterns of various languages . . . But any linguist will tell you that we are nowhere near a complete account of the mental grammar for any language" (1994, p. 26). Linguists could profit by studying the neural bases of motor control.

Take-Home Messages

- The study of linguistics could profit by heeding Dobzhansky: "Nothing in biology makes sense except in the light of evolution." The fruits of biologically motivated studies such as how breathing relates to language, how human beings actually produce and perceive vowels, and the acoustic cues and motor gestures that produce stop consonants, can inform and refine linguistic theories. The study of language would profit if linguists took account of the principles and findings of evolutionary biology.

- Language entails using anatomic systems, physiology, and brain mechanisms that were not "logically" designed. Systems initially adapted for one behavior often are modified to take on new functions and the result is not necessarily simple or elegant.
- The concept of modularity does not appear to characterize the processes involved in comprehending speech and distinctions in meaning conveyed by syntax. There is little evidence to support any version of Chomsky's innate Universal Grammar, and much that refutes this concept.
- If we take into account the findings of current neurophysiologic studies, serial algorithmic processes are not likely candidates for understanding how the human brain works and the consequent linguistic computations of the human mind.

CHAPTER 8

Where We Might Go

As the word "toward" in this book's title indicates, my object is not to provide a definitive account of the evolutionary biology of language. That lies in the future. In this endeavor, Charles Darwin continues to be our guide. Some aspects of the biologic bases of human language and cognition appear to reflect the results of natural selection that enhanced, to a slight degree, aspects of behavior that gradually evolved. Recall Darwin's description of natural selection:

> Owing to this struggle for life, any variation, however slight and from whatever cause proceeding, if it be in any degree profitable to an individual of any species, in its infinitely complex relations to other organic beings, and to external nature, will tend to the preservation of that individual, and will generally be inherited by its offspring. The offspring, also, will have a better chance of surviving. (1859, p. 61)

For example, the descent of the human tongue into the pharynx, as discussed did not yield a profound "leap" in the efficacy of speech as a means of communication. The human tongue and supralaryngeal vocal tract allow us to produce perceptually salient quantal vowels with less effort (we don't have to change the shape of our tongue). The quantal supervowel [i] also enhances the process of vocal tract normalization, contributing to the robustness of human speech as a means of communication. However, we could communicate vocally, albeit with somewhat greater articulatory efforts and perceptual uncertainty, if we were unable to produce [i]'s. The word identification error rate might increase by 5 to 10 percent, and it might be necessary to listen to a longer stretch of speech before deciding what words you actually had heard, but encoded speech communication still would be possible. However, the small selective advantages conferred from the reshaping and descent of the human tongue apparently resulted in natural selection that yielded our peculiar tongues.

But the human tongue increases the risk of choking to death and we have to keep in mind that there would have been no selective advantage for its evolution unless some form of speech were already present in the species ancestral to *Homo sapiens.* Natural selection works by enhancing existing capabilities. The heated debate on Neanderthal speech often misses this point. As Chapter 6 pointed out, there probably was a small, but significant, difference between the speech capabilities of Neanderthals and contemporary *Homo sapiens;* but Neanderthals and most likely forms of *Homo erectus* used some form of speech.

In contrast, the evolution of human reiterative ability that allows us to form a potentially infinite number of words, sentences, and dances from a finite number of sounds, syntactic processes, and steps may have been abrupt. These and other creative human attributes may have derived from chance mutations on the foxp2 gene, yielding FOXP2. We know that FOXP2 is involved in the embryonic development of the basal ganglia and other subcortical structures that play a part in regulating motor control and cognition. Mutations that disrupt FOXP2 result in the verbal apraxia syndrome and speech motor-sequencing and cognitive deficits occur, including deficits in the comprehension of distinctions in meaning conveyed by syntax. The findings of independent studies of the behavioral deficits of Parkinson's disease, hypoxia, and other instances of brain damage, as well as neuroimaging studies of normal human subjects and neurologic patients, show that the basal ganglia constitute a sequencing engine in these seemingly separable domains.

But the result of the intense selection for FOXP2 in *Homo sapiens* also fits the Darwinian paradigm: neural structures adapted for one purpose—motor control—took on "new" cognitive tasks. The studies that I have discussed point to the human basal ganglia sequencing engine being a key, perhaps *the* key, to how the reiterative capacity that marks creative human behavior is conferred. Studies of living human subjects suffering from developmental verbal apraxia may be germane to assessing the linguistic and cognitive capacities of Neanderthals. Neanderthals might have had similar limitations in speech and syntactic abilities, and they may also have lacked a fully human capacity to adapt to changing circumstances.

Moreover, it is probable that the evolution of language has a long history, congruent with hominid evolution. Comparative studies show that chimpanzees have limited syntactic ability and are able to devise solutions to novel problems. If the earliest stages of the evolution of the striatal sequencing engine arose from the demands of upright bipedal locomotion, the cortical-striatal-cortical circuits of the earliest hominids most likely yielded superior cognitive and linguistic abilities to those of chimpanzees. Neanderthals are far removed from these hominids who lived 6 to 7 million years ago, and surely the Neanderthals had advanced linguistic and cognitive capabilities. But we are here and Neanderthals are not; language and cognition may account for this fact.

It is also clear that our knowledge is so imperfect that there can be no certainty concerning the linguistic or cognitive abilities of any extinct hominid species. For example, Marsden and Obeso in 1994 observed that, though we have a reasonable notion of the "local" sequencing operations that the basal ganglia perform, we do not know how they carry out these operations (see Chapter 4). We still do not know *how* the basal ganglia sequencing engine works. If our basal ganglia circuit diagrams were correct, patients undergoing surgically induced lesions to reduce tremor would become totally rigid, but that fortunately is not the case.

We do not really understand how the brain works, and therefore, we cannot say why chimpanzees cannot talk or think as we do. Moreover, there is more to language than speech and syntax. Words, which convey concepts, are critical elements of language; we humans have far greater lexical capabilities than chimpanzees. Humans commonly acquire tens of thousands of concepts that we can attempt to communicate to others, coded as words. In contrast, chimpanzees can produce no more that 150 to 200 words using manual gestures or keyboards. That distinction must be accounted for.

Specialization

An understanding of the biologic bases or evolution of human language cannot be achieved by research confined by the constraints of any single traditional academic field. The "generative" Chomskian school of linguistics provides a signal lesson. Despite the intensive efforts of a generation of many bright people, few insights on the nature of language have occurred. In contrast, consider the discovery of the FOXP2 gene and its neural and behavioral sequelae. Studies that involved several disciplines yielded a major advance. The starting point was a traditional experiment-in-nature, a behavioral assessment of the deficits of an extended family that pointed to a genetic anomaly. MRI and PET imaging studies identified the subcortical neural loci of a sequencing problem that impeded the execution of orofacial gestures, speech production, language comprehension, and cognition. Insights from independent neurophysiologic studies and studies of Parkinson's disease reinforced this view. The techniques of molecular biology identified the gene, the distinctions between the human, chimpanzee, and mouse versions, and the regulatory gene's role in the embryonic development of the subcortical complex that plays a part in regulating these different aspects of human, chimpanzee, and mouse behavior.

Comparative Studies

Intensive studies of the linguistic and cognitive capabilities of other living species are needed. The foxp2 gene provides an entry point for focused comparative studies that may lead to an understanding of the evolution of human language. Different versions of the foxp2 gene are apparent in mice and chimpanzees and we could profit by learning something about whatever functional distinctions may be present in the neural structures that this gene regulates during embryonic development. A skeptic might declare that mice do not talk nor do they really think. Chimpanzees also are not able to talk and scholars too often categorically state that chimpanzees do not command any aspect of human linguistic ability. So what could comparative studies focusing on the behavioral sequelae of somewhat different foxp2 genes possibly establish?

This negative attitude unfortunately is common; a belief in human uniqueness often overrides the findings of studies that document actual behavior. For example, it is not unusual to find assertions that claim that chimpanzees are incapable of acquiring any words, despite the findings of the papers noted in earlier chapters by the Gardners, Savage-Rumbaugh, and others in peer-reviewed journals. The direct evidence of documentary films that demonstrate the lexical abilities of chimpanzees also does not seem to count.

For that matter, evidence that shows that chimpanzees use cognitive-social mechanisms such as joint attention that play a role in the process by which children learn words is disputed. The term "joint attention" denotes the process by which adults or other caretakers interact with a child to direct attention to a referent of a word (some object or activity) through eye contact, gaze, tone of voice, smiles and so on (for example, Tomasello and Farrar 1986; Bloom 2000). A link is thereby established between some object or action and the sound-pattern or gestures that convey the word. As Tomasello (2004a, b) points out, joint attention plays a role in the manner in which humans acquire other aspects of human culture.

But joint attention—a three-way relationship between an instructor, a novice, and an object or activity—is not an exclusive human paradigm. It enters, perhaps in reduced degree, into the social development of chimpanzees. The ongoing research of Tetsuro Matsuzawa (2004) and his colleagues reveals joint attention in chimpanzee mother-infant pairs starting at age 3 months. Chimpanzee infants and mothers can be observed smiling as they look into each other's open eyes. Joint attention can be observed between chimpanzee playmates as early as age 1 year. In a laboratory setting, by age 22 months, chimpanzee infants in a controlled laboratory setting begin to learn to use sticks as probing tools to obtain honey in a manner analogous to that observed in African settings for termite fishing (Goodall 1986). Joint attention appears to enter into their learning process.

Joint imitation also has been documented in motion pictures of chimpanzees in their natural habitat. "Follow-films" of the Gombe chimpanzees studied by Jane Goodall's research group show them engaging in many activities. Trained observers continually follow the chimpanzees, using small hand-held video cameras. The chimpanzees become habituated and essentially ignore the observer. (The procedure is not unlike that practiced by photojournalists who capture the activities of a willing celebrity.) One follow-film—of an engagement in chimpanzee warfare, noted in Goodall (1986), during which a group of chimpanzees are on patrol looking for "enemy" chimpanzees—clearly shows an instance of joint attention. The patrolling chimpanzees hear the vocalizations of another band of chimpanzees at their territory's boundary. Two members of the patrol rush forward to reconnoiter, side by side. Then, some of the chimpanzees exchange looks with the alpha male, and he looks toward the enemy band and decides that a riposte is in order. Those chimpanzees then vocalize. The primatologist Christopher Boehm (1991) describes the episode in detail.

In fact, it has long been evident that various aspects of chimpanzee behavior are learned from their culture rather than acquired genetically. Groups of chimpanzees that are separated by hundreds of miles have different toolkits and food-gathering and food-preparation traditions. In East Africa, termite and ant fishing with tools prepared from sticks of different sizes can be observed (Goodall 1986; McGrew 1993). In West Africa, chimpanzees crack nuts by placing them on anvils and striking them with hammers (Boesch 1993). Wood hammers are used for soft-shelled nuts and stone hammers for harder nuts. These different, culturally transmitted patterns of behavior clearly must be learned. And pedagogy, another trait that often is thought to be exclusively human, can be observed in chimpanzees. Overt pedagogy has been documented in the films made by Boesch and Boesch (1984), which show young chimpanzees being guided by their mothers toward proficient nut-cracking. Christophe Boesch's account of one filmed incident may perhaps convince skeptics:

> Salome [a chimpanzee mother] was cracking a very hard nut species . . . with her son, Sartre. He took 17 of the 18 nuts she opened. Then, taking her stone hammer. Sartre tried to crack some by himself, with Salome still sitting in front of him. Those hard nuts are tricky to open as they consist of three kernels independently embedded in a hard wooden shell, and the partly opened nut has to be positioned precisely each time to gain access to the different kernels without smashing them. After successfully opening a nut, Sartre replaced it haphazardly on the anvil in order to gain access to the second kernel. But before he could strike it, Salome took the piece of nut in her own hand, cleaned the anvil, and replaced the piece carefully in the correct position. Then, with Salome observing him, he successfully opened it and ate the second kernel. (Boesch 1993, pp. 176–177)

Boesch describes another incident of a chimpanzee mother overtly showing her daughter how to grip the nut-cracking hammer. The "mother, seeing the difficulties of her daughter, corrected an error [an incorrect hammer grip] in a very conspicuous way and proceeded to demonstrate how it [the hammer] works with the proper grip." The learning curve seems agonizingly long compared to that of human children, but the chimpanzees learn how to crack nuts. In a laboratory setting in which the infant chimpanzee Loulis "acquired" the chimpanzee version of American Sign Language from the five adult chimpanzees cross-fostered by the Gardners, the adult chimpanzees were observed overtly molding Loulis's hand into the proper ASL signs (Fouts, Fouts, and Van Cantfort 1989).

Joint attention, pedagogy, imitation, statistical induction, and other "general" cognitive processes are involved in the process by which children acquire language. The published studies of Meltzoff, Gopnik, Greenfield, Tomasello, Karmiloff-Smith, and Bates and their colleagues show that this is the case. It is most unlikely that joint attention, or any factor that enters into the acquisition of language by children, is absent in chimpanzees who are so close to us in time and in their genetic endowment. However, the degree to which these processes are operant in chimpanzees is unclear, as are the nature and limits of chimpanzee linguistic abilities. Although evidence is accumulating concerning the referential uses of chimpanzee vocalizations (for example, Slocombe and Zuberbuhler 2005b), the limits of what and how chimpanzees and other nonhuman primate species communicate by vocal signals still remain a mystery.

Jane Goodall (1986) and other acute observers have compiled lists of chimpanzee calls that they believe have particular referents. However, these inventories are based on auditory impressions. Phenomena such as the perceptual magnet effect (Kuhl et al. 1992; see Chapter 6) make objective acoustic analyses imperative as well as "playback" experiments, similar to those first employed by Cheney and Seyfarth (1990) for monkey calls. Some outstanding questions include: What do the long-distance calls of chimpanzees signify? Are we overlooking acoustic cues that differentiate calls which convey "semantic" information to chimpanzees? The situation is similar to that faced by a naive monolingual speaker of English who, listening to Chinese, does not "hear" the distinctions conveyed by lexical tones of the language. Moreover, as Goodall (1986) and other acute observers have noted, the gestures and facial expressions of primates appear to work in concert with their vocal signals to establish communication in social interactions.

Therefore, before we can discern what the genetic distance between human FOXP2 and chimpanzee foxp2 signifies, we first must determine how and what chimpanzees communicate and their inherent cognitive capacities. The fact that chimpanzees can communicate by means of words and simple syntax when they are exposed to manual sign languages (Gardner and Gardner 1969, 1973, 1984, 1994; Savage-Rumbaugh, Rumbaugh, and McDonald 1985; Savage-Rumbaugh et al. 1986; Savage-Rumbaugh and Rumbaugh 1993) indicates that we have a long way to go before we can state that we know how chimpanzees communicate and think. The fact that the biologic substrate exists that permits chimpanzees to acquire these abilities in the appropriate setting (albeit a human cultural setting) suggests that these elements may be present in the chimpanzees' "native" communications systems.

Genes

The unique genes, or gene knockouts (removal or silencing of genes), that underlie the development of the human skull, face, and supralaryngeal vocal tract have yet to be isolated. The ASPM gene may have resulted in the initial expansion of the hominid brain and perhaps overall body size, but it cannot account for the unique morphology of the human face and tongue. Avian foxp2 appears to be related to the ability of songbirds to acquire their songs. What is the role of FOXP2 in human learning? Research along these lines may determine the mechanisms that structure the sensitive period during which a child can effortlessly acquire any language. Do mice have similar sensitive periods in which they can readily "acquire" some aspects of mouseness? Can we isolate the genetic mechanisms and probable neurotransmitters that trigger and terminate these developmental events? And finally, the antiquity and similar neural expression of foxp2 point to both the continuity and tinkerer's logic of evolution—neural systems initially designed for motor control can take on cognitive functions.

The neural bases of the brain's dictionary are becoming somewhat clearer. There does not appear to be any distinction between real-world knowledge and the semantic knowledge inherent in a word, apart from syntactic constraints. Most syntactic constraints are, in a meaningful sense, also semantic. If a word can function as both a verb conveying some action and as a noun, that information is "semantic." Will different patterns of brain activation be seen when a word is being used as a verb or noun? The link between the neural bases of perception, action, and a word is suggestive of a brain organization indicating that other species also have "words"—a suggestion reinforced by the lexical abilities of chimpanzees, dogs, parrots, dolphins, and other species. The difference between the words of humans and of other species may rest in our ability to communicate our words vocally to a vastly greater degree. Can some genetic basis for this distinction be found?

How?

As should be evident from the range of studies and techniques that have been discussed throughout this book, the goal determines the appropriate disciplines. This presents a challenge to academic research, which is usually rigidly compartmentalized. Tenure, rank, and status depend on the number of publications and talks in specialized journals and meetings. But if the goal is to understand how we are able to talk to each other and think, then we must be able somehow to relate the findings of studies in areas of science that have their own techniques and traditions. In fact we must do more than that. Studies of verbal apraxia in children seemingly have little to do with thinking, but they are yielding insights into the genetic bases and evolution of language and cognitive ability. Ongoing studies of verbal apraxia (Young and Lieberman, in submission) point to further progress. New insights often come from fresh minds. MRI and fMRI imaging studies will have to complement these new studies and similar data on nonhuman primates, even dogs, would be helpful. Clearly, studies centered in a single department of any known university will not suffice.

Would the formation of an Institute for the Evolution of Language be the answer? Probably not, because such an institution might exclude fresh minds—"outsiders" who see connections between old problems and new issues that are invisible to their elders. The unpredictability and infinitely complex connections that course through the community of the mind usually result in formal academic liaisons after the fact. Studies of rodent grooming, monkey brain activity during button-pushing, Parkinson's and Alzheimer's diseases, and climbers on Mount Everest might seem to be unconnected, but I hope that the connections are now evident.

The toolkit is available. Studies of humans and of other species exploring genetic, neurophysiologic, and behavioral issues—as well as linguistic studies grounded in science rather than impoverished formal mathematical procedures and introspection—can yield a better understanding of how the human brain works to make language possible, and how human language came to be. The nature and evolution of the biologic bases of language are inextricably one. An evolutionary biology of language will inform us of its particular nature, as well as of our nature.

The evolution of human language cannot be the result of a singular process. There may have been events that are unique to our extinct hominid ancestors or to *Homo sapiens,* but the result clearly is not a language organ. The neural mechanisms regulating human language also are implicated in other aspects of our behavior—cognition, motor control, and emotion. That clearly is also the case for the anatomy implicated in speech, structured by the demands of life-sustaining breathing and eating. And there is a lesson in this: neither intelligent design nor simplicity is evident in the biology of language.

Notes

References

Index

Notes

Chapter 1

1. Pinker and Bloom (1990) and Hauser, Chomsky, and Fitch (2002) imply that the neural basis of human linguistic ability is devoted to language, and language alone, because it is a spandrel. The concept of a spandrel derives from the Gould and Lewontin 1979 paper, "The spandrels of San Marco and the Panglossian program: A critique of the adaptationist programme." The spandrels of the cathedral are fillers between the walls and load-bearing arches. The arches, not the spandrels, support the cathedral. Gould and Lewontin conclude that biologic "spandrels" exist that play no part in any previous behavior. However, as I shall show, evidence from independent studies of the behavior and brains of humans and other species shows that the neural structures that allow us to reiterate syntactic rules continue to play a part in motor control and also confer cognitive flexibility outside the domain of language.
2. Paradoxically, the Chomskian school has seized on a pathologic genetic variation that supposedly demonstrates the innate basis of syntax. Myrna Gopnik and her colleagues in a 1990 paper claim that the members of an extended family (KE) in London who exhibited a genetic anomaly (subsequently traced to an anomalous FOXP2 gene) were unable to acquire the regular plural and past tense forms of English verbs and nouns such as "boxes" or "walked." In contrast, the afflicted individuals supposedly had no difficulty with irregular forms such as "swam" or "geese."

   These observations, according to Gopnik and Pinker (1994), showed that Chomsky was correct—a fragment of the innate, genetically transmitted Universal Grammar was missing in the brains of the afflicted family KE members. Chomsky's 1981 model called for a set of parameters that would be triggered by the language that a child heard. The regular plural and past tense parameters presumably were missing in the afflicted individuals so they could not acquire these rules, though they could master the irregular forms that are not rule-governed. Steven Pinker organized a session at the annual meeting of the American Association for the Advancement of Science devoted to the "language gene" and a chapter in his 1994 book. However, the fact of the matter is that the syntactic deficits of these individuals are not confined to the regular plural and past tense; they have difficulty with all aspects of syntax. They also suffer from oral apraxia and are unable to sequentially or simultaneously stick their tongues out and purse their lips.

They also cannot repeat two consecutive words and tend to show cognitive deficits when they are tested on standardized intelligence tests (Vargha-Khadem et al. 1995).

Other claims for "specific language impairment" in children have been made (Leonard 1998). However, as yet no convincing evidence has been presented that demonstrates deficits limited to language alone; in many instances the problems documented pertain to reading, a recent historical innovation that involves many skills that do not involve linguistic ability itself. We return to these issues and the FOXP2 gene in the chapters that follow.

Chapter 3

1. In contrast to Muller's work, later nineteenth-century descriptive studies of the larynx focused on its anatomy. The term "vocal folds" is often used to refer to the vocal cords, because they looked like folds when viewed from above by means of a mirror placed at the back of a person's mouth.
2. The comparative electrophysiologic studies that we have (no one has yet attempted a monumental comparative anatomic study like that of Negus) suggest that many species have neural perceptual mechanisms that match the acoustic output of their laryngeal source and their supralaryngeal filter. Peterson et al. (1978), for example, showed neural responses in monkeys to stimuli having the formant frequencies of their species-specific vocal sounds. Studies of cats (Whitfield 1967) show that they are equipped with a neural device that tracks the fundamental frequency of phonation, F0. This is no mean achievement, because we have yet to make an electronic device or write a computer algorithm that will accurately track fundamental frequency, though hundreds of attempts have been submitted to the U.S. Patent Office since 1936. Systems for the detection of laryngeal pathologies (Lieberman 1963a), task-related stress (Coster 1986; Kagan, Reznick, and Snidman 1988) and cognitive impairment (Lieberman et al. 1994, 1995, 2005) that depend on accurate measurements of the fundamental frequency of phonation and other acoustic parameters still involve intervention by skilled human operators who must check the measures provided by automated computer-implemented procedures.
3. Chapter 4 focuses on the apparent neural bases of human language and relevant experimental data.
4. Gerstman (1968) developed an algorithm for computer-implemented speech recognition that used normalization coefficients derived from the F1s and F2s of a speaker's [i] and [u]. However, Gerstman's algorithm involves a listener's deferring her or his judgment of a vowel's identity until tokens that are known to be examples of [i] and [u] are heard. The algorithms developed by Nearey (1979) provided a solution that is a better match to the responses of human listeners. Moreover, as we shall see, the vowel [i] also is less susceptible than other vowels to formant frequency variations deriving from differences in tongue placement (Stevens 1972; Beckman et al. 1995). The vocal tract is perturbed to an extreme position (Stevens and House 1955). In con-

trast to other vowels, the intrinsic muscles of the tongue are sometimes employed when producing an [i] (Fujimura and Kakita 1979).

5. The studies discussed in Chapter 4 suggest that vocal tract modeling most likely involves activating the neural substrate that also regulates overt speech. The effects of vocal tract normalization on speech perception have been noted in other psychoacoustic experiments. May (1976), for example, notes a shift in the boundary for the identification of the fricatives [s] and [s] before the vowel [æ]. The boundary shifted to higher frequencies for the stimuli produced with an [æ] vowel corresponding to a shorter supralaryngeal vocal tract. Rand (1971) obtains similar results for the formant transition boundaries for the stop consonant pairs [b] versus [d] and [d] versus [g]. Strange et al. (1976) show that consonant transitions can serve as cues for vocal tract normalization. The vowel [u], as Nearey (1979) points out, also serves as a vocal tract size-calibrating signal; psychoacoustic experiments again consistently show that [u]'s are identified more reliably than other vowels except [i] (Peterson and Barney 1952; Hillenbrand et al. 1995). Human listeners clearly make use of whatever cues are available to infer the probable length of a speaker's supralaryngeal vocal tract. For example, if the identity of a vowel is known, a listener can derive the length of the vocal tract that would have produced it. Thus vowels like [I] or [æ] and [u], which are not as inherently specified for supralaryngeal vocal tract length (Gerstman 1968; Lieberman et al. 1972; Nearey 1979), can yield more reliable identification by serving as known normalization calibration signals (Verbrugge, Strange, and Shankweiler 1976).

   The general occurrence of stereotyped opening phrases like "Ni hao" in Chinese, "Hello" or "Hi" in English, "Hey" (in general use in Sweden and currently in the United States by young, hip speakers), especially in telephone calls, can be viewed as calibration signals providing supralaryngeal vocal tract normalization. A listener hearing these openers knows the intended phonetic targets. Any stereotyped opener can serve as a calibrating signal for vocal tract normalization. However, there seems to be a tendency to use the quantal vowels [i] and [u] in these openers.

6. Riede and his colleagues analyzed one type of vocalization of Diana monkeys and showed formant transitions that appear to reflect either the monkeys' starting the call with their lips protruded and somewhat constricted or the opening of their lips. The formant frequencies of the calls deviate from those of rhesus macaque measured by Lieberman (1968) and approach those of the vowel [a] (the vowel of the word "ma"); but when the length of the monkeys' airways and the measured shape of the supralaryngeal vocal tracts are compared with those of human speakers, it is evident that they are not producing the vowel [a] (c.f. Lieberman, in press a).

Chapter 4

1. Percheron and his colleagues (Percheron et al. 1984; Percheron and Fillon 1991), noting the extensive dendritic arborization in basal ganglia, argue against completely segregated circuits.

2. Broca's aphasics and Parkinson's disease patients differ in this respect from persons suffering from dementia of Alzheimer's type who cannot associate a word with its semantic attributes. They, for example, won't remember the function of a "key." Anomic Broca's subjects may not be able to recall the name "key" when presented with a key, but they are fully aware of its function; they, moreover, usually have no difficulty classifying words along semantic dimensions.

Chapter 5

1. Recursion as defined by Chomsky is a theory-specific form of embedding (see Chapter 1).

Chapter 6

1. The expansion of the human brain also may play a role in the flexing of the cranial base (D. Lieberman, Ross, and Ravosa 2000); but it cannot, in itself, account for all cranial flexure because brain expansion continues for a few years after the cranial base angle stabilizes.
2. Riede et al. (2005) claim that Diana monkeys can produce the abrupt area-function discontinuity necessary to produce an [a], but no actual monkey SVT area-functions were obtained or modeled in their study. The observed monkey formant frequency pattern, which "sounds" similar to an [a], most likely was produced by a derived aspect of the monkeys' anatomy—air sacs. This is interesting in itself, suggesting parallel evolution for vocal communication using formant frequencies and further study is warranted. The Diana monkey vocalizations are discussed in Lieberman (in press a).
3. Previous independent attempts to assess Neanderthal phonetic capabilities have taken account of some of these anatomic features, including the probable length of the Neanderthal SVTh. However, George's (1976, 1978) analysis of the Denver series of radiographs of SVT development in humans led to overemphasis on the cranial base angle. The studies of Laitman and his colleagues bear replication. These studies used a complex metric that factored in the cranial base angle as well as the length of the basicranium (Laitman, Heimbuch, and Crelin 1978, 1979; Laitman and Heimbuch 1982). However, their statistical procedure makes it difficult to untangle the contributions of cranial base angle (which subsequent studies show cannot be used to predict SVTv length). Virtually all other studies have focused on the position of the larynx rather than on the migration of the human tongue down into the pharynx. Although a quick reading of the Lieberman and Crelin (1971) paper could lead to the impression that the position of the larynx, in itself, is the factor that signifies whether a species had a human SVT, the detailed discussion in Lieberman (1984, pp. 276 to 280) unambiguously states that the descent of the tongue is the key to the enhanced speech-producing capabilities of the human SVT.

   Arensburg et al. (1990) argued that it was possible to determine the position of the larynx by examining the hyoid bone of a Neanderthal fossil.

The fossil's skull was missing and the hyoid bone was larger and outside the range of human variation by one of their own metrics (Laitman et al. 1990). However, Arensburg and his colleagues argued that, because the shape of the Neanderthal hyoid bone was somewhat similar to that of some adult human hyoid bones, it must have occupied the same position as an adult human hyoid. This argument does not hold because the human hyoid bone, with the tongue and larynx, migrates downward in the course of human development without any consistent change in hyoid shape (Laitman et al. 1990; Lieberman 1993). Houghton (1993) essentially assumed that Neanderthals were fully human, but he nonetheless found it necessary to provide his Neanderthal reconstruction with an exceedingly small tongue incapable of swallowing food. That was necessary to avoid endowing the Neanderthal with an SVTv equal in length to the SVTh. A larger tongue, capable of swallowing, would place the putative Neanderthal's larynx in an untenably low position.

Houghton's reconstruction, if true, would have accounted for the extinction of Neanderthals because their tongues would not have enabled them to propel food down the length of their long mouths and swallow (Palmer et al. 1992). Schepartz (1993) simply repeated the arguments made by Falk, Arensburg and his colleagues, and Houghton. Other objections to the Lieberman and Crelin (1971) reconstruction, such as DuBrul (1977), amount to the claim that since Neanderthals produced complex artifacts and were hominids, they must have been fully human.

4. The position advocated by Boe and his colleagues in their papers is difficult to comprehend. On the one hand, they claim that an SVT that has a human tongue and pharynx is not necessary to produce the full range of human speech. However, they also claim that Neanderthals possessed modern human SVTs. Their argument is based on the cranial base angle of the La Chapelle-aux-Saints fossil, which they claim is within the human range. However, although they cite the Lieberman and McCarthy (1999) paper, they ignore its principal finding, that the cranial base angle cannot predict the SVT of a fossil. In short, they assume that the morphology of the Neanderthal skull and mandible morphology conform to those of modern human beings, despite evidence to the contrary (e.g., Howells 1976, 1989; Stringer and Andrews 1988; Stringer 1992, 1998, 2002; Lieberman 1995). Moreover, the Neanderthal SVT reconstruction proposed by Boe and his colleagues also avoids having the larynx in an improbably low position by an arithmetical sleight-of-hand or error. They simply reduce the length of the basicranium of the La Chapelle Neanderthal skull. Although Heim, who is one of the coauthors of the Boe et al. (2002) study, in his own reconstruction (Heim 1989) of the La Chapelle fossil measured the length of the Neanderthal basicranium between the incisors and the foramen magnum (which they take as the length of the oral portion—SVTh) as 12.66 cm, this dimension was inexplicably shortened to 10 cm in the paper that was presented at the Conference on the Evolution of Human Language at Harvard University in 2003.

   The Boe et al. (1999, 2002) studies also rely on an incorrect model of a human newborn airway (Goldstein 1980), which never examined any newborn material. Her "newborn" SVT differs materially from those noted by Negus (1949), Lieberman and Crelin (1971), Bosma (1975), Laitman and Crelin (1976), or any other published data. The putative "newborn" human SVTv to SVTh proportions modeled in their studies correspond to those of five- to six-year-old human children in the Lieberman and McCarthy (1999) and Fitch and Giedd (1999) studies. A detailed discussion of the Boe papers is in Lieberman (in press b).

Chapter 7

1. The breath-group theory derives from the work of Stetson (1951), who noted that sentences and other words that were grouped together syntactically could be segmented by means of speech produced on a single expiration. Stetson, however, neglected the role of the elastic recoil of the lungs and the fact that alveolar and subglottal air pressure must necessarily fall to a negative value during inspiration. The breath-group concept is implicit in phonetic analyses like those of Armstrong and Ward (1926) and the phonemic analyses of Pike (1945) and Trager and Smith (1951).
2. It's not clear that this ability is unique to humans because it may derive from the "implicit" contextual learning noted in Chapter 4, which occurs as you walk down a street and unconsciously encode the landmarks that play a part in retracing your path. Rats and other animals in the state of nature commonly retrace their paths and learn to return to likely sources of food, water, or other attractions.
3. The 1970s data have not been replicated because many of the techniques used to obtain accurate measurements of air pressure and muscle tension data during speech are invasive and are not generally acceptable under current protocols for research in human subjects.
4. Unfortunately, a great deal of otherwise insightful work on the production and acoustics of speech has, so to speak, put the cart before the horse—attempting to "prove" that traditional linguistic concepts are correct instead of testing these claims. Stevens, for example, working at MIT, unfortunately structured much of his research toward reifying the linguistic theories advanced by Noam Chomsky. For example, binary feature contrasts are always used in the phonologic studies of the MIT school of linguistics; Stevens attempts to account for the three-way VOT distinctions of stop consonants by means of a binary vocal cord stiffness factor, for which there is no empirical evidence (Stevens 1998, pp. 251–252, 451–456). The studies outlined in Chapter 4 show that VOT deficits derive from impaired sequencing. If stiffness were a significant factor, then the VOTs of all stop consonants would tend to become longer in patients with Parkinson's disease, which contracts muscles (Jellinger 1990). That is not the case; the VOTs of stops such as [p], [t], and [k] become shorter in PD, overlapping with the VOTs of [b], [d], and [g]. Stevens simply ignores the three-way distinction documented by Lisker and Abramson (1964) for the stop consonants that occur in many of the world's languages that involve prevoicing. In a prevoiced consonant, phona-

tion precedes the primary opening of the SVT. These languages use prevoiced stop consonants similar to Spanish as well as short VOT lag consonants such as most English [b]s, and long VOT lag consonants such as the English [p]. No reference to any study focusing on VOT is referenced in his 1998 magnum opus. Stevens (1998, pp. 249–255) also endorses the traditional tongue maneuvers that supposedly generate and differentiate vowels. No references to the findings of the independent studies that refute the traditional theory (e.g., Russell 1928; Nearey 1979) are included in his 1998 book. The interpretations I have put forth of Stevens's (1972) findings concerning quantal sounds also are mine, and any disputes should be directed toward me, not Stevens. His 1998 book, which presents his views on the acoustics and physiology of speech, does not address the evolution of speech or the particular linguistic issues raised in this chapter.

References

Adcock, G. J., E. S. Dennis, S. Easteal, G. A. Huttley, L. S. Jermiin, W. J. Peacock, and A. Thorne. 2001. Mitochondrial DNA sequences in ancient Australians: Implications for modern human origins. *Proceedings of the National Academy of Sciences, USA* 98:537–542.

Alberch, P. 1989. The logic of monsters. *Geobios,* mem. spec. 12:21–57.

Albert, M. L., R. G. Feldman, and A. L. Willis. 1974. The "subcortical dementia" of progressive supranuclear palsy. *Journal of Neurology, Neurosurgery, and Psychiatry* 37:121–130.

Aldridge, J. W., and K. C. Berridge. 2003. Basal ganglia neural coding of natural action sequences. In *The basal ganglia IV,* ed. A. M. Graybiel, S. T. Kitai, and M. DeLong. New York: Plenum, 279–287.

Aldridge, J. W., K. C. Berridge, M. Herman, and L. Zimmer. 1993. Neuronal coding of serial order: Syntax of grooming in the neostratum. *Psychological Science* 4:391–393.

Alexander, G. E., and M. D. Crutcher. 1990. Functional architecture of basal ganglia circuits: Neural substrates of parallel processing. *Trends in Neuroscience* 13:266–271.

Alexander, G. E., M. R. DeLong, and M. D. Crutcher. 1992. Do cortical and basal ganglionic motor areas use "motor programs" to control movement? *Behavioral and Brain Sciences* 15:656–665.

Alexander, G. E., M. R. DeLong, and P. L. Strick. 1986. Parallel organization of segregated circuits linking basal ganglia and cortex. *Annual Review of Neuroscience* 9:357–381.

Alexander, M. P., M. A. Naeser, and C. L. Palumbo. 1987. Correlations of subcortical CT lesion sites and aphasia profiles. *Brain* 110:961–991.

Anderson, J. A. 1988. Concept formation in neural networks: Implications for evolution of cognitive functions. *Human Evolution* 3:83–100.

——. 1995. *An introduction to neural networks.* Cambridge, MA: MIT Press.

Arensburg, B., L. A. Schepartz, A. M. Tiller, B. Vandermeersch, H. Duday, and Y. Rak. 1990. A reappraisal of the anatomical basis for speech in middle palaeolithic hominids. *American Journal of Physical Anthropology* 83:137–146.

Armstrong, L. E., and I. C. Ward. 1926. *Handbook of English intonation.* Leipzig: Teubner.

Arthur, W. 2002. The emerging conceptual framework of evolutionary developmental biology. *Nature* 415:757–764.

Atal, B. S., and S. C. Hanauer. 1971. Speech and synthesis by linear prediction of the speech wave. *Journal of the Acoustical Society of America* 50:637–655.

Atkinson, J. R. 1973. *Aspects of intonation in speech: Implications from an experimental study of fundamental frequency.* Ph.D. diss., University of Connecticut.

——. 1976. Inter and intraspeaker variation in fundamental voice frequency. *Journal of the Acoustical Society of America* 60:440–446.

Awh, E., J. Jonides, R. E. Smith, E. H. Schumacher, R. A. Koeppe, and S. Katz. 1996. Dissociation of storage and rehearsal in working memory: Evidence from positron emission tomography. *Psychological Science* 7:25–31.

Axelrod, R., and W. D. Hamilton. 1981. The evolution of cooperation. *Science* 211:1390–1396.

Azzarelli, B., K. S. Caldemeyer, J. P. Phillips, and W. E. DeMeyer. 1996. Hypoxic-ischemic encephalopathy in areas of primary myelination: A neuroimaging and PET study. *Pediatric Neurology* 14:108–116.

Baddeley, A. D. 1986. *Working memory.* Oxford: Clarendon.

Baer, T., J. C. Gore, L. C. Gracco, and P. W. Nye. 1991. Analysis of vocal tract shape and dimensions using magnetic resonance imaging: Vowels. *Journal of the Acoustical Society of America* 90:799–828.

Bailey, P. J., A. Q. Summerfield, and M. Dorman. 1977. On the identification of sine-wave analogues of certain speech sounds. *Haskins Laboratories Status Report on Speech Research* 51/52:1–25.

Baldo, J. V., N. F. Dronkers, D. Wilkins, C. Ludy, P. Raskin, and J. Kim. 2005. Is problem solving dependent on language? *Brain and Language* 92:240–250.

Barbujani, G. 1991. What do languages tell us about human microevolution? *Trends in Ecology and Evolution* 6:151–156.

Barbujani, G., and R. R. Sokal. 1990. Zones of sharp genetic change in Europe are also linguistic boundaries. *Proceedings of the National Academy of Sciences, USA* 187:1816–1819.

——. 1991. Genetic population structure of Italy. II. Physical and cultural barriers to gene flow. *American Journal of Human Genetics* 48:398–411.

Barinaga, M. 1995. Remapping the motor cortex. *Science* 268:1696–1698.

Baru, A. V. 1975. Discrimination of synthesized vowels [a] and [i] with varying parameters (fundamental frequency, intensity, duration and number of formants) in dog. In *Auditory analysis and perception of speech,* ed. G. Fant and M. A. A. Tatham. New York: Academic Press, 91–101.

Bates, E., and J. C. Goodman. 1997. On the inseparability of grammar and the lexicon: Evidence from acquisition, aphasia, and real-time processing. *Language and Cognitive Processes* 12:507–586.

Bates, E., D. Thal, and J. Janowsky. 1992. Early language development and its neural correlates. In *Handbook of neuropsychology.* Vol. 7, *Child neuropsychology,* ed. I. Rapin and S. Segalowitz. Amsterdam: Elsevier.

Bates, E., S. Vicari, and D. Trauner. 1999. Neural mediation of language development: Perspectives from lesion studies of infants and children. In *Neurodevelopmental disorders: Contributions to a new framework from the cognitive neurosciences,* ed. H. Tager-Flusberg. Cambridge, MA: MIT Press, 533–581.

Bauer, R. H. 1993. Lateralization of neural control for vocalization by the frog *(Rana pipiens). Psychobiology* 21:243–248.

Baum, S. 1989. On-line sensitivity to local and long-distance syntactic dependencies in Broca's aphasia. *Brain and Language* 37:327–328.

Baum, S., S. E. Blumstein, M. A. Naeser, and C. L. Palumbo. 1990. Temporal dimensions of consonant and vowel production: An acoustic and CT scan analysis of aphasic speech. *Brain and Language* 39:33–56.

Bavin, E. C., and T. Shopen. 1985. Children's acquisition of Walpiri. *Journal of Child Language* 12:597–601.

Bear, M. F., B. W. Conners, and M. A. Paradiso. 1996. *Neuroscience: Exploring the brain.* Baltimore: Williams and Wilkins.

Bear, M. F., L. N. Cooper, and F. F. Ebner. 1987. A physiological basis for a theory of synaptic modification. *Science* 237:42–48.

Beckman, M. E., T.-P. Jung, S.-H. Lee, K. de Jong, A. K. Krishnamurthy, S. C. Ahalt, K. B. Cohen, and M. J. Collins. 1995. Variability in the production of quantal vowels revisited. *Journal of the Acoustical Society of America* 97:471–489.

Bell, A. M. 1867. *Visible speech.* London: Simpkin and Marshall.

Bell, C. 1844. *The anatomy and philosophy of expression as connected with the fine arts.* 3rd ed. London: G. Bell and Sons.

Bell, C. G., H. Fujisaki, J. M. Heinz, K. N. Stevens, and A. S. House. 1961. Reduction of speech spectra by analysis-by-synthesis techniques. *Journal of the Acoustical Society of America* 33:1725–1736.

Bell-Berti, F. 1971. The velopharyngeal mechanism: An electromyographic study. *Haskins Laboratories Status Report on Speech Research* 25/26:117–130.

Benson, D. F., and N. Geschwind. 1985. Aphasia and related disorders: A clinical approach. In *Principles of behavioral neurology,* ed. M. M. Mesulam. Philadelphia: F. A. Davis, 193–228.

Beranek, L. L. 1949. *Acoustics.* New York: McGraw-Hill.

Bergland, O. 1963. The bony nasopharynx: A roentgen-craniometric study. *Acta Odontologica Scandinavica* (Oslo) 21, suppl. 35.

Berridge, K. C., and I. Q. Whitshaw. 1992. Cortex, striatum and cerebellum: Control of serial order in a grooming sequence. *Experimental Brain Research* 90:275–290.

Bickerton, D. 1990. *Language and species.* Chicago: University of Chicago Press.

——. 1998. Review of *Eve spoke: Human language and human evolution. The New York Times Book Review,* March 3, 1998.

Bloom, L. 1970. *One word at a time: The use of single word utterances before syntax.* The Hague: Mouton.

Bloom, P. 2000. *How children learn words.* Cambridge, MA: MIT Press.

Blumstein, S. E. 1994. The neurobiology of the sound structure of language. In *The cognitive neurosciences,* ed. M. S. Gazzaniga. Cambridge, MA: MIT Press.

——. 1995. The neurobiology of language. In *Speech, language and communication,* ed. J. Miller and P. D. Eimas. San Diego: Academic Press, 339–370.

Blumstein, S. E., W. E. Cooper, H. Goodglass, S. Statlender, and J. Gottlieb. 1980. Production deficits in aphasia: A voice-onset time analysis. *Brain and Language* 9:153–170.

Blumstein, S. E., E. Isaacs, and J. Mertus. 1982. The role of the gross spectral shape as a perceptual cue to place of articulation in initial stop consonants. *Journal of the Acoustical Society of America* 72:43–50.

Blumstein, S. E., and K. N. Stevens. 1979. Acoustic invariance in speech production: Evidence from measurements of the spectral properties of stop consonants. *Journal of the Acoustical Society of America* 66:1001–1017.

Boe, L.-J., J.-L. Heim, K. Honda, and S. Maeda. 2002. The potential Neanderthal vowel space was as large as that of modern humans. *Journal of Phonetics* 30:465–484.

Boe, L.-J., S. Maeda, and J.-L. Heim. 1999. Neanderthal man was not morphologically handicapped for speech. *Evolution of Communication* 3:49–77.

Boehm, C. 1991. Lower-level teleology in biological evolution: Decision behavior and reproductive success in two species. *Cultural Dynamics* 4:115–134.

Boesch, C. 1993. Aspects of transmission of tool-use in wild chimpanzees. In *Tools, language and cognition in human evolution,* ed. K. R. Gibson and T. Ingold. Cambridge: Cambridge University Press, 171–184.

Boesch, C., and H. Boesch. 1984. Possible causes of sex differences in the use of natural hammers by wild chimpanzees. *Journal of Human Evolution* 13:415–440.

Bogert, C. M. 1960. The influence of sound on the behavior of amphibians and reptiles. In *Animal sounds and communication,* ed. W. E. Lanyon and W. N. Tavolga. Washington, DC: American Institute of Biological Sciences.

Bond, Z. S. 1976. Identification of vowels excerpted from neutral nasal contexts. *Journal of the Acoustical Society of America* 59:1229–1232.

Bosma, J. F. 1975. Anatomic and physiologic development of the speech apparatus. In *Human communication and its disorders,* ed. D. B. Towers. New York: Raven, 469–481.

Bouhuys, A. 1974. *Breathing.* New York: Grune and Stratton.

Boule, M. 1911–1913. L'homme fossile de la Chapelle-aux-Saints. *Annales Paleontologie* 6:109; 7:21, 85; 8:1.

Boutla, M., T. Supalla, E. L. Newport, and D. Bavelier. 2004. Short-term memory span: insights from sign language. *Nature Neuroscience* 7:997–1002.

Bradshaw, J. L., and N. C. Nettleton. 1981. The nature of hemispheric specialization in man. *Behavioral and Brain Sciences* 4:51–92.

Brainard, M. S., and A. J. Doupe. 2000. Interruption of a basal ganglia-forebrain circuit prevents the plasticity of learned vocalizations. *Nature* 404:762–766.

Bramble, D. M., and D. E. Lieberman. 2004. Endurance running and the evolution of *Homo. Nature* 432:345–352.

Brierley, J. B. 1976. Cerebral hypoxia. In *Greenfield's neuropathology,* ed. W. Blackwood and J. A. N. Corsellis. Chicago: Yearbook Medical Publications, 43–85.

Britten, R. J. 2002. Divergence between samples of chimpanzee and human DNA sequences is 5%, counting indels. *Proceedings of the National Academy of Sciences, USA* 99:13633–13635.

Broca, P. 1861. Nouvelle observation d'aphémie produite par une lésion de la moitié postérieure des deuxième et troisième circonvolutions frontales. *Bulletin Societé Anatomie*, 2nd ser., 6:398–407.

Brodmann, K. 1908. Beiträge zur histologischen Lokalisation der Grosshirnrinde. VII. Mitteilung: Die cytoarchitektonische Cortexgleiderung der Halbaffen (Lemuriden). *Journal für Psychologie und Neurologie* 10:287–334.

——. 1909. *Vergleichende Lokalisationslehre der Grosshirnrinde in iheren Prinzipien dargestellt auf Grund des Zellenbaues.* Leipzig: Barth.

——. 1912. Ergebnisse uber die vergleichende histologische Lokalisation der Grosshirnrinde mit besonderer Berucksichtigung des Stirnhirns. *Anatomischer Anzeiger,* suppl., 41:157–216.

Bronowski, J. 1971. *The identity of man.* Garden City, NY: Natural History Press.

——. 1978. *The origins of knowledge and imagination.* New Haven: Yale University Press.

Brown, R. W. 1973. *A first language.* Cambridge, MA: Harvard University Press.

Buhr, R. D. 1980. The emergence of vowels in an infant. *Journal of Speech and Hearing Research* 23:75–94.

Bunge, M. 1984. Philosophical problems in linguistics. *Erkenntnis* 21:107–173.

——. 1985. From mindless neuroscience and brainless psychology to neuropsychology. *Annals of Theoretical Psychology* 3:115–133.

Burke, R. E., S. O. Franklin, and C. E. Inturrisi. 1994. Acute persistent suppression of preproenkephaline mRNA expression in the striatum following developmental hypoxic-ischemic injury. *Journal of Neurochemistry* 62:1878–1886.

Burling, R. 1993. Primate calls, human language, and nonverbal communication. *Current Anthropology* 34:1–37.

Calvin, W. H., and D. Bickerton. 2000. *Lingua ex machina: Reconciling Darwin and Chomsky with the human brain.* Cambridge, MA: MIT Press.

Cantalupo, C., and W. D. Hopkins. 2001. Asymmetric Broca's area in great apes. *Nature* 414:505.

Capranica, R. R. 1965. *The evoked vocal response of the bullfrog.* Cambridge, MA: MIT Press.

Carew, T. J., E. T. Walters, and E. R. Kandel. 1981. Associative learning in *Aplysia:* Cellular correlates supporting a conditioned fear hypothesis. *Science* 211:501–503.

Carmody, F. 1937. X-ray studies of speech articulation. In *University of California publications in modern philology.* Vol. 20. Berkeley: University of California Press, 187–237.

Carre, R., B. Lindblom, and P. MacNeilage. 1995. Acoustic factors in the evolution of the human vocal tract. *Comptes Rendus de l'Académie des Sciences Paris,* ser. 2b, 320:471–476.

Carroll, S. B, J. K. Grenier, and S. D. Weatherbee. 2001. *From DNA to diversity: Molecular genetics and the evolution of animal design.* Malden, MA: Blackwell.

Cassirer, E. 1944. *An essay on man.* New Haven: Yale University Press.

Chapin, C., C. Y. Tseng, and P. Lieberman. 1982. Short-term release cues for stop consonant place of articulation in child speech. *Journal of the Acoustical Society of America* 71:179–186.

Cheney, D. L., and R. M. Seyfarth. 1980. Vocal recognition in free ranging vervet monkeys. *Animal Behavior* 28:362–367.

——. 1990. *How monkeys see the world: Inside the mind of another species.* Chicago: University of Chicago Press.

Chiappe, L. M., L. Salgado, and R. A. Coria. 2001. Embryonic skulls of titanosaur sauropod dinosaurs. *Science* 293:2444–2446.

Chiba, T., and J. Kajiyama. 1941. *The vowel: Its nature and structure.* Tokyo: Tokyo-Kaisekan Publishing.

Chie, U., Y. Inoue, M. Kimura, E. Kirino, S. Nagaoka, M. Abe, T. Nagata, and H. Arai. 2004. Irreversible subcortical dementia following high altitude illness. *High Altitude Medicine and Biology* 5:77–81.

Chomsky, N. 1957. *Syntactic structures.* The Hague: Mouton.

——. 1966. *Cartesian linguistics.* New York: Harper and Row.

——. 1976. On the nature of language. In *Origins and evolution of language and speech,* ed. H. B. Steklis, S. R. Harnad, and J. Lancaster. New York: New York Academy of Sciences, 46–57.
——. 1980a. Initial states and steady states. In *Language and learning: The debate between Jean Piaget and Noam Chomsky,* ed. M. Piattelli-Palmarini. Cambridge, MA: Harvard University Press, 107–130.
——. 1980b. Rules and representations. *Behavioral and Brain Sciences* 3:1–61.
——. 1986. *Knowledge of language: Its nature, origin and use.* New York: Prager.
——. 1995. *The minimalist program.* Cambridge, MA: MIT Press.
Chomsky, N., and M. Halle. 1968. *The sound pattern of English.* New York: Harper and Row.
Chung, S-J. 2000. *L'expression et la perception de l'émotion extraite de la parole spontanée: Évidences du coréen et de l'anglais.* Thesis, Université de la Sorbonne Nouvelle.
Churchland, P. M. 1995. *The engine of reason, the seat of the soul: A philosophical journey into the brain.* Cambridge, MA: MIT Press.
Clark, J. D., Y. Beyene, G. WoldeGabriel, W. Hart, P. R. Renne, H. Gilbert, A. Defleur, G. Suwa, S. Katoh, K. R. Ludwig, J.-R. Boisserie, B. Asfaw, and T. D. White. 2003. Stratigraphic, chronological and behavioral context of Pleistocene *Homo sapiens* from Middle Awash, Ethiopia. *Nature* 423:747–752.
Classen, J., J. Liepert, S. P. Wise, M. Hallet, and L. G. Cohen. 1998. Rapid plasticity of human cortical movement representation induced by practice. *Journal of Neurophysiology* 79:1117–1123.
Clegg, M. 2001. *The comparative anatomy and evolution of the human vocal tract.* Ph.D. diss., University College, London.
Cohen, L. G., P. Celnik, A. Pascual-Leone, B. Corwell, L. Faiz, J. Dambrosia, M. Honda, N. Sadato, C. Gerloff, M. D. Catata, and M. Hallet. 1997. Functional relevance of cross-modal plasticity in blind humans. *Science* 389:180–184.
Cools, R., R. A. Barker, G. J. Sahakian, and T. W. Robbins. 2001. Mechanisms of cognitive set flexibility in Parkinson's disease. *Brain* 124:2503–2512.
Cooper, W. E., and J. M. Sorenson. 1977. Fundamental frequency contours at syntactic boundaries. *Journal of the Acoustical Society of America* 62:682–692.
——. 1981. *Fundamental frequency in sentence production.* New York: Springer.
Corballis, M. 2002. *From hand to mouth: The origins of language.* Princeton: Princeton University Press.
Coster, W. J. 1986. *Aspects of voice and conversation in behaviorally inhibited and uninhibited children.* Ph.D. diss., Harvard University.
Courtney, S. M., L. Petit, M. M. Jose, L. G. Ungerleider, and J. V. Haxby. 1998. An area specialized for spatial working memory in human frontal cortex. *Science* 279:1347–1351.
Crelin, E. S. 1969. *Anatomy of the newborn: An atlas.* Philadelphia: Lea and Febiger.
——. 1973. *Functional anatomy of the newborn.* New Haven: Yale University Press.
Croft, W. 1991. *Syntactic categories and grammatical relations.* Chicago: University of Chicago.
Crosson, B., A. B. Moore, K. Gopinath, K. D. White, C. E. Wierenga, M. E. Gaiefsky, K. S. Fabrizio, K. K. Peck, D. Soltysik, C. Milsted, R. W. Briggs, T. W. Conway, and L. J. G. Rothi. 2005. Role of the right and left hemispheres in recovery of func-

tion during treatment of intention in aphasia. *Journal of Cognitive Neuroscience* 17:392–406.

Cummings, J. L. 1993. Frontal-subcortical circuits and human behavior. *Archives of Neurology* 50:873–880.

Cummings, J. L., and D. F. Benson. 1984. Subcortical dementia: Review of an emerging concept. *Archives of Neurology* 41:874–879.

Cunnington, R., R. Iansek, J. L. Bradshaw, and J. G. Phillips. 1995. Movement-related potentials in Parkinson's disease: Presence and predictability of temporal and spatial cues. *Brain* 118:935–950.

Curtiss, S. 1977. *Genie: A psycholinguistic study of a modern-day "wild child."* New York: Academic Press.

Cymerman A., P. Lieberman, J. Hochstadt, P. B. Rock, G. E. Butterfield, and L. Moore. 1999. Speech motor control and the development of acute mountain sickness. U.S. Army Research Institute of Environmental Medicine, technical report no. T99-5, February 1999, AD A360764. Alexandria, VA: Defense Technical Information Center.

Damasio, H. 1991. Neuroanatomical correlates of the aphasias. In *Acquired aphasia,* 2nd ed., ed. M. T. Sarno. New York: Academic Press.

Damasio, H., T. J. Grabowski, D. Tranel, R. D. Hichwa, and A. R. Damasio. 1996. A neural basis for lexical retrieval. *Nature* 380:499.

Darnton, R. 1985. *The great cat massacre and other episodes in French cultural history.* New York: Vantage.

Darwin, C. [1859] 1964. *On the origin of species.* Cambridge, MA: Harvard University Press.

——. 1872. *The expression of the emotions in man and animals.* London: John Murray.

Deacon, T. W. 1997. *The symbolic species: The co-evolution of language and the brain.* New York: Norton.

Dehaene-Lambertz, C., S. Dehaene, and L. Hertz-Pannier. 2002. Functional neuroimaging of speech perception in infants. *Science* 298:2013–2015.

Dejours, P. 1963. Control of respiration by arterial chemoreceptors. *Annals of the New York Academy of Sciences* 109:682–695.

DeLong, M. R. 1993. Overview of basal ganglia function. In *Role of the cerebellum and basal ganglia in voluntary movement,* ed. N. Mano, I. Hamada, and M. R. DeLong. Amsterdam: Elsevier.

DeLong, M. R., A. P. Georgopoulos, and M. D. Crutcher. 1983. Cortico-basal ganglia relations and coding of motor performance. In *Neural coding of motor performance.* Suppl. 7, *Experimental brain research,* ed. J. Massion, J. Paillard, W. Schultz, and M. Wiesendanger. Berlin: Springer, 30–40.

Denneberg, V. H. 1981. Hemispheric laterality in animals and the effects of early experience. *Behavioral and Brain Sciences* 4:1–50.

d'Errico, F., J. Zilhao, M. Julien, D. Baffier, and J. Pelegrin. 1998. Neanderthal acculturation in western Europe? A critical review of the evidence and its interpretation. *Current Anthropology* 39:S1–S44.

D'Esposito, M., and M. P. Alexander. 1995. Subcortical aphasia: Distinct profiles following left putaminal hemorrhage. *Neurology* 45:38–41.

D'Esposito, M., J. A. Detre, D. C. Alsop, R. K. Shin, S. Atlas, and M. Grossman. 1995. The neural basis of the central executive system of working memory. *Nature* 378:279–281.

Dewey, R., E. Roy, P. Squire-Storer, and D. Hayden. 1988. Limb and oral praxic abilities of children with verbal sequencing deficits. *Developmental Medicine and Child Neurology* 30:743–751.

Dibble, H. 1989. The implications of stone tool types for the presence of language during the lower and middle Palaeolithic. In *The human revolution: Behavioural and biological perspectives in the origins of modern humans,* ed. P. Mellars and C. B. Stringer. Edinburgh: Edinburgh University Press, 415–432.

Dirnberger, G., C. D. Frith, and M. Jahanshahi. 2005. Executive dysfunction in Parkinson's disease is associated with altered pallidal-frontal processing. *Neuroimage* 25:588–599.

Dobzhansky, T. 1973. Nothing in biology makes sense except in the light of evolution. *American Biology Teacher* 35:125–129.

Donald, M. 1991. *Origins of the modern mind.* Cambridge, MA: Harvard University Press.

Donoghue, J. P. 1995. Plasticity of adult sensorimotor representations. *Current Opinion in Neurobiology* 5:749–754.

Dooling, R. J. 1992. Hearing in birds. In *The evolutionary biology of hearing,* ed. D. B. Webster, R. F. Fay, and A. N. Popper. New York: Springer-Verlag, 545–560.

Draper, M. H., P. Ladefoged, and D. Whitteridge. 1960. Expiratory pressure and air flow during speech. *British Medical Journal* 1:1837–1843.

Driver, J. 1996. Enhancement of selective listening by illusory mislocation of speech sounds due to lip-reading. *Nature* 381:66–68.

Dronkers, N. F., J. K. Shapiro, B. Redfern, and R. T. Knight. 1992. The role of Broca's area in Broca's aphasia. *Journal of Clinical and Experimental Neuropsychology* 14:52–53.

DuBrul, E. L. 1977. Origins of the speech apparatus and its reconstruction in fossils. *Brain and Language* 4:365–381.

Dunbar, R. I. M. 1993. Coevolution of neocortical size and language in humans. *Behavior and Brain Sciences* 16:681–735.

Edelman, G. M. 1987. *Neural Darwinism.* New York: Basic Books.

Eimas, P. D. 1974. Auditory and linguistic processing of cues for place of articulation by infants. *Perception and Psychophysics* 16:513–521.

Eimas, P. D., E. R. Siqueland, P. Jusczyk, and J. Vigorito. 1971. Speech perception in early infancy. *Science* 171:304–306.

Elbert, T., C. Pantev, C. Wienbruch, B. Rockstroh, and E. Taub. 1995. Increased cortical representation of the fingers of the left hand in string players. *Science* 270:305–307.

Eldridge, N., and S. J. Gould. 1972. Punctuated equilibria: An alternative to phyletic gradualism. In *Models in paleobiology,* ed. T. J. M. Schopf. San Francisco: Freeman Cooper.

Elman, J., E. Bates, M. Johnson, A. Karmiloff-Smith, D. Parisi, and K. Plunkett. 1997. *Rethinking innateness: A connectionist perspective on development.* Cambridge, MA: MIT Press.

Enard, W., M. Przeworski, S. E. Fisher, C. S. L. Lai, V. Wiebe, T. Kitano, A. P. Monaco, and S. Paabo. 2002. Molecular evolution of FOXP2, a gene involved in speech and language. *Nature* 41:869–872.

Engen, E., and T. Engen. 1983. *Rhode Island test of language structure.* Baltimore: University Park Press.

Evans, P. D., J. R. Anderson, E. J. Vallender, S. L. Malcom, and S. Dorus. 2004. Adaptive value of ASPM, a major determinant of cerebral cortex size in humans. *Human Molecular Genetics* 13:489–494.

Evarts, E. V. 1973. Motor cortex reflexes associated with learned movement. *Science* 179:501–503.

Falk, D. 1975. Comparative anatomy of the larynx in man and chimpanzee: Implications for language in Neanderthal. *American Journal of Physical Anthropology* 43:123–132.

Falk, D. 2003. Motherese and the evolution of language. Symposium: Evolution of Language Reappraised. Primate Research Center (PRI), Kyoto University, Japan.

Fant, G. 1956. On the predictability of formant levels and spectrum envelopes from formant frequencies. In *For Roman Jakobson*, ed. M. Halle, H. Lunt, and H. MacLean. The Hague: Mouton.

——. 1960. *Acoustic theory of speech production.* The Hague: Mouton.

Feinberg, M. J., and O. Ekberg. 1990. Deglutition after near-fatal choking episode: Radiologic evaluation. *Radiology* 176:637–640.

Fernald, A., and P. K. Kuhl. 1987. Acoustic determinants of infant preference for motherese speech. *Infant Behavior and Development* 10:279–293.

Fernald, A., T. Taeschner, J. Dunn, M. Papousek, B. de Boysson-Bardies, and I. Fukui. 1989. A cross-language study of prosodic modifications in mothers' and fathers' speech to preverbal infants. *Journal of Child Language* 16:477–501.

Ferrein, C. J. 1741. *Mémoires de l'Académie des sciences de Paris,* 409–432 (Nov. 15). Noted in Muller (1848, 1002).

Fink, B. R., and R. J. Demarest. 1978. *Laryngeal biomechanics.* Cambridge, MA: Harvard University Press.

Fink, G. R., R. S. Frackowiak, U. Pietrzyk, and R. E. Passingham. 1997. Multiple nonprimary motor areas in the human cortex. *Journal of Neurophysiology* 77:2164–2174.

Fischer, J., D. L. Cheney, and R. M. Seyfarth. 2000. Development of infant baboons' responses to graded bark variants. *Proceedings of the Royal Society of London* 267:2317–2321.

Fischer, J., K. Hammerschmidt, D. L. Cheney, and R. M. Seyfarth. 2002. Acoustic features of male baboon loud calls: Influences of context, age and individuality. *Journal of the Acoustical Society of America* 111:1465–1474.

Fisher, S. E., F. Vargha-Khadem, K. E. Watkins, A. P. Monaco, and M. E. Pembrey. 1998. Localization of a gene implicated in a severe speech and language disorder. *Nature Genetics* 18:168–170.

Fitch, W. T., III. 1993. *Vocal tract length and the evolution of language.* Ph.D. diss., Brown University.

——. 1997. Vocal tract length and formant frequency dispersion correlate with body size in macaque monkeys. *Journal of the Acoustical Society of America* 102:1213–1222.

——. 2000a. The phonetic potential of nonhuman vocal tracts: Comparative cineradiographic observations of vocalizing animals. *Phonetica* 57:205–218.

——. 2000b. The evolution of speech: A comparative view. *Trends in Cognitive Science* 4:258–267.

——. 2000c. Skull dimensions in relation to body size in nonhuman mammals: The causal bases for acoustic allometry. *Zoology* 103:40–58.

Fitch, W. T., III, and J. Giedd. 1999. Morphology and development of the human vocal tract: A study using magnetic resonance imaging. *Journal of the Acoustical Society of America* 106:1511–1522.

Fitch, W. T., III, and J. P. Kelley. 2000. Perception of vocal tract resonances by whooping cranes *Grus americana. Ethology* 106:559–574.

Fitch, W. T., III, and D. Reby. 2001. The descended larynx is not uniquely human. *Proceedings of the Royal Society of London* B 268:1669–1675.

Flowers, K. A., and C. Robertson. 1985. The effects of Parkinson's disease on the ability to maintain a mental set. *Journal of Neurology, Neurosurgery, and Psychiatry* 48:517–529.

Fodor, J. 1983. *Modularity of mind.* Cambridge, MA: MIT Press.

Fouts, R. S. 1975. Capacities for language in the great apes. In *Society and psychology of primates,* ed. R. H. Tittle. The Hague: Mouton, 371–390.

Fouts, R. S., D. H. Fouts, and T. Van Cantfort. 1989. The infant Loulis learns from cross-fostered chimpanzees. In *Teaching sign language to chimpanzees*, ed. R. A. Gardner, B. T. Gardner, and T. Van Cantfort. Albany: State University of New York Press, 280–292.

Fouts, R. S., A. D. Hirsch, and D. H. Fouts. 1982. Cultural transmission of a human language in a chimpanzee mother-infant relationship. In *Child nurturance,* vol. 3, ed. H. E. Fitzgerald, J. A. Mullins, and P. Gage. New York: Plenum.

Friederici, A. D. 2002. Towards a neural basis of auditory sentence processing. *Trends in Cognitive Science* 6:78–84.

Friedman, J. H., P. Lieberman, M. Epstein, K. Cullen, J. N. Sanes, M. D. Lindquist, and M. Daamen. 1996. Gamma knife pallidotomy in advanced Parkinson's disease. *Annals of Neurology* 39:535–538.

Frishkopf, L. S., and M. H. Goldstein Jr. 1963. Responses to acoustic stimuli from single units in the eighth nerve of the bullfrog. *Journal of the Acoustical Society of America* 35:1219–1228.

Fujii, N., and A. M. Graybiel. 2003. Representation of action sequence boundaries by macaque prefrontal cortical neurons. *Nature* 301:1246–1249.

Fujimura, O., and Y. Kakita. 1979. Remarks on quantitative description of lingual articulation. In *Frontiers of speech communication research*, ed. B. Lindblom and S. Ohman. London: Academic Press, 17–24.

Fujisaki, H., and T. Kawashima. 1968. The role of pitch and higher formants in the perception of vowels. *IEEE Transactions on Audio and Electroacoustics* AV-16:73–77.

Fuster, J. M. 1989. *The prefrontal cortex: Anatomy, physiology, and neuropsychology of the frontal lobe.* 2nd ed. New York: Raven.

Gall, F. J. 1809. *Recherches sur le système nerveux.* Paris: B. Bailliere.

Gannon, P. J., R. L. Holloway, D. C. Broadfield, and A. R. Braun. 1998. Asymmetry of chimpanzee planum temporale: Humanlike pattern of Wernicke's brain language area homolog. *Science* 279:220–222.

Gardner, B. T., and R. A. Gardner. 1994. Development of phrases in the utterances of children and cross-fostered chimpanzees. In *The ethological roots of culture,* ed. R. A. Gardner, B. T. Gardner, B. Chiarelli, and R. Plooj. Dordrecht: Kluwer, 223–255.

Gardner, R. A., and B. T. Gardner. 1969. Teaching sign language to a chimpanzee. *Science* 165:664–672.

——. 1971. Two-way communication with an infant chimpanzee. In *Behavior of nonhuman primates*, vol. 4, ed. A. Schrier and F. Stollnitz. New York: Academic Press.

——. 1973. Teaching sign language to the chimpanzee Washoe. 16mm sound film and transcript. State College, PA: Psychological Film Register.

——. 1974. Comparing the early utterances of child and chimpanzee. In *Minnesota symposium on child psychology,* ed. A. Pick. Minneapolis: University of Minnesota Press.

——. 1978. Comparative psychology and language acquisition. *Annals of the New York Academy of Sciences* 309:37–76.

——. 1980. Two comparative psychologists look at language acquisition. In *Children's language*, vol. 2, ed. K. E. Nelson. New York: Halsted.

——. 1984. A vocabulary test for chimpanzees *(Pan troglodytes). Journal of Comparative Psychology* 4:381–404.

——. 1994. Development of phrases in utterances of children and cross-fostered chimpanzees. In *The ethological roots of culture,* eds. R. A. Gardner, B. T. Gardner, B. Chiarelli, and R. Plooj. Dordrecht: Kluwer Academic Publishers, 223–255.

Gardner, R. A., B. T. Gardner, and T. E. Van Cantfort. 1989. *Teaching sign language to chimpanzees.* Albany: State University of New York Press.

Gathercole, S. E., and A. D. Baddeley. 1993. *Working memory and language.* Hillsdale, NJ: Lawrence Erlbaum.

George, S. L. 1976. *The relationship between cranial base angle morphology and infant vocalizations.* Sc.D. diss., University of Connecticut.

——. 1978. A longitudinal and cross-sectional analysis of the growth of the postnatal cranial base angle. *American Journal of Physical Anthropology* 49:171–178.

Gerstman, L. 1968. Classification of self-normalized vowels. *IEEE Transactions on Audio and Electroacoustics* AV-16:78–80.

Geschwind, N. 1970. The organization of language and the brain. *Science* 170:940–944.

Gold, E. M. 1967. Language identification in the limit. *Information and Control* 10:447–474.

Goldin-Meadow, S. 1993. When does gesture become language? A study of gesture used as the primary communication by deaf children of hearing parents. In *Tools, language and cognition in human evolution,* ed. K. R. Gibson and T. Ingold. Cambridge: Cambridge University Press, 63–85.

Goldman-Eisler, F. 1972. Pauses, clauses, sentences. *Language and Speech* 15:103–113.

Goldstein, K. 1948. *Language and language disturbances.* New York: Grune and Stratton.

Goldstein, U. G. 1980. *An articulatory model for the vocal tracts of growing children.* Sc.D. diss., MIT.

Goodall, J. 1986. *The chimpanzees of Gombe: Patterns of behavior.* Cambridge, MA: Harvard University Press.

Gopnik, A., and A. Meltzoff. 1985. From people, to plans, to objects: Changes in the meanings of early words and their relation to cognitive development. *Journal of Pragmatics* 9:495–512.

——. 1987. The development of categorization in the second year and its relation to other cognitive and linguistic developments. *Child Development* 58:1523–1531.

Gopnik, M. 1990. Dysphasia in an extended family. *Nature* 344:715.

Gopnik, M., and M. Crago. 1991. Familial segregation of a developmental language disorder. *Cognition* 39:1–50.

Gotham, A. M., R. G. Brown, and C. D. Marsden. 1988. "Frontal" cognitive function in patients with Parkinson's disease "on" and "off" levadopa. *Brain* 111:199–321.

Gottlieb, G. 1975. Development of species identification in ducklings: Nature of perceptual deficits caused by embryonic auditory deprivation. *Journal of Comparative and Physiological Psychology* 89:387–389.

Gould, S. J. 1977. *Ontogeny and phylogeny*. Cambridge, MA: Harvard University Press.

Gould, S. J., and N. Eldridge. 1977. Punctuated equilibria: The tempo and mode of evolution reconsidered. *Paleobiology* 3:115–151.

Gould, S. J., and R. C. Lewontin. 1979. The spandrals of San Marco and the Panglossian program: A critique of the adaptationist programme. *Proceedings of the Royal Society of London* 205:281–288.

Grafman, J. 1989. Plans, actions and mental sets: The role of the frontal lobes. In *Integrating theory and practice in clinical neuropsychology*, ed. E. Perecman. Hillsdale, NJ: Lawrence Erlbaum.

Graybiel, A. M. 1995. Building action repertoires: Memory and learning functions of the basal ganglia. *Current Opinion in Neurobiology* 5:733–741.

——. 1997. The basal ganglia and cognitive pattern generators. *Schizophrenia Bulletin* 23:459–469.

——. 1998. The basal ganglia and chunking of action repertoires. *Neurobiology of Memory and Learning* 70:119–136.

Graybiel, A. M., T. Aosaki, A. W. Flaherty, and M. Kimura. 1994. The basal ganglia and adaptive motor control. *Science* 265:1826–1831.

Greenberg, B. D., D. L. Murphy, and S. A. Rasmussen. 2000. Neuroanatomically based approaches to obsessive-compulsive disorder: Neurosurgery and transcranial magnetic stimulation. *Psychiatric Clinics North America* 23:671–685.

Greenberg, J. 1963. *Universals of language*. Cambridge, MA: MIT Press.

Greenewalt, C. A. 1968. *Bird song: Acoustics and physiology*. Washington, DC: Smithsonian Institution Press.

Greenfield, P. M. 1991. Language, tools and brain: The ontogeny and phylogeny of hierarchically organized sequential behavior. *Behavioral and Brain Sciences* 14:531–577.

Grice, P. 1989. *Studies in the way of words*. Cambridge, MA: Harvard University Press.

Grieser, D. L., and P. K. Kuhl. 1988. Maternal speech to infants in a tonal language: Support for universal prosodic features in motherese. *Developmental Psychology* 24:14–20.

——. 1989. Categorization of speech by infants: Support for speech-sound prototypes. *Developmental Psychology* 25:577–588.

Grosmangin, C. 1979. *Base du crane et pharynx dans leur rapports avec l'appareil du language articule*. Mémoires du laboratoire d'anatomie de la faculté de medécine de Paris, no. 40–1979.

Gross, M. 1979. On the failure of generative grammar. *Language* 55:859–885.

Grossman, M., S. Carvell, S. Gollomp, M. B. Stern, M. Reivich, D. Morrison, A. Alavi, and H. I. Hurtig. 1993. Cognitive and physiological substrates of impaired sentence processing in Parkinson's disease. *Journal of Cognitive Neuroscience* 5:480–498.

Grossman, M., S. Carvell, S. Gollomp, M. B. Stern, G. Vernon, and H. I. Hurtig. 1991. Sentence comprehension and praxis deficits in Parkinson's disease. *Neurology* 41:1620–1628.

Grossman, M., S. Carvell, M. B. Stern, S. Gollomp, and H. I. Hurtig. 1992. Sentence comprehension in Parkinson's disease: The role of attention and memory. *Brain and Language* 42:347–384.

Grossman, M., J. Glosser, J. Kalmanson, M. B. Morris, H. Stren, and H. I. Hurtig. 2001. Dopamine supports sentence comprehension in Parkinson's disease. *Journal of the Neurological Sciences* 184:123–130.

Haeckel, E. 1866. *Generelle morphologie der organismen: Allgemeine Grundzüge der organischen formen-wissenschaft, mechanisch begründet durch die von Charles Darwin reformirte descendenz-theorie,* 2 vols. Berlin: Reimer.

——. 1895. *The evolution of man: A popular exposition of the principal points of human ontogeny and phylogeny.* New York: Appleton.

Haesler, S., K. Wada, A. Nshdejan, E. E. Morrisey, T. Lints, E. D. Jarvis, and C. Scharff. 2004. FoxP2 expression in avian learners and non-learners. *Journal of Neuroscience* 31:64–75.

Harrington, D. L., and K. Y. Haaland. 1991. Sequencing in Parkinson's disease: Abnormalities in programming and controlling movement. *Brain* 114:99–115.

Harris, C. M. 1953. A study of the building blocks of speech. *Journal of the Acoustical Society of America* 25:962–969.

Harris, K. S. 1977. The study of articulatory organization: Some negative progress. *Haskins Laboratories Status Report on Speech Research* 50:13–20.

Hata, Y., and M. P. Stryker. 1994. Control of thalamocortical afferent rearrangement by postsynaptic activity in developing visual cortex. *Science* 263:1732–1735.

Hauser, M. D. 1996. *The evolution of communication.* Cambridge, MA: MIT Press.

Hauser, M. D., N. Chomsky, and W. T. Fitch. 2002. The faculty of language: What is it, who has it, and how did it evolve? *Science* 298:1569–1579.

Hauser, M. D., and C. Fowler. 1991. Declination in fundamental frequency is not unique to human speech. *Journal of the Acoustical Society of America* 91:363–369.

Hayes, K. J., and C. Hayes. 1951. The intellectual development of a home-raised chimpanzee. *Proceedings of the American Philosophical Society* 95:105–109.

Hebb, D. O. 1949. *The organization of behavior: A neuropsychological theory.* New York: Wiley.

Heffner, R., and H. Heffner. 1980. Hearing in the elephant *(Elephas maximus). Science* 208:518–520.

Heim, J.-L. 1976. *Les hommes fossiles de la Ferrassie.* Paris: Masson.

——. 1989. La nouvelle reconstitution du crâne néanderthalien de la Chapelle-aux-Saints. Méthode et résultats. *Bulletin et Mémoires de la Société d'Anthropologie de Paris,* n.s., 1:95–118.

Heinz, R. D., M. B. Sachs, and J. M. Sinnott. 1981. Discrimination of steady state vowels by blackbirds and pigeons. *Journal of the Acoustical Society of America* 70:699–706.

Henke, W. L. 1966. *Dynamic articulatory model of speech production using computer simulation.* Ph.D. diss., MIT.

Herman, L. M., and W. N. Tavolga. 1980. The communication systems of cetaceans. In *Cetacean behavior: Mechanisms and functions.* New York: John Wiley and Sons.

Hermann, L. 1894. Nachtrag zur Untersuchung der Vocalcurven. *Archiv fur der Geschichte des Physiologie* 58:264–279.

Herrnstein, R. J. 1979. Acquisition, generalization, and discrimination of a natural concept. *Journal of Experimental Psychology and Animal Behavioral Processes* 5:116–129.

Herrnstein, R. J., and P. A. de Villiers. 1980. Fish as a natural category for people and pigeons. In *The psychology of learning and motivation,* vol. 14, ed. G. H. Bower. New York: Academic Press, 59–95.

Hewes, G. W. 1973. Primate communication and the gestural origin of language. *Current Anthropology* 14:5–24.

Hiiemae, K. M., J. B. Palmer, A. W. Crompton, J. Liu, D. Sapper, K. Hamblett, and F. Dodge. 2003. Tongue palate, hyoid and the capacity for modern speech.

Hiiemae, K. M., J. B. Palmer, S. W. Medicis, J. Hegener, B. S. Jackson, and D. E. Lieberman. 2002. Hyoid and tongue movements in speaking and eating. *Archives of Oral Biology* 47:11–27.

Hillenbrand, J. L., A. Getty, M. J. Clark, and K. Wheeler. 1995. Acoustic characteristics of American English vowels. *Journal of the Acoustical Society of America* 97:3099–3111.

Hillenbrand, L. 2002. *Seabiscuit: An American legend.* New York: Ballantine.

Hochstadt, J. 2004. *The nature and causes of sentence comprehension deficits in Parkinson's disease: Insights from eye tracking during sentence-picture matching.* Ph.D. diss., Brown University.

Hochstadt, J., and P. Lieberman. 2000. Eye tracking of sentence-picture matching: Comparisons between normal and Parkinsonian subjects. Poster. 13th Annual CUNY Conference on Human Sentence Processing, La Jolla, CA, March 30–April 1, 2000.

Hockett, C. F. 1960. Logical considerations in the study of animal communication. In *Animal sounds and communication,* ed. V. E. Lanyon and W. N. Tavolga. Washington, DC: American Institute of Biological Sciences, 392–430.

Hoehn, M. M., and M. D. Yahr. 1967. Parkinsonism: Onset, progression and mortality. *Neurology* 17:427–442.

Holloway, R. L. 1995. Evidence for POT expansion in early *Homo:* A pretty theory with ugly (or no) paleoneurological facts. *Behavioral and Brain Sciences* 18:191–193.

Hoover, J. E., and P. L. Strick. 1993. Multiple output channels in the basal ganglia. *Science* 259:819–821.

Horning, J. J. 1969. *A study of grammatical inference.* Ph.D. diss., Stanford University.

Houghton, P. 1993. Neanderthal supralaryngeal vocal tract. *American Journal of Physical Anthropology* 90:139–146.

Howard, L. A., M. G. Binks, A. P. Moore, and J. R. Playfer. 2001. The contribution of apraxic speech to working memory deficits in Parkinson's disease. *Brain and Language* 74:269–288.

Howells, W. W. 1976. Neanderthal man: Facts and figures. In *Proceedings of the Ninth International Congress of Anthropological and Ethnological Sciences, Chicago 1973.* The Hague: Mouton.

——. 1989. *Skull shapes and the map: Craniometric analyses in the dispersion of modern* Homo. Cambridge, MA: Peabody Museum of Archaeology and Ethnology, Harvard University.

Hoy, R. R., and R. C. Paul. 1973. Genetic coding of song specificity in crickets. *Science* 180:82–83.

Hublin, J.-J., F. Spoor, M. Braun, F. Zonneveld, and S. Condemi. 1996. A late Neanderthal associated with Upper Paleolithic artifacts. *Nature* 381:224–226.

Huettel, S. A., P. B. Mack, and G. McCarthy. 2002. Perceiving patterns in random series: Dynamic processing of sequence in prefrontal cortex. *Nature Neuroscience* 5:485–490.

Hulme, C., N. Thomson, C. Muir, and A. Lawrence. 1984. Speech rate and the development of short-term memory span. *Journal of Experimental Child Psychology* 47:241–253.

Illes, J., E. J. Metter, W. R. Hanson, and S. Iritani. 1988. Language production in Parkinson's disease: Acoustic and linguistic considerations. *Brain and Language* 33:146–160.

Ingman, M., H. Kaessmann, S. Paabo, and U. Gyliensten. 2000. Mitochondrial genome variation and the origin of modern humans. *Nature* 408:708–713.

Inoue, T., H. Kato, T. Araki, and K. Kogure. 1992. Emphasized selective vulnerability after repeated nonlethal cerebral ischemic insults in rats. *Stroke* 23:739–745.

International Phonetic Association. 1949. *The principles of the International Phonetic Association: Being a description of the International Phonetic Alphabet and the manner of using it.* London: Department of Phonetics, University College.

Irwin, O. C. 1948. Infant speech: Development of vowel sounds. *Journal of Speech and Hearing Disorders* 13:31–34.

Ivry, R. B., and H. S. Gopal. 1992. Speech production and perception in patients with cerebellar lesions. In *Attention and performance XIV: Synergies in experimental psychology, artificial intelligence and cognitive neuroscience*, ed. D. E. Meyer and S. Kornblum. Cambridge, MA: MIT Press, 772–803.

Ivry, R. B., and S. W. Keele. 1989. Timing functions of the cerebellum. *Journal of Cognitive Neuroscience* 1:134–150.

Jackendoff, R. 1994. *Patterns in the mind: Language and human nature.* New York: Basic Books.

Jackson, J. H. 1915. On affectations of speech from diseases of the brain. *Brain* 38:106–174.

Jacob, F. 1977. Evolution and tinkering. *Science* 196:1161–1166.

Jakobson, R. 1940. Kindersprache, aphasie, und allgemeine lautgesetze. In *Selected writings.* The Hague: Mouton. Trans. A. R. Keiler. 1968. *Child language, aphasia, and phonological universals.* The Hague: Mouton.

Jakobson, R., C. G. M. Fant, and M. Halle. 1952. *Preliminaries to speech analysis.* Cambridge, MA: MIT Press.

Jansen, E. M., L. Solberg, S. Underhill, S. Wilson, C. Cozzari, B. K. Hartman, P. L. Faris, and W. C. Low. 1997. Transplantation of fetal neocortex ameliorates sensorimotor and locomotor deficits following neonatal ischemic-hypoxic brain injury in rats. *Experimental Neurology* 147:487–497.

Jean, A. 1990. Brainstem control of swallowing. In *Neurophysiology of the jaws and teeth*, ed. A. Taylor. London: MacMillan, 294–321.

Jeffrey, N. In submission. Cranial base angulation and growth of the human fetal pharynx. *Anatomical Record.*

Jeffrey, N., and C. Spoor. 2002. Brain size and the human cranial base: A prenatal perspective. *American Journal of Physical Anthropology* 118:324–340.

Jellinger, K. 1990. New developments in the pathology of Parkinson's disease. In *Advances in neurology.* Vol. 53, *Parkinson's disease: Anatomy, pathology and therapy,* ed. M. B. Streifler, A. D. Korezyn, J. Melamed, and M. B. H. Youdim. New York: Raven, 1–15.

Jeong, J. H., J. C. Kwon, J. H. Chin, S. J. Yoon, and D. L. Na. 2002. Globus pallidus lesions associated with high mountain climbing. *Journal of Korean Medical Science* 17:861–863.

Jerison, H. J. 1970. Brain evolution: New light on old principles. *Science* 170:1224–1225.

——. 1973. *Evolution of the brain and intelligence.* New York: Academic Press.

Johnson, J. S., and E. L. Newport. 1989. Critical period effects in second language learning: The influence of maturational state on the acquisition of English as a second language. *Cognitive Psychology* 21:60–99.

Johnson, K. forthcoming. Massive reduction in conversational American English.

Jones, D. 1932. *An outline of English phonetics.* 3rd ed. New York: Dutton.

Just, M. A., P. A. Carpenter, T. A. Keller, W. F. M. Eddy, and K. R. Thulborn. 1996. Brain activation modulated by sentence comprehension. *Science* 274:114–116.

Kagan, J., J. S. Reznick, and N. Snidman. 1988. Biological bases of childhood shyness. *Science* 240:167–171.

Kaminski, J., J. Call, and J. Fisher. 2004. Word learning in a domestic dog: Evidence for "fast mapping." *Science* 304:1682–1683.

Kant, I. [1785] 1981. *Groundings for the metaphysics of morals.* Trans. J. W. Ellington. Indianapolis: Hackett.

Karmiloff, K., and A. Karmiloff-Smith. 2001. *Pathways to language: From fetus to adolescent.* Cambridge, MA: Harvard University Press.

Karni, A., G. Meyer, P. Jezzard, M. M. Adams, R. Turner, and L. G. Ungerleider. 1995. Functional MRI evidence for adult motor cortex plasticity during motor skill learning. *Nature* 377:155–158.

Karni, A., G. Meyer, C. Rey-Hipolito, P. Jezzard, M. M. Adams, and L. G. Ungerleider. 1998. The acquisition of skilled motor performance: Fast and slow experience-driven changes in primary motor cortex. *Proceedings of the National Academy of Sciences, USA* 953:861–868.

Katz, W., J. Machetanz, U. Orth, and P. Schonle. 1990. A kinematic analysis of anticipatory coarticulation in the speech of anterior aphasic subjects using electromagnetic articulography. *Brain and Language* 38:555–575.

Katz, W. F. 1988. Anticipatory coarticulation in aphasia: Acoustic and perceptual data. *Brain and Language* 35:340–368.

Kay, R. F., M. Cartmill, and M. Balow. 1998. The hypoglossal canal and the origin of human vocal behavior. *Proceedings of the National Academy of Sciences, USA* 95:5417–5419.

Keating, P. J., and R. Buhr. 1978. Fundamental frequency in the speech of infants and children. *Journal of the Acoustical Society of America* 63:567–571.

Kegl, J., A. Senghas, and M. Coppola. 1999. Creation through contact: Sign language emergence and sign language change in Nicaragua. In *Language creation and language change: Creolization, diachrony, and development*, ed. M. Degraff. Cambridge, MA: MIT Press, 179–237.

Kimura, D. 1979. Neuromotor mechanisms in the evolution of human communication. In *Neurobiology of social communication in primates,* ed. H. D. Steklis and M. J. Raleigh. New York: Academic Press.

Kimura, D. 1993. *Neuromotor mechanisms in human communication.* Oxford: Oxford University Press.

Kimura, M., T. Aosaki, and A. Graybiel. 1993. Role of basal ganglia in the acquisition and initiation of learned movement. In *Role of the cerebellum and basal ganglia in voluntary movements,* ed. N. Mano, I. Hamada, and M. R. DeLong. Amsterdam: Elsevier, 83–87.

Kimura, D., and N. Watson. 1989. The relation between oral movement and speech. *Brain and Language* 37:565–590.

Kirchoff, B. A., A. D. Wagner, A. Naril, and C. E. Stern. 2000. Prefrontal-temporal circuitry for episodic encoding and subsequent memory. *Journal of Neuroscience* 20:6173–6180.

Klatt, D. H. 1976. Linguistic uses of segmental duration in English: Acoustic and perceptual evidence. *Journal of the Acoustical Society of America* 59:1208–1221.

Klatt, D. H., and R. A. Stefanski. 1974. How does a mynah bird imitate human speech? *Journal of the Acoustical Society of America* 55:822–832.

Klatt, D. H., K. N. Stevens, and J. Mead. 1968. Studies of articulatory activity and airflow during speech. *Annals of the New York Academy of Sciences* 155:42–54.

Klein, D., B. Milner, R. J. Zatorre, E. Meyer, and A. C. Evans. 1995. The neural substrates underlying word generation: A bilingual functional imaging study. *Proceedings of the National Academy of Sciences, USA* 92:2899–2903.

Klein, D., R. J. Zatorre, B. Milner, E. Meyer, and A. C. Evans. 1994. Left putaminal activation when speaking a second language: Evidence from PET. *NeuroReport* 5:2295–2297.

Klein, R. G. 1999. *The human career.* 2nd ed. Chicago: Chicago University Press.

Kluender, K. R., R. L. Diehl, and P. R. Killeen. 1987. Japanese quail can learn phonetic categories. *Science* 237:1195–1197.

Knowlton, B. J., J. A. Mangels, and L. R. Squire. 1996. A neostratal learning system in humans. *Science* 273:1399–1402.

Kohler, E., C. Keysers, M. A. Umilta, L. de Fogassi, V. Gallese, and G. Rizzolatti. 2002. Hearing sounds, understanding actions: Action representation in mirror neurons. *Science* 297:846–848.

Kohonen, T. 1984. *Self-organization and associative memory.* New York: Springer-Verlag.

Kosslyn, S. M., A. Pascual-Leone, O. Felician, S. Camposano, J. P. Keenan, W. L. Thompson, G. Ganis, K. E. Sukel, and N. M. Alpert. 1999. The role of area 17 in visual imagery: Convergent evidence from PET and rTMS. *Science* 284:167–170.

Kotz, S. A., S. Frisch., D. Y. von Cramon, and A. D. Friederici. 2003a. Syntactic language processing: ERP lesion data on the role of the basal ganglia. *Journal of the International Neuropsychological Society* 9:1053–1060.

Kotz, S. A., M. Meyer, K. Alter, M. Besson, D. Y. von Cramon, and A. D. Friederici.

2003b. On the lateralization of emotional prosody: An fMRI investigation. *Brain and Language* 86:366–376.

Krakauer, J. 1997. *Into thin air.* New York: Villard.

Krams, M., M. F. Rushworth, M. P. Deiber, R. S. Frackowiak, and R. E. Passingham. 1998. The preparation, execution and suppression of copied movements in the human brain. *Experimental Brain Research* 120:386–398.

Kratzenstein, C. G. [1780] 1782. Sur la naissance de la formation des voyelles. *Journal of Physiology* 21:358–381. Trans. from *Acta Academie Petrograd.*

Krings, M., A. Stone, R. W. Schmitz, H. Krainitzky, M. Stoneking, and S. Paabo. 1997. Neanderthal DNA sequences and the origin of modern humans. *Cell* 90:19–30.

Kubaska, C. A., and P. Keating. 1981. Word duration in early child speech. *Journal of Speech and Hearing Research* 24:614–621.

Kuhl, P. K. 1978. Speech perception by the chinchilla: Identification functions for synthetic VOT stimuli. *Journal of the Acoustical Society of America* 63:905–916.

———. 1981. Discrimination of speech by nonhuman animals: Basic auditory sensitivities conducive to the perception of speech-sound categories. *Journal of the Acoustical Society of America* 70:340–349.

———. 1988. Auditory perception and the evolution of speech. *Human Evolution* 3:21–45.

Kuhl, P. K., and A. N. Meltzoff. 1996. Infant vocalizations in response to speech: Vocal imitation and developmental change. *Journal of the Acoustical Society of America* 100:2425–2438.

Kuhl, P. K., K. A. Williams, F. Lacerda, K. N. Stevens, and B. Lindblom. 1992. Linguistic experience alters phonetic perception in infants by 6 months of age. *Science* 255:606–608.

Kuoppamaki, M., K. P. Bhatia, and N. Quinn. 2002. Progressive delayed-onset dystonia after cerebral anoxic insult in adults. *Movement Disorders* 17:1345–1349.

Kutas, M. 1997. Views on how the electrical activity that the brain generates reflects the functions of different language structures. *Psychophysiology* 34:383–398.

Labov, W., M. Yaeger, and R. Steiner. 1972. *A quantitative study of sound change in progress.* Philadelphia: U.S. Regional Survey.

Ladefoged, P., and D. E. Broadbent. 1957. Information conveyed by vowels. *Journal of the Acoustical Society of America* 29:98–104.

Ladefoged, P., J. De Clerk, M. Lindau, and G. Papcun. 1972. An auditory-motor theory of speech production. *UCLA Working Papers in Phonetics* 22:48–76.

Lai, C. S., D. Gerrelli, A. P. Monaco, S. E. Fisher, and A. J. Copp. 2003. FOXP2 expression during brain development coincides with adult sites of pathology in a severe speech and language disorder. *Brain* 126:2455–2462.

Lai, S. J., S. E. Fisher, J. A. Hurst, F. Vargha-Khadem, and A. P. Monaco. 2001. A forkhead-domain gene is mutated in a severe speech and language disorder. *Nature* 413:519–523.

Laitman, J. T., and E. S. Crelin. 1976. Postnatal development of the basicranium and vocal tract region in man. In *Symposium on development of the basicranium*, ed. J. Bosma. Washington DC: U.S. Government Printing Office, 206–219.

Laitman, J. T., and R. C. Heimbuch. 1982. The basicranium of Plio-Pleistocene hominids as an indicator of their upper respiratory systems. *American Journal of Physical Anthropology* 59:323–344.

Laitman, J. T., R. C. Heimbuch, and E. S. Crelin. 1978. Developmental changes in a basicranial line and its relationship to the upper respiratory system in living primates. *American Journal of Anatomy* 152:467–482.

——. 1979. The basicranium of fossil hominids as an indicator of their upper respiratory systems. *American Journal of Physical Anthropology* 51:15–34.

Laitman, J. T., J. S. Reidenberg, P. J. Gannon, B. Johansson, K. Landahl, and P. Lieberman. 1990. The Kebara hyoid: What can it tell us about the evolution of the hominid vocal tract? *American Journal of Physical Anthropology* 81:254.

LaMettrie, J. O. [1747] 1960. *De l'homme machine.* Ed. A. Vartanian. Princeton: Princeton University Press.

Landahl, K. L. 1982. *The onset of structural discourse: A developmental study of the acquisition of language.* Ph.D. diss., Brown University.

Lange, K. W., T. W. Robbins, C. D. Marsden, M. James, A. M. Owen, and G. M. Paul. 1992. L-dopa withdrawal in Parkinson's disease selectively impairs cognitive performance in tests sensitive to frontal lobe dysfunction. *Psychopharmacology* 107:394–404.

Langlois, A., R. J. Baken, and C. N. Wilder. 1980. Pre-speech respiratory activity in the first year of life. In *Infant communication, cry and early speech,* ed. T. Mury and J. Mury. Houston: College-Hill Press.

Laplane, D., M. Baulac, D. Widlocher, and B. Dubois. 1984. Pure psychic akinesia with bilateral lesions of basal ganglia. *Journal of Neurology, Neurosurgery and Psychiatry* 47:377–385.

Laplane, D., M. Levasseur, B. Pillon, B. Dubois, M. Baulac, B. Mazoyer, S. Tran Dinh, G. Sette, F. Danze, and J. C. Baron. 1989. Obsessive-compulsive and other behavioural changes with bilateral basal ganglia lesions. *Brain* 112:699–725.

Larkin, M. 2003. Detection of cognitive impairment: The final frontier. *The Lancet—Neurology* 2:590–591.

Lartet, E. 1868. De quelques cas de progression organique vérifiables dans la succession des temps, géologiques sur des mammifères de même famille et de même genre. *Comptes Rendus de l'Académie des Sciences Paris* 66:1119–1122.

Lashley, K. S. 1951. The problem of serial order in behavior. In *Cerebral mechanisms in behavior,* ed. L. A. Jeffress. New York: Wiley, 112–146.

Lattimore, E. [1934] 1994. *Turkestan reunion.* New York: Kodansha America.

Lee, S., A. Potamianos, and S. Narayanan. 1999. Acoustics of children's speech: Developmental changes of temporal and spectral parameters. *Journal of the Acoustical Society of America* 105:1455–1468.

Lehericy, S., M. Ducros, P.-F. Van de Moortele, C. Francois, L. Thivard, C. Poupon, N. Swindale, K. Ugurbil, and D.-S. Kim. 2004. Diffusion tensor tracking shows distinct corticostriatal circuits in humans. *Annals of Neurology* 55:522–529.

Leibniz, G. W. von. 1949. *Nouveaux essais sur l'entendement humain.* Trans. A. G. Langley. LaSalle, IL: Open Court.

Leiner, H. C., A. L. Leiner, and R. S. Dow. 1993. Cognitive and language functions of the human cerebellum. *Trends in Neuroscience* 16:444–447, 453–454.

Lenneberg, E. H. 1967. *Biological foundations of language.* New York: Wiley.

le Normand, M.-T., L. Vaivre-Douret, C. Payan, and H. Cohen. 2000. Neuromotor development and language processing in developmental dyspraxia: A follow-up case study. *Journal of Clinical and Experimental Neuropsychology* 22:408–417.

Leonard, L. E. 1998. *Children with specific language impairment.* Cambridge, MA: MIT Press.

Levelt, W. J. 1989. *Speaking: From intention to articulation.* Cambridge, MA: MIT Press.

Lewis, M. M. 1936. *Infant speech: A study of the beginnings of language.* New York: Harcourt Brace.

Lewis, S. J. G., A. Slabosz, T. W. Robbins, R. A. Barker, and A. M. Owen. 2005. Dopaminergic basis for deficits in working memory but not attentional set-shifting in Parkinson's disease. *Neuropsychologia* 43:823–832.

Liberman, A. M., F. S. Cooper, D. P. Shankweiler, and M. Studdert-Kennedy. 1967. Perception of the speech code. *Psychological Review* 74:431–461.

Liberman, M. Y. 1978. *The intonational system of English.* Bloomington: Indiana University Linguistics Club.

Lichtheim, L. 1885. On aphasia. *Brain* 7:433–484.

Lieberman, D. E. 1995. Testing hypotheses about recent human evolution from skulls. *Current Anthropology* 36:159–198.

——. 1998. Sphenoid shortening and the evolution of modern human cranial shape. *Nature* 393:158–162.

Lieberman, D. E., B. M. McBratney, and G. Krovitz. 2002. The evolution and development of cranial form in *Homo sapiens. Proceedings of the National Academy of Sciences, USA* 99:1134–1139.

Lieberman, D. E., and R. C. McCarthy. 1999. The ontogeny of cranial base angulation in humans and chimpanzees and its implications for reconstructing pharyngeal dimensions. *Journal of Human Evolution* 36:487–517.

Lieberman, D. E., R. C. McCarthy, K. M. Hiiemae, and J. B. Palmer. 2001. Ontogeny of postnatal hyoid and laryngeal descent: Implications for deglutition and vocalization. *Archives of Oral Biology* 46:117–128.

Lieberman, D. E., C. F. Ross, and M. J. Ravosa. 2000. The primate cranial base: Ontogeny, function and integration. *Yearbook of Physical Anthropology* 43:117–169.

Lieberman, D. E., and J. J. Shea. 1994. Behavioral differences between archaic and modern humans in the Levantine Mousterian. *American Anthropologist* 96:300–332.

Lieberman, M. R., and P. Lieberman. 1973. Olson's "projective verse" and the use of breath control as a structural element. *Language and Style* 5:287–298.

Lieberman, P. 1961. Perturbations in vocal pitch. *Journal of the Acoustical Society of America* 33:597–603.

——. 1963a. Some measures of the fundamental periodicity of normal and pathologic larynges. *Journal of the Acoustical Society of America* 35:344–353.

——. 1963b. Some effects of semantic and grammatical context on the production and perception of speech. *Language and Speech* 6:172–187.

——. 1965. On the acoustic basis of the perception of intonation and stress by linguists. *Word* 21:40–54.

——. 1967. *Intonation, perception and language.* Cambridge, MA: MIT Press.

——. 1968a. Primate vocalizations and human linguistic ability. *Journal of the Acoustical Society of America* 44:1157–1164.

——. 1968b. Direct comparison of subglottal and esophageal pressure during speech. *Journal of the Acoustical Society of America* 43:1157–1164.
——. 1975. *On the origins of language: An introduction to the evolution of speech.* New York: Macmillan.
——. 1980. On the development of vowel production in young children. In *Child phonology, perception and production,* ed. G. Yeni-Komshian and J. Kavanagh. New York: Academic Press, 113–142.
——. 1982. Can chimpanzees swallow or talk? A reply to Falk. *American Anthropologist* 84:148–152.
——. 1984. *The biology and evolution of language.* Cambridge, MA: Harvard University Press.
——. 1985. On the evolution of human syntactic ability: Its pre-adaptive bases—motor control and speech. *Journal of Human Evolution* 14:657–668.
——. 1991. *Uniquely human: The evolution of speech, thought and selfless behavior.* Cambridge, MA: Harvard University Press.
——. 1992. On Neanderthal speech and Neanderthal extinction. *Current Anthropology* 33:409–410.
——. 1993. The Kebara KMH-2 hyoid and Neanderthal speech. *Current Anthropology* 34:172–175.
——. 1994a. Functional tongues and Neanderthal vocal tract reconstruction: A reply to Houghton (1993). *American Journal of Physical Anthropology* 95:443–452.
——. 1994b. Biologically bound behavior, free will, and human evolution. In *Conflict and cooperation in nature,* ed. J. I. Casti. New York: John Wiley and Sons, 133–163.
——. 1998. *Eve spoke: Human language and human evolution.* New York: Norton.
——. 2000. *Human language and our reptilian brain: The subcortical bases of speech, syntax, and thought.* Cambridge, MA: Harvard University Press.
——. 2002. On the nature and evolution of the neural bases of human language. *Yearbook of Physical Anthropology* 45:36–62.
——. In press a. Limits on tongue deformation—Diana monkey formants and the impossible vocal tract shapes proposed by Riede et al. (2005). *Journal of Human Evolution.*
——. In press b. Current views on Neanderthal speech capabilities: A reply to Boe et al. (2002). *Journal of Phonetics.*
Lieberman, P., and S. E. Blumstein. 1988. *Speech physiology, speech perception, and acoustic phonetics.* Cambridge: Cambridge University Press.
Lieberman, P., and E. S. Crelin. 1971. On the speech of Neanderthal man. *Linguistic Inquiry* 2:203–222.
Lieberman, P., E. S. Crelin, and D. H. Klatt. 1972. Phonetic ability and related anatomy of the newborn, adult human, Neanderthal man, and the chimpanzee. *American Anthropologist* 74:287–307.
Lieberman, P., L. S. Feldman, S. Aronson, and B. Engen. 1989. Sentence comprehension, syntax and vowel duration in aged people. *Clinical Linguistics and Phonetics* 3:299–311.
Lieberman, P., J. Friedman, and L. S. Feldman. 1990. Syntactic deficits in Parkinson's disease. *Journal of Nervous and Mental Disease* 178:360–365.
Lieberman, P., K. S. Harris, P. Wolff, and L. H. Russell. 1972. Newborn infant

cry and nonhuman primate vocalizations. *Journal of Speech and Hearing Research* 14:718–727.

Lieberman, P., E. T. Kako, J. Friedman, G. Tajchman, L. S. Feldman, and E. B. Jiminez. 1992. Speech production, syntax comprehension, and cognitive deficits in Parkinson's disease. *Brain and Language* 43:169–189.

Lieberman, P., B. G. Kanki, and A. Protopappas. 1995. Speech production and cognitive decrements on Mount Everest. *Aviation, Space and Environmental Medicine* 66:857–864.

Lieberman, P., B. G. Kanki, A. Protopappas, E. Reed, and J. W. Youngs. 1994. Cognitive defects at altitude. *Nature* 372:325.

Lieberman, P., W. Katz, A. Jongman, R. Zimmerman, and M. Miller. 1984. Measures of the sentence intonation of read and spontaneous speech in American English. *Journal of the Acoustical Society of America* 77:649–657.

Lieberman, P., D. H. Klatt, and W. H. Wilson. 1969. Vocal tract limitations on the vowel repertoires of rhesus monkey and other nonhuman primates. *Science* 164:1185–1187.

Lieberman, P., and S. B. Michaels. 1962. Some aspects of fundamental frequency and envelope amplitude as related to the emotional content of speech. *Journal of the Acoustical Society of America* 34:922–927.

Lieberman, P., A. Morey, J. Hochstadt, M. Larson, and S. Mather. 2005. Mount Everest: A space-analogue for speech monitoring of cognitive deficits and stress. *Aviation, Space and Environmental Medicine* 76:198–207.

Lieberman, P., J. Ryalls, and S. Rabson. 1982. On the early imitation of intonation and vowels. *Handbook of the Seventh Annual Boston University Conference on Language Development,* 34–35.

Liegeois, F., T. Baldeweg, A. Connelly, D. G. Gadian, M. Mishkin, and F. Vargha-Khadem. 2003. Language fMRI abnormalities associated with FOXP2 gene mutation. *Nature Neuroscience* 6:1230–1237.

Lindblom, B. 1988. Models of phonetic variation and selection. In *Language change and biological evolution.* Torino, Italy: Institute for Scientific Interchange.

——. 1996. Role of articulation in speech perception: Clues from production. *Journal of the Acoustical Society of America* 99:1683–1692.

Linebarger, M., M. Schwartz, and E. Saffran. 1983. Sensitivity to grammatical structure in so-called agrammatic aphasics. *Cognition* 13:361–392.

Lisker, L., and A. S. Abramson. 1964. A cross-language study of voicing in initial stops: Acoustical measurements. *Word* 20:384–442.

Logothetis, N. K., J. Pauls, M. Augath, T. Trinath, and A. Oeltermann. 2001. Neurophysiological investigation of the basis of the fMRI signal. *Nature* 412:150–157.

Long, C. A. 1969. The origin and evolution of mammary glands. *Biological Sciences* 19:519–523.

Longworth, C. E., S. E. Keenan, R. A. Barker, W. D. Marslen-Wilson, and L. K. Tyler. 2005. The basal ganglia and rule-governed language use: Evidence from vascular and degenerative conditions. *Brain* 128:584–596.

Lubker, J., and T. Gay. 1982. Anticipatory labial coarticulation: Experimental, biological, and linguistic variables. *Journal of the Acoustical Society of America* 71:437–438.

MacDonald, M. C. 1994. Probabilistic constraints and syntactic ambiguity resolution. *Language and Cognitive Processes* 9:157–201.

MacLean, P. D. 1986. Neurobehavioral significance of the mammal-like reptiles (therapsids). In *The ecology and biology of mammal-like reptiles,* ed. N. Hotton III, J. J. Roth, and E. C. Roth. Washington, DC: Smithsonian Institution Press, 1–21.

MacLean, P. D., and J. D. Newman. 1988. Role of midline frontolimbic cortex in the production of the isolation call of squirrel monkeys. *Brain Research* 450:111–123.

MacNeilage, P. F. 1991. The "postural origins" theory of primate neurobiological asymmetries. In *Biological foundations of language development,* ed. N. Krasnegor, D. Rumbaugh, M. Studdert-Kennedy, and R. Schiefelbusch. Hillsdale, NJ: Lawrence Erlbaum, 165–188.

Maddieson, I. 1984. *Patterns of sounds.* Cambridge: Cambridge University Press.

Maeda, S. 1976. *A characterization of American English intonation.* Ph.D. diss., MIT.

——. 1989. Compensatory articulation during speech: Evidence from the analysis and synthesis of vocal-tract shapes using an articulatory model. In *Speech production and modelling,* ed. W. J. Hardcastle and A. Marchal. Kluwer: Academic Publishers.

Maess, B., S. Koelsch, T. C. Gunter, and A. D. Friederici. 2001. Music syntax is processed in the area of Broca: An MEG study. *Nature Neuroscience* 4:540–545.

Magee, J. C., and D. Johnston. 1997. A synaptically controlled associative signal for Hebbian plasticity in hippocampal neurons. *Science* 275:209–213.

Mahajan, P. V., and B. A. Bharucha. 1994. Evaluation of short neck: Percentiles and linear correlations with height and sitting height. *Indian Pediatrics* 31:1193–1203.

Manley, R. S., and L. C. Braley. 1950. Masticatory performance and efficiency. *Journal of Dental Research* 29:314–321.

Manley, R. S., and F. R. Shiere. 1950. The effect of dental efficiency on mastication and food preference. *Oral Surgery, Oral Medicine and Oral Pathology* 3:674–685.

Manning, C. D., and H. Schutze. 1999. *Foundations of statistical natural language processing.* Cambridge, MA: MIT Press, 386–387.

Maresh, M. M. 1948. Growth of the heart related to bodily growth during childhood and adolescence. *Pediatrics* 2:382–402.

Maresh, M. M., and A. H. Washburn. 1938. Size of the heart in healthy children. *American Journal of Disease in Children* 56:33–60.

Marie, P. 1926. *Traveaux et mémoires.* Paris: Masson.

Marin, O., W. J. Smeets, and A. Gonzalez. 1998. Evolution of the basal ganglia in tetrapods: A new perspective based on recent studies in amphibians. *Trends in Neurosciences* 21:487–494.

Markram, H., J. Lubke, M. Frotscher, and B. Sakmann. 1997. Regulation of synaptic efficacy by coincidence of postsynaptic APs and EPSPs. *Science* 275:213–215.

Marsden, C. D., and J. A. Obeso. 1994. The functions of the basal ganglia and the paradox of sterotaxic surgery in Parkinson's disease. *Brain* 117:877–897.

Martin, A., and L. L. Chao. 2001. Semantic memory and the brain: Structure and processes. *Current Opinion in Neurobiology* 11:194–201.

Martin, A., J. V. Haxby, F. M. Lalonde, C. L. Wiggs, and L. G. Ungerleider. 1995a. Discrete cortical regions associated with knowledge of color and knowledge of action. *Science* 270:102–105.

Martin, A., C. L. Wiggs, L. G. Ungerleider, and J. V. Haxby. 1995b. Neural correlates of category-specific knowledge. *Nature* 379:649–652.

Massaro, D. W., and M. M. Cohen. 1995. Perceiving talking faces. *Current Directions in Psychological Science* 4:104–109.

Matsuzawa, T. 2004. Chimpanzee Ai homepage. Primate Research Institute, Kyoto University, Japan. www.pri.kyoto-u.ac.jp/ai.

May, J. 1976. Vocal tract normalization for /s/ and /ŝ/. *Haskins Laboratories Status Report on Speech Research* 48:67–73.

Mayr, E. 1982. *The growth of biological thought.* Cambridge, MA: Harvard University Press.

McBrearty, S., and A. S. Brooks. 2000. The revolution that wasn't: A new interpretation of the origin of modern human behavior. *Journal of Human Evolution* 39:453–563.

McCammon, R. 1970. *Human growth and development.* Springfield, IL: Thomas.

McCarthy, R. C., and D. Lieberman. 2001. The posterior maxillary (PM) plane and anterior cranial architecture in primates. *Anatomical Record* 264:247-260.

McCarthy, R. C., D. S. Strait, F. Yates, and P. Lieberman. Forthcoming. The origin of human speech.

McGrew, W. C. 1993. The intelligent use of tools: Twenty propositions. In *Tools, language and cognition in human evolution,* ed. K. R. Gibson and T. Ingold. Cambridge: Cambridge University Press, 151–170.

McGurk, H., and J. MacDonald. 1976. Hearing lips and seeing voices. *Nature* 263:747–748.

McNeill, D. 1985. So you think gestures are nonverbal? *Psychological Review* 92:350–371.

Mead, J., A. Bouhuys, and D. F. Proctor. 1968. Mechanisms generating subglottic pressure. *Annals of the New York Academy of Sciences* 155:177–181.

Mega, M. S., and M. P. Alexander. 1994. Subcortical aphasia: The core profile of capsulostriatal infarction. *Neurology* 44:1824–1829.

Meltzoff, A. N., and M. K. Moore. 1977. Imitation of facial and manual gestures by human neonates. *Science* 198:75–78.

———. 1983. Newborn infants imitate adult facial gestures. *Child Development* 54:702–709.

Merzenich, M. M. 1987a. In *The neural and molecular bases of learning: Report of the Dahlem Workshop on the Neural and Molecular Bases of Learning*, ed. J.-P. Changeux and M. Konishi. Chichester: Wiley, 337–358.

———. 1987b. Cerebral cortex: A quiet revolution in thinking. *Nature* 328:572–573.

Merzenich, M. M., R. J. Nelson, M. P. Stryker, M. S. Cynader, A. Schoppmann, and J. M. Zook. 1984. Somatosensory cortical map changes following digit amputation in adult monkeys. *Journal of Comparative Neurology* 224:591–605.

Mesulam, M. M. 1985. Patterns in behavioral neuroanatomy: Association areas, the limbic system and hemispheric specialization. In *Principles of behavioral neurology*, ed. M. M. Mesulam. Philadelphia: F. A. Davis, 1–70.

———. 1990. Large-scale neurocognitive networks and distributed processing for attention, language, and memory. *Annals of Neurology* 28:597–613.

Metter, E. J., D. Kempler, C. Jackson, W. R. Hanson, J. C. Mazziotta, and M. E.

Phelps. 1989. Cerebral glucose metabolism in Wernicke's, Broca's, and conduction aphasia. *Archives of Neurology* 46:27–34.

Meyer-Luehmann, M., J. F. Thompson, K. C. Berridge, and J. W. Aldridge. 2002. Substantia nigra pars reticulata neurons code initiation of a serial pattern: Implications for neural action sequences and sequential disorders. *European Journal of Neuroscience* 16:1599–1608.

Middleton, F. A., and P. L. Strick. 1994. Anatomical evidence for cerebellar and basal ganglia involvement in higher cognition. *Science* 266:458–461.

Miles, H. L. 1976. *Conversations with apes: The use of sign language by two chimpanzees.* Ph.D. diss., University of Connecticut.

Miles, H. L. 1983. Apes and language: The search for communicative competence. In *Language and primates: Perspectives and implications*, ed. J. deLucas and H. Wilder. New York: Springer-Verlag, 43–61.

Miller, G. A. 1956. The magical number seven, plus or minus two: Some limits on our capacity for processing information. *Psychological Review* 63:81–97.

Mirenowicz, J., and W. Schultz. 1996. Preferential activation of midbrain dopamine neurons by appetitive rather than aversive stimuli. *Nature* 379:449–451.

Miyai, I., A. D. Blau, M. J. Reding, and B. T. Volpe. 1997. Patients with stroke confined to basal ganglia have diminished response to rehabilitation efforts. *Neurology* 48:95–101.

Monchi, O., P. Petrides, V. Petre, K. Worsley, and A. Dagher. 2001. Wisconsin card sorting revisited: Distinct neural circuits participating in different stages of the task identified by event-related functional magnetic resonance imaging. *Journal of Neuroscience* 21:7733–7741.

Montague, R. 1974. *Formal philosophy: Selected papers of Richard Montague*, ed. R. Thomason. New Haven: Yale University Press.

Moore, L. G., S. Niermeyer, and S. Zamudio. 1998. Human adaptation to high altitude: Regional and life-cycle perspectives. *Yearbook of Physical Anthropology* 41:25–61.

Moore, L. G., S. Zamudio, J. Zhuang, S. Sun, and T. Droma. 2001. Oxygen transport in Tibetan women during pregnancy at 3,658 m. *American Journal of Physical Anthropology* 114:42–53.

Morris, R. G., J. J. Downes, B. J. Sahakian, J. L. Evenden, A. Heald, and T. W. Robbins. 1988. Planning and spatial working memory in Parkinson's disease. *Journal of Neurology, Neurosurgery, and Psychiatry* 51:757–766.

Mouse Genome Sequencing Consortium. 2002. Initial sequencing and comparative analysis of the mouse genome. *Nature* 420:520–562.

Muller, J. 1848. *The physiology of the senses, voice and muscular motion with the mental faculties.* Trans. W. Baly. London: Walton and Maberly.

Naeser, M. A., M. P. Alexander, N. Helms-Estabrooks, H. L. Levine, S. A. Laughlin, and N. Geschwind. 1982. Aphasia with predominantly subcortical lesion sites: Description of three capsular/putaminal aphasia syndromes. *Archives of Neurology* 39:2–14.

National Transportation Safety Board. 2002. Aircraft accident brief: EgyptAir flight 990. Washington, DC: National Transportation Safety Board.

Natsopoulos, D., G. Grouios, S. Bostantzopoulou, G. Mentenopoulos, Z. Katsarou, and J. Logothetis. 1993. Algorithmic and heuristic strategies in comprehension of complement clauses by patients with Parkinson's disease. *Neuropsychologia* 31:951–964.

Nearey, T. 1979. *Phonetic features for vowels.* Bloomington: Indiana University Linguistics Club.

Negus, V. E. 1928. *The mechanism of the larynx.* London: Heinemann.

Negus, V. E. 1949. *The comparative anatomy and physiology of the larynx.* New York: Hafner.

Newman, A. J., D. Bavelier, D. Corina, P. Jezzard, and H. J. Neville. 2002. A critical period for right hemisphere recruitment in American Sign Language processing. *Nature Neuroscience* 5:76–80.

Newman, J. D. 1985. The infant cry of primates: An evolutionary perspective. In *Infant crying: Theoretical and research perspectives,* ed. B. M. Lester and C. F. Zachariah Boukydis. New York: Plenum.

Newman, J. D., and P. D. MacLean. 1982. Effects of tegmental lesions on the isolation call of squirrel monkeys. *Brain Research* 232:317–329.

Nishimura, H., K. Hashikawa, K. Doi, T. Iwaki, Y. Watanabe, H. Kusuoka, T. Nishimura, and T. Kubo. 1999. Sign language "heard" in the auditory cortex. *Nature* 397:116.

Nishimura, T., A. Mikami, J. Suzuki, and T. Matsuzawa. 2003. Descent of the larynx in chimpanzee infants. *Proceedings of the National Academy of Sciences, USA* 100:6930–6933.

Nittrouer, S., S. Estee, J. H. Lowenstein, and J. Smith. 2005. The emergence of mature gestural patterns in the production of voiceless and voiced word-final stops. *Journal of the Acoustical Society of America* 117:361–364.

Nowak, M. A., and D. C. Krakauer. 1999. The evolution of language. *Proceedings of the National Academy of Sciences, USA* 96:8028–8033.

Nowak, M. A., J. B. Plotkin, and V. A. A. Janson. 2000. The evolution of syntactic communication. *Nature* 404:495–498.

Nudo, R. J., G. W. Milliken, W. M. Jenkins, and M. M. Merzenich. 1996. Use-dependent alterations of movement representations in primary motor cortex of adult squirrel monkeys. *Journal of Neuroscience* 16:785–807.

Oelz, O., and M. Regard. 1988. Physiological and neurophysiological characteristics of world-class extreme-altitude climbers. *American Alpine Journal* 83–86.

Ohala, J. 1970. Aspects of the control and production of speech. In *UCLA Working Papers in Phonetics* 15. Los Angeles: UCLA Phonetics Laboratory.

Ohman, S. E. G. 1966. Coarticulation in VCV utterances: Spectrographic measurements. *Journal of the Acoustical Society of America* 39:151–168.

Ojemann, G. A., and C. Mateer. 1979. Human language cortex: Localization of memory, syntax, and sequential motor-phoneme identification systems. *Science* 205:1401–1403.

Ojemann, G. A., F. Ojemann, E. Lettich, and M. Berger. 1989. Cortical language localization in left dominant hemisphere: An electrical stimulation mapping investigation in 117 patients. *Journal of Neurosurgery* 71:316–326.

Olmsted, D. L. 1971. *Out of the mouth of babes.* The Hague: Mouton.

Olson, C. 1959. *Projective verse.* New York: Totem.

Ovchinnikov, I. V., A. Gotherstrom, G. P. Romanova, V. M. Kharitonov, K. Liden, and W. Goodwin. 2000. Molecular analysis of Neanderthal DNA from the northern Caucasus. *Nature* 404:453–454.

Owren, M. J. 1990. Acoustic classification of alarm calls by vervet monkeys *(Cercopithecus aethiops)* and humans *(Homo sapiens)*: II. Synthetic calls. *Journal of Comparative Psychology* 104:29–40.

Owren, M. J., and R. Bernacki. 1988. The acoustic features of vervet monkey alarm calls. *Journal of the Acoustical Society of America* 83:1927–1935.

Pagani, M., R. Ansjon, F. Lind, J. Uusijarvi, G. Sumen, C. Jonsson, D. Salmaso, H. Jacobsson, and S. A. Larsson. 2000. Effects of acute hypobaric hypoxia on regional cerebral blood flow: A single photon computed tomography study in humans. *Acta Pysiologica Scandanavica* 168:377–383.

Palmer, J. B., N. J. Rudin, G. Lara, and A. W. Crompton. 1992. Coordination of mastication and swallowing. *Dysphagia* 7:187–200.

Pantev, C., R. Oostenveld, A. Engelien, B. Ross, L. E. Roberts, and M. Hoke. 1998. Increased auditory cortical representation in musicians. *Nature* 392:811–814

Parent, A. 1986. *Comparative neurobiology of the basal ganglia.* New York: John Wiley.

Parkinson's Study Group. 1989. The DATATOP series. *Archives of Neurology* 46:1052–1060.

Pascual-Leone, A., D. Nguyet, L. G. Cohen, J. P. Brasil-Neto, A. Cammaroya, and M. Hallet. 1995. Modulation of motor responses evoked by transcranial magnetic stimulation during the acquisition of new fine motor skills. *Journal of Neurophysiology* 74:1037–1045.

Paulesu, E., C. Frith, and R. Frackowiak. 1993. The neural correlates of the verbal component of working memory. *Nature* 362:342–345.

Paus, T. D., W. Perry, R. A. Zatorre, K. J. Worsley, and A. C. Evans. 1996. Modulation of cerebral blood flow in the human auditory cortex during speech: role of motor-to-sensory discharges. *European Journal of Neuroscience* 8:2236–2246.

Pena, M., L. L. Bonatti, M. Nespor, and J. Mehler. 2002. Signal-driven computations in speech processing. *Science* 298:604–607.

Pepperberg, I. M. 1981. Functional vocalizations by an African grey parrot *(Psittacus erithacus). Zeitschrift fur Tierpsychologie* 55:139–160.

——. 2002. *The Alex studies.* Cambridge, MA: Harvard University Press.

Percheron, G., and M. Fillon. 1991. Parallel processing in the basal ganglia up to a point (letter comment). *Trends in Neuroscience* 14:55–59.

Percheron, G., J. Yelnick, and C. A. Francois. 1984. Golgi analysis of the primate globus pallidus. III. Spatial organization of the striato-pallidal complex. *Journal of Comparative Neurology* 227:214–227.

Perkell, J. S. 1969. *Physiology of speech production: Results and implications of a quantitative cineradiographic study.* Cambridge, MA: MIT Press.

Perkell, J. S., and W. L. Nelson. 1982. Articulatory targets and speech motor control: A study of vowel production. In *Speech motor control,* ed. S. Grillner, A. Persson, B. Lindblom, and J. Lubker. New York: Pergamon, 187–204.

Peterson, G. E., and H. L. Barney. 1952. Control methods used in a study of the vowels. *Journal of the Acoustical Society of America* 24:175–184.

Peterson, G. E., W. Wang, and E. Sivertsen. 1958. Segmentation technique in speech synthesis. *Journal of the Acoustical Society of America* 30:739–742.

Peterson, M. R., M. D. Deecher, S. R. Zolith, D. B. Moody, and W. C. Stebbens. 1978. Species-specific perceptual processing of vocal sounds by monkeys. *Science* 202:324–326.

Peterson, S. E., P. T. Fox, M. I. Posner, M. Minton, and M. E. Raichle. 1988. Positron emission tomographic studies of the cortical anatomy of single-word processing. *Nature* 331:585-589.

Pfungst, O. 1907. *Das Pferd des Herrn von Osten: Der kluge Hans.* Leipzig: Barth.

Piatelli-Palmarini, M. 1989. Evolution, selection, and cognition: From "learning" to parameter-setting in biology and the study of language. *Cognition* 31:1–44.

Pickett, E. R. 1998. *Language and the cerebellum.* Ph.D. diss., Brown University.

Pickett, E. R., E. Kuniholm, A. Protopappas, J. Friedman, and P. Lieberman. 1998. Selective speech motor, syntax and cognitive deficits associated with bilateral damage to the putamen and the head of the caudate nucleus: A case study. *Neuropsychologia* 36:173–188.

Pierrehumbert, J. 1979. The perception of fundamental frequency declination. *Journal of the Acoustical Society of America* 66:363–369.

Pike, K. E. 1945. *The intonation of American English.* Ann Arbor: University of Michigan Press.

Pinker, S. 1994. *The language instinct: How the mind creates language.* New York: William Morrow.

——. 1998. *How the mind works.* New York: Norton.

——. 2002. *The blank slate.* New York: Norton.

Pinker, S., and P. Bloom. 1990. Natural selection and natural language. *Behavioral and Brain Sciences* 13:707–784.

Pirozzolo, F. J., E. C. Hansch, J. A. Mortimer, D. D. Webster, and M. A. Kuskowski. 1982. Dementia in Parkinson's disease: A neuropsychological analysis. *Brain and Cognition* 1:71–83.

Pisoni, D. B., and C. S. Martin. 1989. Effects of alcohol on the acoustic-phonetic properties of speech: Perceptual and acoustic analyses. *Alcoholism: Clinical and Experimental Research* 13:577–587.

Pitt, M. A., and A. G. Samuel. 1995. Lexical and sublexical feedback in auditory word recognition. *Cognitive Psychology* 29:149–188.

Poizner, H., E. S. Klima, and U. Bellugi. 1987. *What the hands reveal about the brain.* Cambridge, MA: MIT Press.

Polich, L. 2006. *The emergence of the deaf community in Nicaragua: With sign language you can learn so much.* Washington, DC: Gallaudet University Press.

Polit, A., and E. Bizzi. 1978. Processes controlling arm movements in monkeys. *Science* 201:1235–1237.

Pollack, I., and J. M. Pickett. 1963. The intelligibility of excerpts from conversation. *Language and Speech* 6:165–171.

Ponce de Leon, M. S., and C. P. Zollikofer. 2001. Neanderthal cranial ontogeny and its implications for late hominid diversity. *Nature* 412:534–538.

Poremba, A., M. Malloy, R. C. Saunders, R. E. Carlson, P. Herscovitch, and M. Miskin. 2004. Species-specific calls evoke asymmetric activity in the monkey's temporal poles. *Nature* 429:448–451.

Prinzo, O. V., P. Lieberman, and E. Pickett. 1998. *An acoustic analysis of ATC communication.* Washington, DC: Office of Aviation Medicine. DOT/FAA/AM-98/20.

Protopappas, A., and P. Lieberman. 1997. Fundamental frequency of phonation and perceived emotional stress. *Journal of the Acoustical Society of America* 101:267–277.

Rabiner, L. R., and J. Shafter. 1979. *Digital processing of speech.* New York: McGraw-Hill.

Rand, T. C. 1971. Vocal tract normalization in the perception of stop consonants. *Haskins Laboratories Status Report on Speech Research* 25/26:141–146.

Rauschecker, J. P., and M. Korte. 1993. Auditory compensation for early blindness in cat cerebral cortex. *Journal of Neuroscience* 18:4538–4548.

Regard, M., O. Oelz, P. Brugger, and T. Landis. 1989. Persistent cognitive impairment in climbers after repeated exposure to altitude. *Neurology* 39:210–213.

Remez, R. E., P. E. Rubin, D. B. Pisoni, and T. O. Carrell. 1981. Speech perception without traditional cues. *Science* 212:947–950.

Rendall, D., S. Kollias, C. Ney, and P. Loyd. Forthcoming. Pitch (Fo) and formant profiles of human and vowel-like baboon grunts: The role of vocalizer body size and voice-acoustic allometry. *Journal of the Acoustical Society of America.*

Richardson, M. K. 1999. Vertebrate evolution: The developmental origins of adult variation. *Bioessays* 21 (7):604–612.

Riede, T., E. Bronson, H. Hatzikirou, and K. Zuberbuhler. 2005. Vocal production in a non-human primate: Morphological data and a model. *Journal of Human Evolution* 48:85–96.

Rilling, J. K., and R. A. Seligman. 2002. A quantitative morphometric comparative analysis of the primate temporal lobe. *Journal of Human Evolution* 42:505–533.

Rissman, J., J. C. Eliassen, and S. E. Blumstein. 2003. An event-related fMRI investigation of implicit semantic priming. *Journal of Cognitive Neuroscience* 15:1160–1175.

Rizzolatti, G., and M. A. Arbib. 1998. Language within our grasp. *Trends in Neuroscience* 21:188–194.

Rizzolatti, G., L. Fadiga, V. Galiese, and L. Fogassi. 1996. Premotor cortex and the recognition of motor actions. *Cognitive Brain Research* 3:131–141.

Robbins, T. W. 2000. Chemical neuromodulation of frontal-executive functions in humans and other animals. *Experimental Brain Research* 133:130–138.

Robertson, R. L., L. Ben-Sira, P. D. Barnes, R. V. Mulkern, C. D. Robson, S. E. Maier, M. J. Rivkin, and A. J. du Plessis. 1999. MR line-scan diffusion-weighted imaging of term neonates with perinatal brain ischemia. *American Journal of Neuroradiology* 20:1658–1670.

Roche, H., A. Delagnes, J.-P. Brugal, C. Feibel, M. Kibunjia, V. Mourre, and P.-J. Texier. 1999. Early hominid stone tool production and technical skill 2.34 Myr ago in West Turkana, Kenya. *Nature* 399:57–60.

Ruhlen, M. 1994. *On the origin of language: Tracing the evolution of the mother tongue.* New York: John Wiley.

Russell, B. R. 1967. *The autobiography of Bertrand Russell, 1872–1913.* Boston: Little, Brown.

Russell, G. O. 1928. *The vowel.* Columbus: Ohio State University Press.

Ryalls, J. 1981. Motor aphasia: Acoustic correlates of phonetic disintegration in vowels. *Neuropsychologia* 20:355–360.

Ryalls, J. 1986. An acoustic study of vowel production in aphasia. *Brain and Language* 29:48–67.

Ryalls, J., and P. Lieberman. 1982. Fundamental frequency and vowel perception. *Journal of the Acoustical Society of America* 72:1631–1634.

Sachs, J., P. Lieberman, and D. Erickson. 1972. Anatomical and cultural determinants of male and female speech. In *Language attitudes: Current trends and prospects.* Monograph no. 25. Ed. R. W. Shuy and R. W. Fasold. Washington, DC: Georgetown University Press, 74–84.

Sadato, N., A. Pascual-Leone, J. Grafman, V. Ibanez, M.-P. Deiber, G. Dold, and M. Hallett. 1996. Activation of the primary visual cortex by Braille reading in blind subjects. *Nature* 380:526–528.

Saffran, J. R., R. N. Aslin, and E. L. Newport. 2001. Statistical learning by 8-month-old infants. *Science* 274:1926–1928.

Samuel, A. G. 1996. Phoneme restoration. *Language and Cognitive Processes* 11:647–653.

——. 1997. Lexical activation produces potent phonemic percepts. *Cognitive Psychology* 32:97–127.

——. 2001. Knowing a word affects the fundamental perception of the sounds within it. *Psychological Science* 12:348–351.

Sandner, G. W. 1981. Communications with a three-month-old baby. In *Proceedings of the Thirteenth Annual Child Language Research Forum.* Stanford, CA: Child Language Project, Stanford University.

Sanes, J. N., and J. P. Donoghue. 1994. Plasticity of cortical representations and its implication for neurorehabilitation. In *Principles and practice of rehabilitation medicine,* ed. B. T. Dhanai. Baltimore: Williams and Wilkins.

——. 1996. Static and dynamic organization of motor cortex. *Advances in Neurology, Brain Plasticity* 73:277–296.

——. 1997. Dynamic motor cortical organization. *The Neuroscientist* 3:158–165.

Sanes, J. N., J. P. Donoghue, V. Thangaraj, R. R. Edelman, and S. Warach. 1995. Shared neural substrates controlling hand movements in human motor cortex. *Science* 268:1775–1777.

Sarich, V. M. 1974. Just how old is the hominid line? In *Yearbook of physical anthropology, 1973.* Washington, DC: American Association of Physical Anthropologists.

Saussure, F. de. 1959. *Course in general linguistics.* Trans. W. Baskin. New York: McGraw-Hill.

Savage-Rumbaugh, S., K. McDonald, R. A. Sevcik, W. D. Hopkins, and E. Rubert. 1986. Spontaneous symbol acquisition and communicative use by pygmy chimpanzees *(Pan paniscus). Journal of Experimental Psychology, General* 115:211–235.

Savage-Rumbaugh, S., and D. Rumbaugh. 1993. The emergence of language. In *Tools, language and cognition in human evolution,* ed. K. R. Gibson and T. Ingold. Cambridge: Cambridge University Press, 86–100.

Savage-Rumbaugh, S., D. Rumbaugh, and K. McDonald. 1985. Language learning in two species of apes. *Neuroscience and Biobehavioral Reviews* 9:653–665.

Savariaux, C., P. Perrier, and J. P. Orliaguet. 1995. Compensation strategy for the production of the round vowel [u] using a lip tube: A study of the control space in speech production. *Journal of the Acoustical Society of America* 98:2428–2242.

Sayigh, L. S., P. L. Tyack, R. S. Wells, and M. D. Scott. 1990. Signature whistles of free-ranging bottlenose dolphins *Tursiops truncatus:* Stability and mother-offspring comparisons. *Behavioral Ecology and Sociobiology* 26:247–260.

Schepartz, L. A. 1993. Language and modern human origins. *Yearbook of Physical Anthropology* 36:91–126.

Schmahmann, J. D. 1991. An emerging concept: The cerebellar contribution to higher function. *Archives of Neurology* 48:1178–1186.

Schoenemann, P. T. Forthcoming. Conceptual complexity and the brain: Understanding language origins. In *Language acquisition, change and emergence: Essays in evolutionary linguistics*, ed. W. S-Y Wang and J. M. Minett. Hong Kong: City University of Hong Kong Press.

Schoenemann, P. T., T. F. Budinger, V. M. Sarich, and W. Wang. 2000. Brain size does not predict general cognitive ability within families. *Proceedings of the National Academy of Sciences, USA* 97:4932–4937.

Scott, R. B., J. Harrison, C. Boulton, J. Wilson, R. Gregory, S. Parkin, P. G. Bain, C. Joint, J. Stein, and T. Z. Aziz. 2002. Global attentional-executive sequelae following surgical lesions to globus pallidus interna. *Brain* 125:562–574.

Seebach, B. S., N. Intrator, P. Lieberman, and L. N. Cooper. 1994. A model of prenatal acquisition of speech parameters. *Proceedings of the National Academy of Sciences, USA* 91:7473–7476.

Seidenberg, M., and L. Petitto. 1979. Signing behavior in apes: A critical review. *Cognition* 7:177–215.

Sejnowski, T. J. 1997. The year of the dendrite. *Science* 275:178–179.

Sejnowski, T. J., C. Koch, and P. S. Churchland. 1988. Computational neuroscience. *Nature* 241:1299–1306.

Semendeferi, K., and H. Damasio. 2000. The brain and its main anatomical subdivisions in living hominoids using magnetic resonance imaging. *Journal of Human Evolution* 38:317–332.

Semendeferi, K., H. Damasio, R. Frank, and G. W. Van Hoesen. 1997. The evolution of the frontal lobes: A volumetric analysis based on three-dimensional reconstructions of magnetic resonance scans of human and ape brains. *Journal of Human Evolution* 32:375–378.

Semendeferi, K., A. Lu, N. Schenker, and H. Damasio. 2002. Humans and apes share a large frontal cortex. *Nature Neuroscience* 5:272–276.

Semino, O., G. Passarino, P. J. Oefner, A. A. Lin, S. Arbuzova, L. E. Beckman, G. de Benedictis, P. Francalacci, A. Kouvatsi, S. Limborska, M. Marcikiae, A. Mika, B. Mika, D. Primorac, A. S. Santachiara-Benerecetti, L. L. Cavalli-Sforza, and P. A. Underhill. 2000. The genetic legacy of Paleolithic *Homo sapiens sapiens* in extent Europeans: A Y chromosome perspective. *Science* 290:1155–1159.

Sereno, J., S. R. Baum, G. C. Marean, and P. Lieberman. 1987. Acoustic analyses and perceptual data on anticipatory labial coarticulation in adults and children. *Journal of the Acoustical Society of America* 81:512–519.

Sereno, J., and P. Lieberman. 1987. Developmental aspects of lingual coarticulation. *Journal of Phonetics* 15:247–257.

Shephard, J., and S. M. Kosslyn. 2005. The MiniCog rapid assessment battery: Developing a "blood pressure" cuff for the mind. *Aviation Space and Environmental Medicine* 76:192–197

Sherrington, C. S. 1947. *The integrative action of the nervous system*. New York: Yale University Press.
Shockey, L. 2003. *Sound patterns of spoken English*. Oxford: Blackwell.
Simpson, G. G. 1966. The biological nature of man. *Science* 152:472–478.
Singleton, J. L., and E. L. Newport. 1989. When learners surpass their models: The acquisition of American Sign Language from impoverished input. In *Proceedings of the 14th Annual Boston University Conference on Language Development*, vol. 15. Boston: Program in Applied Linguistics, Boston University.
Skinner, B. F. 1957. *Verbal behavior*. New York: Appleton-Century-Crofts.
Slocombe, K., and K. Zuberbuhler. 2005a. Agonistic screams in wild chimpanzees *(Pan trogloydytes schweinfurthii)* vary as a function of social life. *Journal of Comparative Psychology* 119:66–77.
Slocombe, K., and K. Zuberbuhler. 2005b. Functionally referential communication in a chimpanzee. *Current Biology* 15:1–6.
Slotnick, B. M. 1967. Disturbances of maternal behavior in the rat following lesions of the cingulate cortex. *Behavior* 24:204–236.
Smith, A., and L. Goffman. 1998. Stability of speech movement sequences in children and adults. *Journal of Speech, Language and Hearing Research* 41:18–30.
Smith, B. L. 1978. Temporal aspects of English speech production: A developmental perspective. *Journal of Phonetics* 6:37–68.
Smith, W. J. 1977. *The behavior of communicating*. Cambridge, MA: Harvard University Press.
Solecki, R. S. 1971. *Shanidar, the first flower people*. New York: Knopf.
Solzhenitsyn, A. 1965. *The first circle*. Moscow: Tvardovsky.
Spurzheim, J. K. 1815. *The physiognomical system of Drs. Gall and Spurzheim*. London: Printed for Baldwin, Cradock, and Joy.
Stamm, J. S. 1955. The function of the medial cerebral cortex in maternal behavior of rats. *Journal of Comparative Physiology and Psychology* 48:347–356.
Stephan, H., H. Frahm, and G. Baron. 1981. New and revised data on volumes of brain structures in insectivores and primates. *Folia Primatologia* 35:1–29.
Stetson, R. H. 1951. *Motor phonetics: A study of speech movements in action*. Amsterdam: North-Holland.
Stevens, K. N. 1972. Quantal nature of speech. In *Human communication: A unified view*, ed. E. E. David Jr. and P. B. Denes. New York: McGraw Hill, 51–66.
———. 1998. *Acoustic Phonetics*. Cambridge, MA: MIT Press.
Stevens, K. N., R. P. Bastide, and C. P. Smith. 1955. Electrical synthesizer of continuous speech. *Journal of the Acoustical Society of America* 27:207.
Stevens, K. N., and S. E. Blumstein. 1978. Invariant cues for place of articulation in stop consonants. *Journal of the Acoustical Society of America* 64:1358–1368.
Stevens, K. N., and A. S. House. 1955. Development of a quantitative description of vowel articulation. *Journal of the Acoustical Society of America* 27:484–493.
Stokoe, W. 1978. Sign language versus spoken language. *Sign Language Studies* 18:69–90.
Stone, M., and A. Lundberg. 1996. Three dimensional tongue surface shapes of English consonants and vowels. *Journal of the Acoustical Society of America* 99:3728–3736.
Story, B. H. 2004. On the ability of a physiologically constrained area function

model of the vocal tract to produce normal formant patterns under perturbed conditions. *Journal of the Acoustical Society of America* 115:1760–1770.
Story, B. H., and I. R. Titze. 1998. Paramaterization of vocal tract area functions by empirical orthogonal modes. *Journal of Phonetics* 26:223–260.
Story, B. H., I. R. Titze, and E. A. Hoffman. 1996. Vocal tract area functions from magnetic resonance imaging. *Journal of the Acoustical Society of America* 100:537–554.
Stowe, L. A., A. M.-J. Paans, A. A. Wijers, and F. Zwarts. 2004. Activations of "motor" and other non-language structures during sentence comprehension. *Brain and Language* 89:290–299.
Strange, W., R. R. Verbrugge, D. P. Shankweiler, and T. R. Edman. 1976. Consonantal environment specifies vowel identity. *Journal of the Acoustical Society of America* 60:213–224.
Strauss, W. L. Jr., and A. J. E. Cave. 1957. Pathology and posture of Neanderthal man. *Quarterly Review of Biology* 32:348–363.
Stringer, C. B. 1992. Evolution of early humans. In *The Cambridge encyclopedia of human evolution,* ed. S. Jones, R. Martin, and D. Pilbeam. Cambridge: Cambridge University Press, 241–251.
——. 1998. Chronological and biogeographic perspectives on later human evolution. In *Neanderthals and modern humans in western Asia*, ed. T. Akazawa, K. Akoi, and O. Bar-Yosef. New York: Plenum, 29–38.
——. 2002. Modern human origins: Progress and prospects. *Philosophical Transactions of the Royal Society of London* B 357:563–579.
Stringer, C. B., and P. Andrews. 1988. Genetic and fossil evidence for the origin of modern humans. *Science* 239:1263–1268.
Stringer, C. B., R. Grun, H. P. Schwarcz, and P. Goldberg. 1989. ESR dates for the hominid burial site of Es Skhul in Israel. *Nature* 338:756–758.
Stromswold, K., D. Caplan, N. Alpert, and S. Rausch. 1996. Localization of syntactic processing by positron emission tomography. *Brain and Language* 51:452–473.
Strub, R. L. 1989. Frontal lobe syndrome in a patient with bilateral globus pallidus lesions. *Archives of Neurology* 46:1024–1027.
Studdert-Kennedy, M. 2000. Implications of the particulate principle. In *The evolutionary emergence of language: Social function and the origins of linguistic form*, ed. C. Knight, M. Studdert-Kennedy, and J. A. Hurford. New York: Cambridge University Press, 161–176.
Stuss, D. T., and D. F. Benson. 1986. *The frontal lobes.* New York: Raven.
Susman, R. L. 1994. Fossil evidence for early hominid tool use. *Science* 265:1570–1573.
Sutton, D., and U. Jurgens. 1988. Neural control of vocalization. In *Comparative primate biology*, vol. 4, ed. H. D. Steklis and J. Erwin. New York: Arthur D. Liss, 625–647.
Swift, J. [1726] 1970. *Gulliver's travels.* New York: Norton.
Takahashi, K., F. C. Liu, K. Hirokawa, and H. Takahashi. 2003. Expression of FoxP2, a gene involved in speech and language in the developing and adult striatum. *Journal of Neuroscience Research* 73:62–72.
Taylor, A. E., J. A. Saint-Cyr, and A. E. Lang. 1990. Memory and learning in early

Parkinson's disease: Evidence for a "frontal lobe syndrome." *Brain and Cognition* 13:211–232.

Templeton, A. R. 2002. Out of Africa again and again. *Nature* 416:45–51.

Terrace, H. S. 1979. *Nim*. New York: Knopf.

Terrace, H. S., L. A. Petitto, R. J. Sanders, and T. G. Bever. 1979. Can an ape create a sentence? *Science* 206:821–901.

Thach, W. T. 1996. On the specific role of the cerebellum in motor learning and cognition: Clues from PET activation and lesion studies in man. *Behavioral and Brain Sciences* 19:411–431.

Thach, W. T., J. W. Mink, H. P. Goodkin, and J. G. Keating. 1993. Combining versus gating motor programs: Differential roles for cerebellum and basal ganglia. In *Role of the cerebellum and basal ganglia in voluntary movement*, ed. N. Mano, I. Hmada, and M. R. DeLong. Amsterdam: Elsevier.

Thelen, E. 1984. Learning to walk: Ecological demands and phylogenetic constraints. In *Advances in infancy research*, vol. 3, ed. L. Lipsitt. Norwood, NJ: Ablex, 213–250.

Thieme, H. 1997. Lower Paleolithic hunting spears from Germany. *Nature* 385:807–810.

Timcke, R., H. von Leden, and P. Moore. 1958. Laryngeal vibrations: Measurements of the glottic wave. *American Medical Association Archives of Otolaryngology* 68:1–19.

Tinbergen, N. 1953. *Social behavior in animals*. London: Methuen.

Tomasello, M. 2004a. *Constructing a language: A usage-based theory of language acquisition*. Cambridge, MA: Harvard University Press.

———. 2004b. *The cultural origins of human cognition*. Cambridge, MA: Harvard University Press.

Tomasello, M., M. Davis-Dasilva, L. Camak, and K. Bard. 1987. Observational learning of tool-use by young chimpanzees. *Human Evolution* 2:175–183.

Tomasello, M., and M. J. Farrar. 1986. Joint attention and early language. *Child Development* 57:1454–1463.

Toth, N., and K. Schick. 1993. Early stone industries. In *Tools, language and cognition in human evolution*, ed. K. R. Gibson and T. Ingold. Cambridge: Cambridge University Press, 346–362.

Trager, G. L., and H. L. Smith. 1951. *Outline of English structure*. Norman, OK: Battenburg.

Truby, H. L., J. F. Bosma, and J. Lind. 1965. *Newborn infant cry*. Uppsala: Almquist and Wiksell.

Tseng, C. Y. 1981. An acoustic study of tones in Mandarin. Ph.D. diss., Brown University.

Tyson, E. 1699. *Orang-outang, sive, Homo sylvestris; or, The anatomy of a pygmie compared with that of a monkey, an ape, and a man*. London: Printed for Thomas Bennet and Daniel Brown.

Ullman, M. T. 2004. Contributions of memory circuits to language: The declarative/procedural model. *Cognition* 92:311–270.

Ungerleider, L. G. 1995. Functional brain imaging studies of cortical mechanisms for memory. *Science* 270:769–775.

Utman, J. A., S. E. Blumstein, and K. Sullivan. 2001. Mapping from sound to meaning: Reduced lexical activation in Broca's aphasia. *Brain and Language* 79:444–472.

Vallar, G., A. M. D. Betta, and M. C. Silveri. 1997. The phonological short-term store-rehearsal system. *Neuropsychologia* 35:795–812.

Van den Berg, J. 1958. Myoelastic-aerodynamic theory of voice production. *Journal of Speech and Hearing Research* 1:227–244.

Vargha-Khadem, F., D. G. Gadian, A. Copp, and M. Mishkin. 2005. FOXP2 and the neuroanatomy of speech and language. *Nature Reviews, Neuroscience* 6:131–138.

Vargha-Khadem, F., K. Watkins, K. Alcock, P. Fletcher, and R. Passingham. 1995. Praxic and nonverbal cognitive deficits in a large family with a genetically transmitted speech and language disorder. *Proceedings of the National Academy of Sciences, USA* 92:930–933.

Vargha-Khadem, F., K. E. Watkins, C. J. Price, J. Ashburner, K. J. Alcock, A. Connelly, R. S. Frackowiak, K. J. Friston, M. E. Pembrey, M. Mishkin, D. G. Gadian, and R. E. Passingham. 1998. Neural basis of an inherited speech and language disorder. *Proceedings of the National Academy of Sciences, USA* 95:12695–12700.

Velanova, K., L. I. Jacoby, M. E. Wheeler, M. A. McAvoy, S. E. Peterson, and R. L. Buckner. 2003. Functional-anatomic correlates of sustained and transient processing components engaged during controlled retrieval. *Journal of Neuroscience* 23:8460–8470.

Verbrugge, R., W. Strange, and D. Shankweiler. 1976. What information enables a listener to map a talker's vowel space? *Haskins Laboratories Status Report on Speech Research* 37/38:199–208.

von Baer, K. E. 1828. *Über entwickelungsgeschichte der thiere: Beobachtung und reflexion.* Königsberg: Bei den Gebrüdern Bornträger.

——. 1876. *Studien aus dem gebiete der naturwissenschaften.* St. Petersburg: H. Schmitzdorff.

von Kempelen, W. R. 1791. *Le Mechanisme de la parole suivi de la description d'une machine parlant.* Vienna: J. V. Degen.

Vorperian, H. K., R. D. Kent, M. Lindstrom, C. M. Kalina, L. R. Gentry, and B. S. Yandell. 2005. Development of vocal tract length during early childhood: A magnetic resonance imaging study. *Journal of the Acoustical Society of America* 117:338–350.

Wallace, D. C. 2004. Mitochondrial DNA sequence variation in human evolution and disease. *Proceedings of the National Academy of Sciences, USA* 91:8736–8746.

Walters, T., T. J. Carew, and E. R. Kandel. 1981. Associative learning in *Aplysia:* Evidence for conditioned fear in an invertebrate. *Science* 211:404–506.

Warden, C. J., and L. H. Warner. 1928. The sensory capacities and intelligence of dogs, with a report on the ability of the noted dog "Fellow" to respond to verbal stimuli. *Quarterly Review of Biology* 3:1–28.

Warrington, E., V. Logue, and R. Pratt. 1971. The anatomical localization of the selective impairment of auditory short-term memory. *Neuropsychologia* 9:377–387.

Watkin, K. L., and D. Fromm. 1984. Labial coordination in children: Preliminary considerations. *Journal of the Acoustical Society of America* 75:629–632.

Watkins, K. E., A. P. Strafella, and T. Paus. 2003. Seeing and hearing speech excites the motor system involved in speech production. *Neuropsychologia* 41:989–994.

Watkins, K. E., F. Vargha-Khadem, J. Ashburner, R. E. Passingham, A. Connelly, K. J. Friston, R. S. J. Frackiwiak, M. Miskin, and D. G. Gadian. 2002. MRI analysis of an inherited speech and language disorder: Structural brain abnormalities. *Brain* 125:465–478.

Weisengrubber, G. E., G. Forstenpointner, G. Peters, A. Kubber-Heiss, and W. T. Fitch. 2002. Hyoid apparatus and pharynx in the lion *(Panthera leo)*, jaguar *(Panthera onca)*, tiger *(Panthera tigris)*, cheetah *(Acinonyx jubatus)* and domestic cat *(Felis silvestris f. catus)*. *Journal of Anatomy* 201:195–201.

Wernicke, C. [1874] 1967. The aphasic symptom complex: A psychological study on a neurological basis. In *Proceedings of the Boston Colloquium for the Philosophy of Science,* vol. 4, ed. R. S. Cohen and M. W. Wartofsky. Dordrecht: Reidel.

Wheeler, M. F., and R. J. Buckner. 2003. Functional dissociation among components of remembering: Control, perceived oldness, and content. *Journal of Neuroscience* 23:3869–3880.

White, T. D., B. Asfaw, D. DeGusta, H. Gilbert, G. D. Richards, G. Suwa, and F. Clark Howell. 2003. Pleistocene *Homo sapiens* from Middle Awash, Ethiopia. *Nature* 423:742–747.

Whitfield, 1967. *The auditory pathway.* London: Arnold.

Wilkins, W. K., and J. Wakefield. 1995. Brain evolution and neurolinguistic preconditions. *Behavioral and Brain Sciences* 18:161–162.

Wilson, S. M., A. P. Saygin, M. I. Sereno, and M. Iacoboni. 2004. Listening to speech activates motor areas involved in speech production. *Nature Neuroscience* 7:701–702.

Wimsatt, W. 1985. Developmental constraints, generative entrenchment and the innate-acquired distinction. In *Integrating scientific disciplines,* ed. W. Bechtel. Dordrect: Martinus Nijhoff, 185–208.

Wood, B., and M. Collard. 1999. The human genus. *Science* 284:65–71.

Wood, B. A. 1992. Evolution of australopithecines. In *The Cambridge encyclopedia of human evolution,* ed. S. Jones, R. Martin, and D. Pilbeam. Cambridge: Cambridge University Press, 231–240.

Wood, E. R., P. A. Dudchenko, and H. Eichenbaum. 1999. The global record of memory in hippocampal neuronal activity. *Nature* 397:613–616.

Wuethrich, B. 2000. Learning the world's languages before they vanish. *Science* 288:1156–1159.

Xuerob, J. H., B. E. Tomlinson, D. Irving, R. H. Perry, G. Blessed, and E. K. Perry. 1990. Cortical and subcortical pathology in Parkinson's disease: Relationship to Parkinsonian dementia. In *Advances in neurology.* Vol. 53, *Parkinson's disease: Anatomy, pathology and therapy,* ed. M. B. Streifler, A. D. Korezyn, J. Melamed, and M. B. H. Youdim. New York: Raven, 35–39.

Young, B., and P. Lieberman. forthcoming. Corticostriatal circuit function and developmental verbal apraxia.

Zhang, J. 2003. Evolution of the human ASPM gene, a major determinant of brain size. *Genetics* 165:2063–2070.

Zhu, R. X., K. A. Huffman, R. Potts, C. L. Deng, Y. X. Pair, B. Guo, C. D. Shi, Z. T. Guo, B. Y. Yuan, Y. M. Hou, and W. W. Huang. 2001. Earliest presence of humans in northeast Asia. *Nature* 413:413–417.

Ziles, K., G. Schlaug, M. Matelli, G. Luppino, A. Schleicher, M. Qu, A. Dabringhaus,

R. Seitz, and P. E. Roland. 1995. Mapping of human and macaque sensorimotor areas by integrating architectonic, transmitter receptor, MRI, and PET data. *Journal of Anatomy* 187:515–537.

Zinkin, N. I. 1968. *Mechanisms of speech.* The Hague: Mouton.

Zuberbuhler, K. 2002. A syntactic rule in forest monkey communication. *Animal Behavior* 63:293–299.

Zubrow, E. 1990. The demographic modeling of Neanderthal extinction. In *The human revolution: Behavioral and biological perspectives on the origin of modern humans,* vol. 1, ed. P. Mellars and C. B. Stringer. Edinburgh: Edinburgh University Press, 212–231.

Zurif, E. B., A. Caramazza, and R. Meyerson. 1972. Grammatical judgments of agrammatic aphasics. *Neuropsychologia* 10:405–418.

Index